Structural Integrity

Volume 31

Shun-Peng Zhu · Behrooz Keshtegar

Structural Reliability Analysis

Analytical Methods

Springer

Shun-Peng Zhu
School of Mechanical and Electrical Engineering
University of Electronic Science and Technology of China
Chengdu, China

Behrooz Keshtegar
Department of Civil Engineering
University of Zabol
Zabol, Iran

ISSN 2522-560X ISSN 2522-5618 (electronic)
Structural Integrity
ISBN 978-3-031-75196-7 ISBN 978-3-031-75197-4 (eBook)
https://doi.org/10.1007/978-3-031-75197-4

This Springer imprint is published by the registered company Springer Nature Switzerland AG
The registered company address is: Gewerbestrasse 11, 6330 Cham, Switzerland

Foreword

With the rapid advancement of design and manufacturing technologies in engineering, major equipment and engineering structures are evolving toward higher reliability and safety. The uncertainties encountered during the design, manufacturing, and service phases significantly impact the stability of structural performance and its service life. Today, reliability analysis that thoroughly incorporates structural uncertainties has become one of the most prominent research topics in reliability engineering. However, the complexity of extreme states, the coupling of multi-physical fields, and the multi-source nature of variables present significant challenges to reliability analysis. Consequently, developing efficient, accurate, and robust structural reliability analysis methods is one of the key scientific problems that must be addressed. Within this framework, Professors Shun-Peng Zhu and Behrooz Keshtegar have authored the monograph "*Structural Reliability Analysis - Analytical Methods*" in response to modern high-performance engineering structure design and manufacturing.

This monograph centers on the First-Order Reliability Method (FORM), providing a detailed discussion of various iterative forms, convergence control conditions, and their improved applications, exploring how to balance accuracy and efficiency in complex structural analyses. Specifically, it delves into the stability issues arising from iterative methods and convergence control conditions within FORM analysis, particularly in highly nonlinear problems. The content includes algorithmic formulas, the framework for computer program implementation, and a thorough comparison of the advantages and disadvantages of different iterative methods and convergence control conditions, alongside their applicable scenarios. Furthermore, the monograph comprehensively evaluates the contributions of FORM-based methods, such as the steepest descent method, directional stability transformation method, and conjugate gradient method, in addressing the challenge of efficient reliability analysis in the context of multi-physical field coupling. An innovative search formula suitable for interval variable problems is also introduced, effectively overcoming the challenges posed by the multi-source nature of variable types. Overall, the book focuses on efficient, accurate, and robust reliability analysis and deeply explores the key role and application value of advanced analytical algorithms in high-performance engineering structure design and manufacturing.

This monograph first presents the basic principles of reliability analysis in a clear and accessible manner, systematically comparing the performance, strengths, and weaknesses of various methods, providing a solid foundation for beginners. By integrating theory with practice, the book also outlines the future research directions for analytical methods, offering valuable insights and guidance for further study in the field, thus becoming a classic educational reference. Furthermore, the book thoroughly summarizes the classic reliability analysis methods, providing detailed explanations for each step, from theoretical derivation to algorithm implementation, ensuring that engineering designers can easily understand and apply these methods. As an essential resource in the field of reliability analysis, it serves not only as a valuable academic reference but also as a practical guide for engineering design, significantly enhancing design efficiency and accuracy. Therefore, whether in advancing academic research or driving innovation in engineering applications, this monograph injects new momentum into the development of related fields, making an important contribution to scientific progress and advancement of engineering technologies.

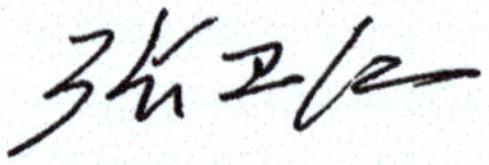

January 2025

Professor Wei-Hong Zhang
Member of Chinese Academy of Sciences
Northwestern Polytechnical University, Xi'An, China

Preface

The primary goal of the reliability methods is to approximate the structural failure probability and to solve a multi-dimensional integration in engineering practice. For this purpose, Monte Carlo simulation, first-order/second-order reliability methods, and moment methods have been commonly used in reality. The main process of first-order reliability method (FORM) is to determine the MPP (which is applied for computing the reliability index) in reliability analysis. The second-moment method is established based on the mean and standard deviation (std) of the limit state function (LSF) that the mean and std of LSF cannot confirm an accurate joint distribution for performance function with complex simulated responses. On the other hand, the moment methods are strongly dependent on sampling approach, number of samples, evaluating method for approximation of performance function that these factors are directly affected on the joint distribution function applied for detecting the failure domain. Consequently, for a complex engineering problem with implicit LSF, the moment method may provide inaccurate results. In approximating the failure probability of structural components, the simulation methods have shown high accuracy compared to analytical approaches, but for an LSF with implicit nonlinear dynamic performance, the use of simulation methods but, application of simulation methods for engineering problems analyzed by a finite element model with nonlinear behavior of dynamical response and problems with low-failure probability is generally impossible. For the simulation methods, it requires a high number of simulations for approximating a very low-failure probability (for example, of $P_f = 10^{-5}$), thus computational cost/time increases. Especially, in applicable engineering problems simulated in finite element methods for obtaining their responses by a dynamic analysis, contact behavior, interaction problems such as soil (water)—structure interaction, nonlinear buckling problems, fatigue evaluations and cracking detections, nonlinear problems as large deformations or material nonlinearity, the simulation methods are not computational efficient approaches. Therefore, as an efficient reliability method, the first-order reliability methods are offered by providing a good balancing between accuracy and efficiency.

The FORM procedure requires a formulation with complex reactions than the moment method. Nevertheless, it takes much less time for analysis than the simulation methods. The accuracy of failure probability in FORM depends on the curvature of LSF at MPP that its approximation can be improved by using the second-order reliability method. The basic issue of the FORM method is to find a reliability index with instable iterative formulations as periodic (oscillation), bifurcation or chaos for highly nonlinear problems. Since engineering problems have different performances with various degrees of non-linearity, in this book, the main attempt is to present the novel analytical methods-based FORM for controlling the instability of iterative formulation that the iteration formulas have been shown a high-efficiency as low computational burden. Also, the FORM formulations are improved with suitable adjustment inexact conditions to provide the stable convergence for high nonlinear problems. Accordingly, we have investigated the various instability of FORM formula for estimating the failure probability and have improved the efficiency and robustness of FORM. For these purposes, there have been developed several convergence criteria for determining the stability of FORM formulation, and have been established a novel search formulation for problems with interval variables named original probabilistic method. Based on the presented main efforts, this book is written based on seven main chapters that each chapter involves of several iterative formulations of FORM to search MPP and discussions on their advantages and limitations by numerical examples and related suitable references to help the reader. The briefly details of these seen chapter are described as below:

Chapter 1 introduces the basic concepts of SRA. In general, several main defeminations in the SRA and the role of SRA are also briefly discussed. Uncertainty is fundamental to SRA, which can be divided into natural uncertainties and human uncertainties. Based on this, the SRA framework can be developed including the LSF, the concept of failure and safety, failure probability, reliability index as well as the distribution function. Besides, the development and prospects of SRA are also explored including system reliability, time-variant reliability and machine learning.

Chapter 2 briefly introduces several typical SRA methods which can be mainly divided into analytical methods and simulation methods. Specifically, the first-order reliability method, second-order reliability method, moment method, first-order saddle point approximation method, non-probabilistic reliability model and dimension reduction method are six classical analytical methods. Similarly, the Monte Carlo simulation, importance sampling, uniform importance sampling, enhanced Monte Carlo simulation, line sampling, subset simulation, directional sampling as well as direct probabilistic integral method are the classical simulation methods introduced in this chapter.

Chapter 3 presents the FORM-based steepest descent search direction. The iterative formulation of FORM is established based on the steepest descent sensitivity vectors that almost analytical methods of FORM are enhanced by a step size less than 1. In this chapter, all collected the FORM formulas are developed as an improved version of the HL-RF method named as steepest descent method. By using HL-RF method, we try to collect several approaches of FORM named as (i) improve HL-RF,

(ii) chaos control method, (iii) relaxed HL-RF, (iv) non-negative constraint method that they are formulated for improving gradient line search direction based on a step size.

Chapter 4 introduces the convergence control of FORM-based steepest formulation given by HL-RF formulated using gradient sensitivity vector. Three conditions as (i) merit condition applied in improved HL-RF method, (ii) angle condition, and (iii) steepest descent condition is presented for controlling instability of the FORM-based improved HL-RF and relaxed FORM approach. The adjusting step size given by inexact line is computed based on an inner loop applied in main loop of FORM.

Chapter 5 presents another type of FORM-based steepest descent condition as FORM formulated by the directional steepest decent sensitivity vector. In this section, several formulations computed by gradient method are presented as direction FORM. The main format of these approaches is developed by finite step length method. And then, several approaches including (i) directional stability transformation method, (ii) adaptive directional stability transformation method, (iii) enhanced finite step length method, (iv) dynamical relaxed directional method have been introduced.

Chapter 6 introduces directional conjugate gradient approaches. The basic formulation of conjugate FORM-based FR is presented, and then several improvements in conjugate search direction vectors are established. These methods are named as (i) limited FR, (ii) enhanced FR-based variant type in gradient vector, (iii) chaotic conjugate chaos control method, (iv) chaotic conjugate stability transformation method. The step size is adaptively adjusted based on the sufficient decent condition and the Armijo-type condition in these formulations.

Chapter 7 presents the extended different types of conjugate FORM which are formulated with different scalar factors by comparing with FR. Several conjugate FORM formulations are established as (1) hybrid conjugate finite-step length method; (2) hybrid self-adaptive directional conjugate method; and (3) three-term conjugate gradient method. These three methods are investigated for convergence performances for both accuracy and efficiency in this current chapter. The parameters applied in computing the line search including (i) step length and (ii) conjugate gradient vectors are discussed through several examples.

Chapter 8 presents a novel optimization model for MPP search in original space for interval basic random variables named original probabilistic model (OPM). The novel MPP search approach is formulated by using joint cumulative distribution function (CDF) of basic random variables in X-space. In novel searching MPP, an individual MPP can be provided for multi-MPP problems. Then, harmony search (HS) optimization as a music-inspired using human-based optimization method is applied for searching MPP by presented OPM in this chapter. The modified version of HS optimization is provided for accurate MPP results.

Chengdu, China
Zabol, Iran
August 2024

Shun-Peng Zhu
Behrooz Keshtegar

Acknowledgments

This book is based on jointed research conducted by the authors between 2016 to the present, especially for Dr. Behrooz Keshtegar working as research associate at the University of Electronic Science and Technology of China during 2018–2024. We are grateful for a number of friends and colleagues encouraging us to start the work, persevere with it, and finally to publish it.

For this book, we acknowledge the support and funding provided by Key Program of the National Natural Science Foundation of China (Grant No. 12232004), Sichuan Science and Technology Program (Grant No. 2022JDJQ0024), University of Zabol (Grant No. IR-UOZ-GR-2117), and Iran National Science Foundation (INSF) (Grant No. 4012606).

Additionally, we would like to offer our most sincere and profound acknowledgment to Prof. Hong-Zhong Huang, founder of the reliability team at the University of Electronic Science and Technology of China. Without Prof. Huang's steadfast devotion to the advancement of our field and his relentless pursuit of knowledge, this team would not have achieved what it has achieved, and this book would never have been existed.

Finally, we would like to acknowledge with gratitude, the support and love of our families, they all kept us going, and this book would not have been possible without them. Specifically, Dr. Behrooz Keshtegar thank his wife, Ms. Zienab Sarani, for her best support on him to complete this book. The authors are grateful to several people who have contributed to this book in various ways: Debiao Meng, Changqi Luo, Yanjing Lv, Xinya You, Mengli Yan, Hao Ma, Yibo Wang, and Xingchan Zhang they are membered in the *Structural Integrity and Intelligent Design Group* at School of Mechanical and Electrical Engineering, University of Electronic Science and Technology of China, China.

Contents

About the Authors

Shun-Peng Zhu is a Professor in the School of Mechanical and Electrical Engineering at University of Electronic Science and Technology of China, China. He received his Ph.D. in Mechanical Engineering from UESTC in 2011, worked as a PIF Fellow of Politecnico di Milano, Italy, from 2016 to 2018 and was a research associate at the University of Maryland, United States, from 2010 to 2011. His research, published in scholarly journals and edited volumes, over 150 peer-reviewed book chapters, journals, and proceedings papers, explores the aspects of fatigue design, structural reliability, probabilistic failure modeling, multi-physics damage modeling, and life prediction for engineering structures/components under uncertainty.

He received 2nd Prizes of Natural Science Awards of the Sichuan Province and Chinese Society of Aeronautics and Astronautics in 2022, four Most Cited Paper Awards of *FFEMS Journal* since 2018, Award of Merit (2019) and Robert Moskovic Award (2024) of ESIS TC12, FEUP Medal of University of Porto in 2024, Top 100 Most Influential International Academic Paper Award of China in 2018, Polimi International Fellowship in 2016, Hiwin Doctoral Dissertation Award in 2012, Best Paper Awards of 5 international conferences. He has been featured as the Elsevier's Most Cited Chinese Researcher in the field of Safety, Risk, Reliability, and Quality since 2018, World's Top 2% Scientists since 2020, and serves as Co-Editor-in-Chief of the *International Journal of Structural Integrity*, guest editor, editorial board member of 7 international journals and Springer, Taylor & Francis book series.

Behrooz Keshtegar is a staff member in Department of Civil Engineering at University of Zabol, Zabol, Iran. He received his Ph.D. from Sistan and Baluchestan, Zahedan, Iran, in 2013. He established a research lab named Modeling and Simulation-based Optimization and Reliability (MASRO) in University of Zabol in 2023. His research fields focused on the multidisciplinary topics such as; structural reliability analysis, reliability-based design optimization, machine learning methods, and optimization-based population approaches. He established several data-driven models, extended several analytical reliability methods applied in engineering problems and reliability-based design optimization and also enhanced several optimization methods. His relative works extremely focused on the structural reliability analysis and reliability-based design optimization coupled with an improved first-order reliability method given from the conjugate sensitivity vector. He has the vest contributions with well-known researchers in the words from Iran, China, Portugal, Australia, Georgia, and Canada. He has published more than 150 peer-reviewed papers in scientific journal. He received a prize of University of Zabol for four most cited papers published in *JOF* and *CMAME* journals in 2022. He got distinguished top researcher from University of Zabol at 2018 to 2020. He received young researcher of civil engineering from The Academy of Science of Iran, Iran, at 2022 and World's Top 2% Scientists since 2019.

Abbreviations

AC	Adaptive conjugate
AMV	Advanced mean value
bw	Bandwidth
CC	Chaos control
CCC	Chaotic conjugate chaos control method
CCSTM	Chaotic conjugate stability transformation method
CD	Conjugate descent
CDF	Cumulative distribution function
CF	Call function
CFORM	Conjugate first-order reliability method
CFSL	Conjugate finite-step length
CFSL-H	Hybrid conjugate finite-step length
CG	Conjugate gradient
CGF	Cumulant generating function
CHL-RF	Conjugate Hasofer–Lind and Rackwitz–Fiessler
COV	Coefficient of variation
CPU	Central processing unit
CS	Conditional sampling
CSDV	Conjugate search direction vector
CSF	Conjugate scalar factor
CSTM	Conjugate stability transformation method
DPIM	Direct probabilistic integral method
DRD	Dynamical relaxed directional method
DS	Directional sampling
DSD	Directional steepest decent
DSTM	Directional stability transformation method
EDSTM	Enhance directional stability transformation method
EFSL	Enhance finite-step length
EMCS	Enhanced Monte Carlo simulation
ESTM	Enhanced stability transformation method
FORM	First-order reliability method

FORM-SQP	First-order reliability method with sequence quadratic program
FOSAM	First-order saddle point approximation method
FR	Fletcher-Reeves
FSL	Finite step length
FSS	Finite step size
GFD	Generalized F-discrepancy-based point selection approach
GGHS	Gaussian global-best harmony search
GHS	Global-best harmony search
HCLS+	HSAC without finite step size
HL-RF	Hasofer and Lind-Rackwitz and Fiessler
HM	Harmony memory
HMCR	Harmony memory considering rate
HS	Harmony search
HSAC	Hybrid self-adaptive conjugate
HSAC+	Hybrid self-adaptive conjugate without step length
IEF	Improved efficiency factor
IFR	Improved FR
iHL-RF	Improve Hasofer–Lind and Rackwitz–Fiessler
IHM	Improved harmony memory
IHS	Improved harmony search
IS	Importance sampling
Iter	Iteration
LFR	Limited FR
LS	Liu–Storey
LSF	Limit state function
MC	Merit condition
MCMC	Markov Chain Monte Carlo
MCS	Monte Carlo simulation
MFR	Modified FR
MH	Metropolis-Hastings
MICS	Multiple indicator cluster survey
MMH	Modified Metropolis-Hastings
MPP	Most probable point
MS	Memory size
NCSV	Normalized conjugate sensitivity vector
NI	Number of iterations
NNCM	Non-negative constraint method
NPR	Non-probabilistic reliability
NRBDO	Non-probabilistic reliability-based design optimization
NSD	Normalized search direction
NSDV	Normalized search direction vector
OPM	Original probabilistic model
OPM-GGHS	Original probabilistic model with Gaussian global-best harmony search
OPM-SQP	Original probabilistic model with Sequence quadratic program

PAR	Pitch adjusting rate
pCN	Preconditioned Crank–Nicolson
PDF	Probability density function
PDIE	Probability density integral equation
QMC	Quasi-Monte Carlo approach
RBDO	Reliability-based design optimization
RHL-RF	Relaxed Hasofer and Lind-Rackwitz and Fiessler
RMIL	Rivaie, Mustafa, Ismail and Leong
RN	Ranked number
RV	Numbers of random variable
SAC	Self-adaptive CSDV
SCEI	Scenario 1
SCEII	Scenario 2
SCEIII	Scenario 3
SDC	Sufficient descent condition
SDSD	Steepest descent search direction
SDV	Search direction vector
SF	Safety factor
SGNI	Sparse grid numerical integration approach
SIS	Standardized importance sampling
SORM	Second-order reliability method
SQP	Sequence quadratic program
SRA	Structural reliability analysis
SS	Subset simulation
STD	Standard deviation
STM	Stability transformation method
SVFR	Spectral-variant FR
TCFS	Three-term conjugate finite-step length
UIS	Uniform importance sampling
WUS	Simulation-based weighted uniform

Chapter 1
Basic Concepts

Abstract In this chapter, basic concepts of structural reliability analysis are briefly introduced mainly including the uncertainty, limit state function and distribution function. Uncertainties are fundamental to structural reliability analysis (SRA), which can be divided into natural uncertainties and human uncertainties. Besides, the development and prospects of SRA are also discussed including system reliability, time-variant reliability and machine learning-based reliability analysis. The basic varibles that they are followed by the various distribution function are described based on normal and well-known non-normal probability density functions.

Keywords Structural reliability analysis · Uncertainty · Limit state function · Distribution function

1.1 Introduction

In the structural design/assessment, multi-source uncertainties may be provided for elevating the performances of structural systems. These uncertainties are defined based on two main performance factors which are related to applied loads and strength capacities of structural members. In general, the uncertainty sources can be defined based on random variables that are introduced using statistical properties including mean, standard divisions and distributions. Therefore, the structural design procedure cannot be analyzed based on a deterministic analysis which is applied to elevate the performances of structures in the design phase. Therefore, in the performance level applied in designed structural members, there is a limited probability of failure which is applied for evaluating the safety of structural members under external loads. The vehicle load imposed on a bridge structure is a suitable example to show the randomness of an input design variable. Vehicle load depends on several factors such as the number of vehicles, weight of the vehicle and speed of the vehicle. Also, the weight of vehicles depends on size, shape, and carried load properties by the vehicle. On the other hand, the number of vehicles is also changing on the bridge which depends on the length of the bridge and distance between the vehicles. The speed of

S.-P. Zhu and B. Keshtegar, *Structural Reliability Analysis*, Structural Integrity 31,
https://doi.org/10.1007/978-3-031-75197-4_1

the truck also provides a dynamical load which can be defined based on an impact load, which depends on the roughness of the bridge, loading conditions of a vehicle, and rotational speed of the vehicle. Since there is not enough information about the exact simulation of a bridge load given from a truck. The number of crossing vehicles on the bridge at the same time is also not shown as well as the dimensions, size and loads. Therefore, the load on the bridge is an uncertain process. Consequently, loads of structures including live loads (e.g. human, equipment and separating walls, …), dead loads (e.g. roof, walls, the body weight of structures, …), thermal loads, weather loads (e.g. ice, flood, rain, …), vehicle loads, dynamic loads (e.g. earthquake, impact, cyclic compact, blast, …) as well as dimensional sizes and topology of structures and material properties including mechanical characteristics, thermal capacities, optical performances and physical performances can be considered based on random variables in the design process. There should be a probabilistic model for evaluating the uncertainties-based random design domain. The probabilistic performances of structures are evaluated by a limit state function which involves several random variables as loads and strengths of engineering systems. The structural reliability analysis (SRA) is a useful tool for evaluating the probabilistic performances that there is approximated in the safe design domain by using the failure probability. The process for evaluating multi-source uncertainties by reliability analysis is presented in Fig. 1.1 where the main procedure is to find multi-source uncertainties which are defined based on basic random variables applied for definitions of probabilistic performance function in safe domain estimation. Finally, structural reliability analysis can use probabilistic performance functions to define the safe region and derive the probability of failure.

In this chapter, an introduction to uncertainties and the probabilistic analysis applied in SRA has been presented, including various types of uncertainties, and basic concepts of reliability including reliability index, failure probability, and performance function (limit state functions).

1.2 Categories of Uncertainties

In most engineering systems, it is necessary to consider a probabilistic model using mathematical functions in the design procedure [1, 2]. In reality, this model consists of a complex system that the engineer has decided about reliable conditions. Accordingly, a structural component may be considered or omitted from the system during the stages of a project including design, construction and serviceability. As mentioned, it can be stated that an engineering system includes various uncertainties which can be classified into (i) natural and (ii) human resources as follows,

(i) Natural uncertainties

These uncertainties for applied loads, size and material properties are given as:

(1) Applied loads such as wind load, snow, ice, water pressure and live loads, etc.

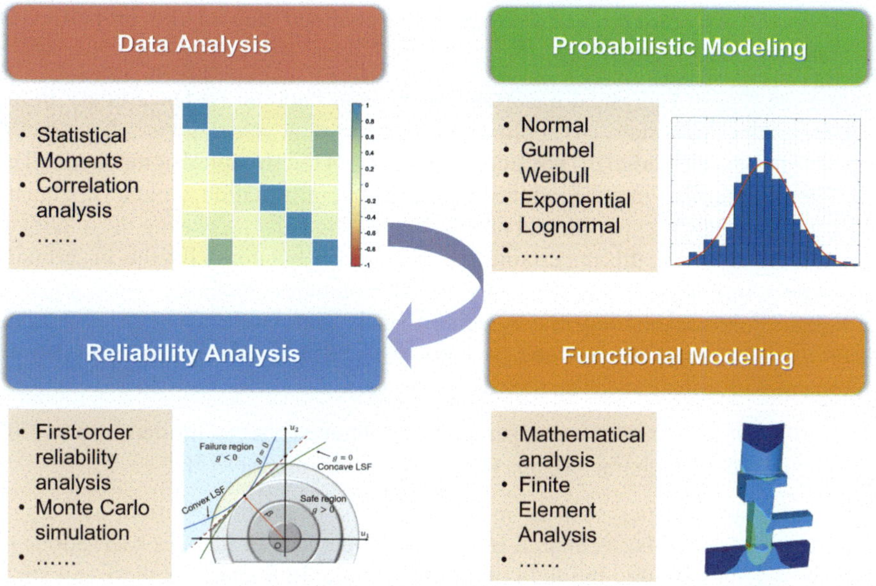

Fig. 1.1 The framework of structural reliability analysis

(2) Mechanical behavior of the materials as material properties applied for building the structures including the density of mass, Poisson's ratio, ultimate strength, yield strength, homogeneity of materials, modulus of elasticity and non-linear behavior related to the relation between strain and stress.

(3) Sizes and dimensions that it related to the shape and topology of structures and their members which are connected together for building a system. The size uncertainties include cross-section, the moment of inertia, element patterns, diameter, angle, thickness, length, width and depth.

(4) Human uncertainties

In the design process, it may raise human uncertainties due to simplifications of the future, material and boundary conditions of structure given by the user for computing its response. For example, human uncertainties can refer to estimations, analysis and modeling simulation errors in the design phase that may be provided based on a lack of knowledge or simplifications applied by the engineers. In addition, based on the knowledge about the engineering system, some characteristics of the system may be unknown thus, it can be increased the level of uncertainty in computing the system performances. Some of these uncertainties can be considered randomly with the help of probability theory.

Sometimes in engineering systems, by definition of statistical properties for variables, the uncertainty may be increased for probability analysis of a system or it is may be simplified the relationship or correlations between variables. Therefore, the uncertainties can be divided into three types of categories as another definition presented as,

(a) *Physical uncertainty:* variables that are obtained due to the properties of materials and dimensions.

Among the types of physical uncertainties, we can mention (i) Uncertainty in the yield strength of materials, (ii) applied loads such as rain, vehicle, wind, floor, walls, snow or earthquake loads, (iii) physical dimensions of the structural components. The physical uncertainties may be reduced by increasing information about a variable by collecting data-based experimental results. For example, by increasing the quality of strength for a material, this uncertainty can be reduced, but normally, this uncertainty cannot be completely eliminated. Thus, for environmental loads e.g. wind, snow or earthquake loads, evaluating the accurate load is impossible. Generally, physical uncertainty for a random variable is unknown and it can be approximated from excremental observations or technical reports.

(b) *Statistical uncertainty*: variables that are obtained due to limited information about statistical characteristics of basic random variables [3].

Statistical characteristics such as mean, moments, standard deviation, and distribution functions can be estimated by statistical analysis of actual observations of random variables. Within the field of structural reliability analysis, these statistical analyses are widely used, with the mean, standard deviation, and persistence function based on the distribution of the data being incorporated into the probabilistic analysis framework. It is the ability to calculate the statistical properties of a random variable based on its actual data points and to use these properties in probabilistic models for structural reliability assessment.

Generally, observations as accrual information for a random variable cannot be a complete statement of the probability characteristics for random variables. Since it is possible that an error occurred when recorded the data or data may be followed to a complex distribution. Therefore, the error of recording equipment for the dataset and the complex distributions of variables in statistical analysis are the main sources of statistical uncertainty.

(c) *Uncertainty in the model*: due to the simplified assumptions of real engineering systems applied in the analytical model or simplification applied for methods of system analysis. Model uncertainty is related to the simulated approach for computing results of structural systems, the definition for simulating the analytical model, damage or safe region applied in structure [4, 5], dimensions, skill and experience of engineers, environmental conditions on the project and existing buildings [6]. Human errors and internal communications between members of engineering systems can also be classified in this category.

Uncertainties in the model due to simplifying assumptions in analytical or predicted models, simplifications of simulation methods, methods for analysis are created. Finally, it is relationship between actual response and predicted results given from a model. The actual behavior of the engineering system for a performance is not accurately predicted based on models with a lack of knowledge. By increasing the obtained data or complex analysis, the uncertainty in the model can be reduced.

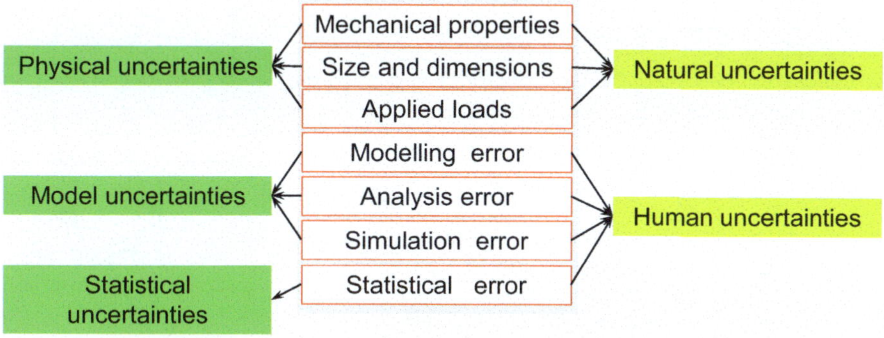

Fig. 1.2 The schematic view of the various uncertainties

As a fast review, the various uncertainties are summarized in Fig. 1.2. As seen, we have a statistical problem involving various uncertainties. Therefore, it is important to incorporate reliability analysis throughout the structural design process to clarify its safe domain. In reliability analysis, the types of uncertainty factors can be introduced to consider the safety of random variables and approximate distributions and designs.

1.3 The Role of Structural Reliability Analysis

As aforementioned, the structural reliability analysis (SRA) can be used for the reliable design of new structures and SRA can be applied for the evaluation of the safety conditions of old structures under multi-source uncertainties in load and strength. Basically, SRA led to vest roles in several directions like,

(i) Estimate the safety (failure probability) of structures accordingly. The designer can evaluate the current performance of the structure by assessing the probability of failure derived from the reliability analysis. This probability of failure can then be compared with the finite probability of failure provided by the specifications of the design. By comparing the failure probability obtained from SRA, a suitable condition of design is provided when it is computed a low failure probability, while for a high computed failure probability, the structures should be refined or retrofitted to satisfy the limit conditions of failure probability.

(ii) Reliability analysis is an important design tool in future decisions of new structures. The failure probability is giving from probabilistic models which are defined by load and resistance random variables. In general, the design-based reliability analysis is a more effective tool for evaluating the new structures and the SRA can be provided for designers with a perspective view about the safety of new structures subjected to carried loads. Consequently, it can achieve the following goals by SRA: (1) The safety factors of structural members can

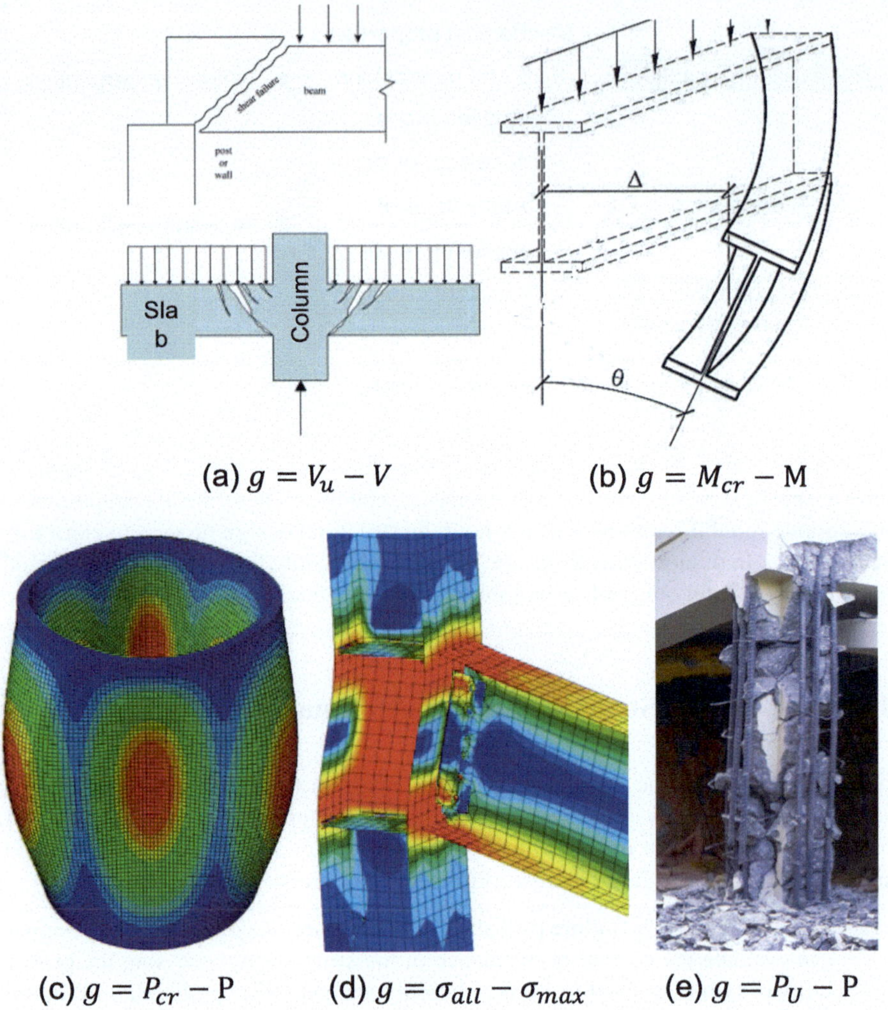

Fig. 1.3 Ultimate limit sate functions for **a** shear failure for beam (slab) concrete; **b** lateral torsional buckling of steel beam; **c** buckling of a shell; **d** beam to column connection region; **e** concrete column

be evaluated based on multi-source uncertainties. Therefore, by considering a limited failure probability in the structural analysis procedure, the confidence levels for a new design are increased thus a high-safety condition is considered for the new structures. (2) A more economical plan can be presented based on a failure probability. In general, SRA can be applied as a comparative factor for increasing the confidence level to obtain a safe condition of structure at being built for an optimum economical design. By adjusting the random design variables applied in performance conditions of a structure, it controls the safety of structure at an acceptable level and then an economical design is provided for a new engineering system.

(iii) In addition, reliability analyses based on approximating the failure probability provide good decision-making about failure and safety, capability and incapability of existing structures. SRA can be considered as a critical criterion for elevating the safety of current structures in which their strengths are reduced by consuming times form their serviceability. Consequently, it can provide an acceptable view to make future decisions including repairing, destroying a built structure, or providing a reliable design condition.

1.4 Basic Definitions in Structural Reliability Analyses

The goal of structural reliability analysis is to calculate the probability of failure by probabilistic modeling of the limit state function. This section briefly describes important definitions in structural reliability analysis.

1.4.1 Limit State Function (Performance Function)

A relationship which requires for description of performances a structural system or a part of the structure is called a limit state function (LSF). The LSF may be devised in accordance with the performance specifications of the structure, and the performance specifications may be estimated based on the available data pertaining to the structure's constraints. In general, there is a boundary condition between safety region and failure region that a structural member is performed in the probabilistic phase. The LSF is mathematically expressed in the form of below equation,

$$g(R, Q) = R - Q \tag{1.1}$$

in which, $g(R, Q)$ is limit state surface as a relation between load and resistance of the structure. R is related to a function for evaluating the structural strength (capacity) and Q is referred to a function of the effect of loads of structure. In Eq. 1.1, The resistance (R) and the load (Q) are each function of multiple random variables that all have different probability density functions which are given based on the dimensions, sizes and mechanical characteristics of materials. Types of limit state functions for structural problems can be presented as follows:

(i) Ultimate limit conditions: It often expresses a relationship between the resistance and the load of a structure. This condition is performed by almost main design condition of structure for evaluating the maximum stress under external loads. For example, damage can be caused due to the increased load thus it reduces the capacity of a structural member. The created plastic joints in a bending member, and instability of beam elements under moment load, the collapse of concrete column under compressive force, shear failure of a welded connection in steel structures or lateral torsional, local and global buckling of mechanical problems are kind of various ultimate failure conditions. Several failure modes and related LSF have presented in Fig. 1.3 for ultimate conditions of structures.

(ii) *Serviceability limit conditions*: It states the relationship between gradual deterioration, consumer comfort and maintenance costs. It can be referred to increase in deformation (strain) of elements, displacement of components, fluctuation of floor and these LSFs are described by vibration of components also local conditions such as crack width in concrete beam, development length of embedded bars of concrete structure and also crack propagation for the structural components. Figure 1.4 presents several failure models and related LSF for Serviceability conditions.

(iii) *Fatigue limit conditions:* It is stated to be a loss of resistance of the structure under cyclic loads. It is observed that the members of the structure are damaged under a small periodic load. The failure mechanism is based on the shape and growth of the crack during destruction, therefore, two factors of number of cycles and the load range, play a critical role in the failure of the structure under fatigue.

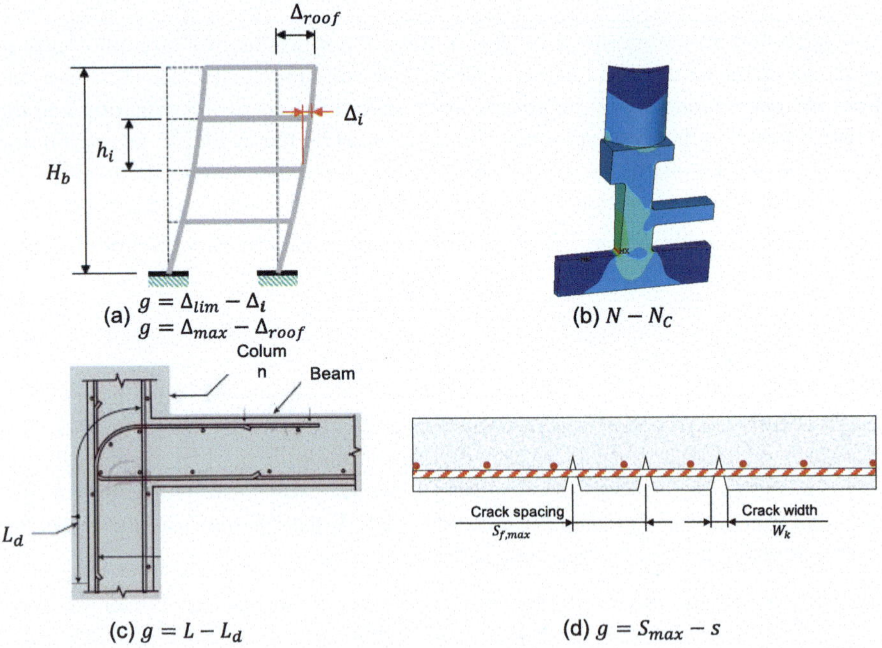

Fig. 1.4 Serviceability limit state functions for **a** drift of story; **b** frequency; **c** crack width in concrete bema; **d** development length of embedded bars

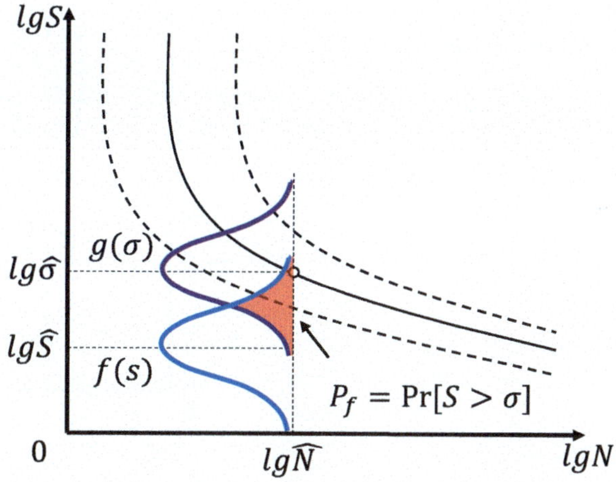

Fig. 1.5 Stress-strength interference model

The stress-strength interference theory is currently widely utilized for SRA in practical engineering due to its inherent simplicity, economic viability, and the fact that strength and stress are statistically independent. $g(\sigma)$ and $f(s)$ denote the probability distributions of strength and stress, respectively. The stress-strength interference theory, as it pertains to normal random loads and strengths, can be employed to ascertain the probability of failure, as shown in Fig. 1.5.

$$P_f = P_r[S > \sigma] = \phi\left(-\frac{\sigma_{mean} - S_{mean}}{\sqrt{\sigma_{std}^2 + S_{std}^2}}\right) \tag{1.2}$$

The properties of the normal distribution of strength and stress can be utilized to calculate $[\sigma_{mean}, S_{mean}, \sigma_{std}, S_{std}]$.

Given the inherent uncertainty and complexity of stress distribution, it is reasonable to assume that the strength and stress exhibit similar dispersions. The mean value is calculated using the following equation,

$$S_{mean} = \frac{\sigma_{mean} - \sigma_{std}}{SF} \tag{1.3}$$

The safety factor (SF) represents the focal point of the relationship between the allowable stress/design stress intensity and the test value.

The probability distribution of the fatigue life of a material is usually regarded as a lognormal distribution and its life can be validated by fatigue testing. Accordingly, the load-life interference theory may be employed to ascertain the fatigue reliability of a material. A schematic of the load-life interference model is provided in Fig. 1.6 for reference. The LSF g_1 is given by,

$$g_1 = lgN - lgN_d \tag{1.4}$$

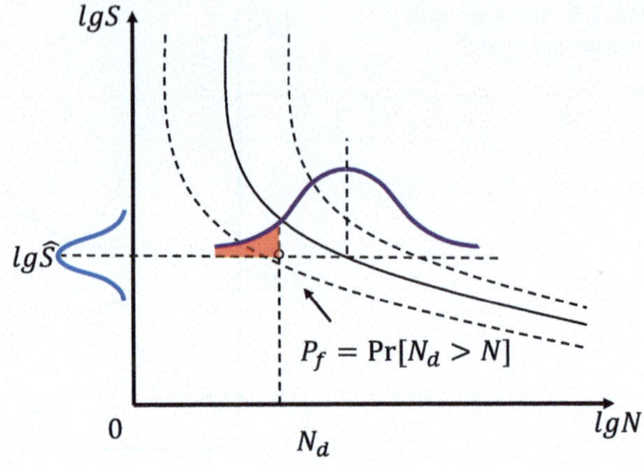

Fig. 1.6 Load-life interference model

The number of failure cycles, N, can be obtained through the fatigue test and is distributed according to a lognormal distribution. The design life of the fatigue design curve, N_d, is also a function of this distribution. The design life is calculated using the mean S–N curve and the adjustment factor,

$$\begin{cases} lgN_d = \mu_{lgN} - lg\gamma_N \\ lgN_d = \mu_{lgN} + \frac{lg\gamma_S}{\kappa} \end{cases} \tag{1.5}$$

where μ_{lgN} represents the mean number of failure cycles derived from the average S–N curve. γ denotes the life or pressure adjustment factor, while κ signifies the slope of the average S–N curve.

Equation 1.2 gives the fatigue life distribution at stress S. The red area indicates the failure probability P_f. The structure is safe when the design life N_d is less than the fatigue-tested life. The failure probability P_f can be given by the following equation,

$$P_f = P[g_1 \leq 0] = \phi\left(\frac{lgN_d - \mu_{lgN}}{\sigma_{lgN}}\right) \tag{1.6}$$

The fatigue life dispersion of the nuclear material is expressed in terms of the standard deviation σ_{lgN} at stress S, by quantifying the equation based on the P-S-N curve fitted from fatigue tests,

$$\sigma_{lgN} = -\left[\frac{lgN_{95\%} - lgN_{5\%}}{\Phi^{-}(0.95) - \Phi^{-}(0.05)}\right] \tag{1.7}$$

If the measured survival rates are 95% and 5% after a given stress level, the life of the S–N curve can be expressed as $lgN_{95\%}$ and $lgN_{5\%}$, respectively; where Φ^{-} is the inverse function of the standard normal distribution function Φ. The fatigue life of a material usually depends upon the manner in which the material changes state throughout the fatigue damage process. This change in state plays an integral role in fatigue life prediction of engineering components. From the perspective of damage accumulation, the fatigue reliability assessment method based on strength-damage interference can be expressed as follows,

$$g_2 = lgD_c - lgD_0 \tag{1.8}$$

$$P_f = P(g_2 \leq 0) = \phi\left(-\frac{\mu_{lgD_c} - \mu_{lgD_c}}{\sqrt{\sigma^2_{lgD_c} + \sigma^2_{lgD_0}}}\right) \tag{1.9}$$

where D_0 and D_c are the cumulative and critical damage, respectively, and can be determined through the application of the following equation,

$$D_0 = \frac{N_d}{N} \tag{1.10}$$

$$D_C = \frac{N_d}{N} \tag{1.11}$$

where N_d is the designed life, N denotes the tested fatigue life, and the median fatigue life, designated as $\overline{N}$, is derived from the S–N curve. The schematic representation of the strength-damage interference model, as previously described, is presented in Fig. 1.7. The test values D_C and D_0 are random variables that follow a lognormal distribution. In the event that the critical damage, D_C, is less than the cumulative damage D_0, the structure is deemed unsafe. In accordance with the principles set forth in Eq. 1.5 and the average S–N curve, the logarithmic fatigue life distribution is as follows. By using Eqs. 1.4, 1.10 and 1.11 and analyzing the experimental data, the distributions of D_C and D_0 used in the analysis can be given as [7],

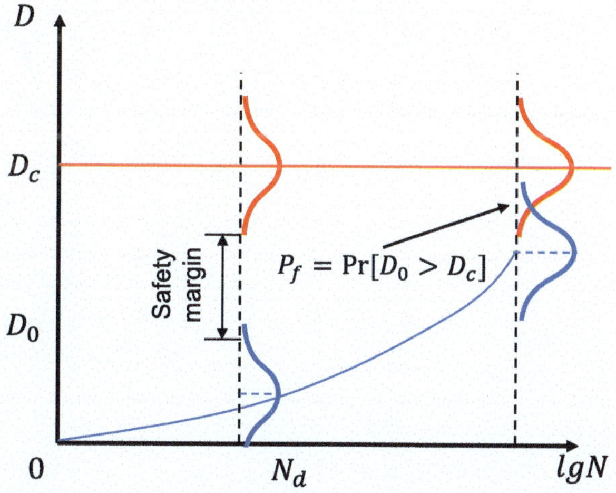

Fig. 1.7 Damage-strength interference model

$$\begin{cases} lgD_c \sim Normal(0, lgN) \\ lgD_0 \sim Normal(-lg\gamma_N, \sigma_{lgN}) \\ lgD_0 \sim Normal\left(\frac{lg\gamma_S}{\kappa}, \sigma_{lgN}\right) \end{cases} \tag{1.12}$$

1.4.2 *Failure and Safety Regions*

The definition of the LSF allows for the consideration of three aspects, each representing a specific state of the structure in terms of failure or safety. These three states are defined as follows:

(i) $R\text{-}Q = 0$ i.e. $g = 0$, the boundary between failure and safety of the structure that it represents to the LSF as a relation between load and resistance of structure;
(ii) $R\text{-}Q < 0$ i.e. $g < 0$, this region denotes the failure domain of structure;
(iii) $R\text{-}Q > 0$ i.e. g > 0, this region denotes the safety domain of structure.

In simple terms, it can be stated that the structure failed when the load exceeded its resistance. Three regions LSF, failure and safe domains are schematically plotted in Fig. 1.8. As seen, we can define two LSFs as convex and concave LSFs. In convex LSF, the main curvatures are negative while the main curvatures are positive in convex LSF.

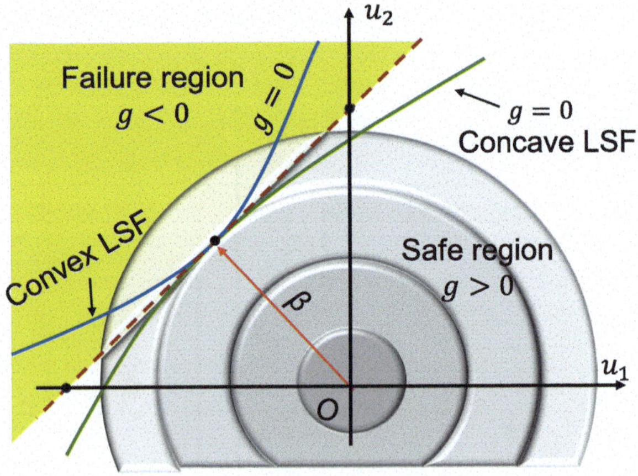

Fig. 1.8 Schematic view of the limit state function and failure (safe) domain

1.4.3 *Probability of Failure*

The failure probability of a structure can be expressed as the overlapping region of two probability distribution functions for the resistance (S) and load (Q) random variables, as shown by the red region in Fig. 1.9. Note that, the load distribution function has a mode and mean values less than the resistance distribution function. At the same time, the failure probability is followed to 1 when load and resistance distribution functions have the same parameters in shape and location. By using LSF and applying independent random variables of R and Q in interval $q < \infty$ and $r \geq 0$, the overlap region for these two probability distribution functions is computed as below,

$$P_f[Q > R] = \iint\limits_{q-r>0} f_{Q,R}(q, r)dqdr \tag{1.13}$$

where P_f represents the failure probability of the structure; $f_{Q,R}(q, r)$ is the joint probability density function for variables R and Q.

The primary objective of structural reliability analysis is to ascertain the probability of failure, denoted as P_f, through the integration over the failure region defined by the LSF $g(\mathbf{X})$, where $\mathbf{X} = (x_1, x_2, \ldots, x_n)^T$ [8, 9],

$$P_f = \int\limits_{g(X)\leq 0} \cdots \int f_X(x_1, \ldots, x_n)dx_1 \ldots dx_n \tag{1.14}$$

In this instance, f_X denotes the joint probability density function of the random variables under consideration.

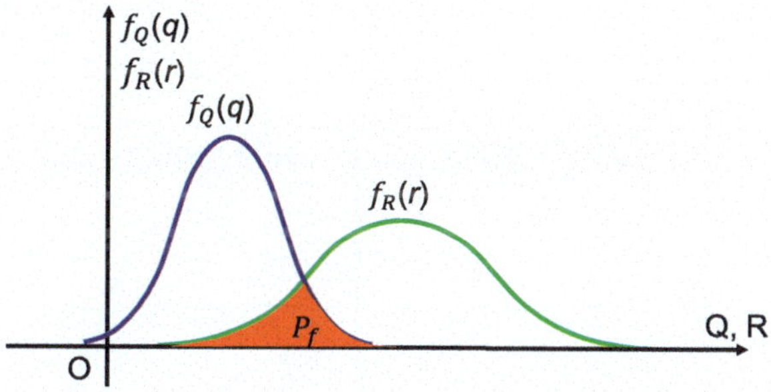

Fig. 1.9 Failure probability based on two basic distribution functions of load and resistance

1.4.4 Reliability Index

According to the LSF, it is possible to define the margin of safety between the resistance and the loading of the structure at $g = 0$. In this case, the failure probability can be expressed as [10],

$$P_f = P[g < 0] \tag{1.15}$$

in which $g = 0$ represents the boundary between the regions of failure and safety. As illustrated in Fig. 1.10, the probability of failure for a given structure can be determined by calculating the area within the failure region where
$g < 0$.

The mean and standard deviation of the PDF $f_G(g)$ can be derived indirectly from the random variables R and Q i.e. load and resistance as follows [11],

$$\mu_g = \mu_R - \mu_Q \tag{1.16}$$

$$\sigma_g = \sqrt{\sigma_R^2 + \sigma_Q^2 - 2\rho_{RQ}\sigma_R\sigma_Q} \tag{1.17}$$

where μ_R and μ_Q are the means of the resistance and load variables, respectively, ρ_{RQ} is the correlation coefficient with the random variable, and σ_R and σ_Q are the standard deviations of the resistance and load, respectively. In order to ascertain the reliability index β, it is necessary to calculate the mean, standard deviation, and correlation coefficient associated with the random variables in question, as outlined in the relevant methodology, according to,

$$\beta = \frac{\mu_g}{\sigma_g} = \frac{\mu_R - \mu_Q}{\sqrt{\sigma_R^2 + \sigma_Q^2 - 2\rho_{RQ}\sigma_R\sigma_Q}} \tag{1.18}$$

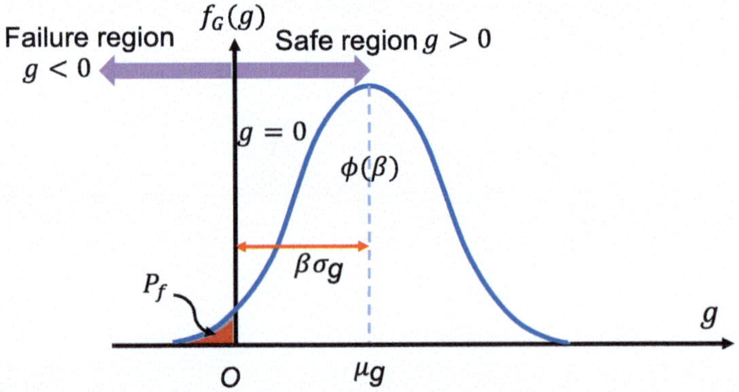

Fig. 1.10 Reliability index and failure probability distribution

According to Eq. 1.18, the reliability index β is correlated with the mean and standard deviation of resistance and load.

For a specific state of resistance and load with a normal and independent probability distribution function, the LSF distribution can be also considered normal, thus $g = 0$ for Eq. 1.1, is a linear function. In this instance, the probability density function of the LSF is given as,

$$f_G(g) = \frac{1}{\sigma_g\sqrt{2\pi}}\exp\left[-\frac{1}{2}\left(\frac{g-\mu_g}{\sigma_g}\right)^2\right] \tag{1.19}$$

The probability of failure can be calculated by [12],

$$\begin{aligned} \mathrm{P}_f &= \int_{-\infty}^{0} f_G(g)dg = \int_{-\infty}^{0} \frac{1}{\sigma_g\sqrt{2\pi}}\exp\left[-\frac{1}{2}\left(\frac{g-\mu_g}{\sigma_g}\right)^2\right]dg \\ &= \int_{-\infty}^{-\beta} \frac{1}{\sqrt{2\pi}}\exp\left[-\frac{1}{2}\right] = \Phi(-\beta) \end{aligned} \tag{1.20}$$

In this case, the reliability index β is represented by the variable R, while the standard normal cumulative distribution function is represented by the variable Φ. According to Eq. 1.14, the failure probability of the LSF $g(X)$ defined by the random variables X can be calculated by Eq. 1.21 that is rewritten as below [13],

$$\mathrm{P}_f = \Phi[\mathrm{g(X)} \leq 0] = \int \ldots \int_{g(X)\leq 0} f_X(x_1, x_2, \ldots, x_n)dx_1dx_2\ldots dx_n \approx \Phi(-\beta) \tag{1.21}$$

For the general case of engineering problems with nonlinear LSFs or implicit performance functions [14], the integrals in Eq. 1.21 cannot be solved in closed form because these problems involve many underlying random variables described by non-normal distribution functions. Different approximation methods can be used to compute the aforementioned multidimensional integrals with complex failure regions, and several of these methods for structural reliability analysis are briefly described below.

1.4.5 Distribution Function

The design, construction, and operation of engineering structures are inherently uncertain processes, given the inherent variability of mechanical properties, geometric dimensions, and external loads [8]. These uncertainty parameters, which can be denoted by random variable $\boldsymbol{X} = (x_1, x_2, \ldots, x_n)$. Uncertainty quantification is often used in practical engineering structural problems to determine the distribution function. Six common distribution functions are described as below:

(1) *Uniform Distribution*

The uniform distribution, also referred to as the rectangular distribution, represents a symmetric probability distribution exhibiting the equal probability of occurrence across intervals of equal length. A uniform distribution is characterized by two parameters, a and b, which correspond to the minimum and maximum values, respectively, on the numerical axis, which can be denoted as $X \sim U(a, b)$. In accordance with the definition of a uniform distribution, it is established that the probability density function (PDF) can be expressed as follows,

$$f(x) = \begin{cases} \frac{1}{b-a} & a < x < b \\ 0 & \text{else} \end{cases} \tag{1.22}$$

And the schematic diagrams of PDF of the uniform distributed random variables are plotted in Fig. 1.11. The cumulative distribution function (CDF) is,

$$F(x) = \begin{cases} 0, & x < a \\ \frac{x-a}{b-a}, & a < x < b \\ 1, & x > b \end{cases} \tag{1.23}$$

To calculate the mean μ and standard deviation σ, the following equation can be used,

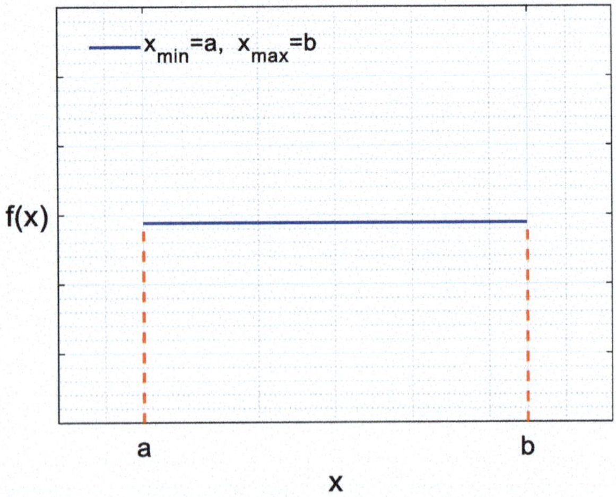

Fig. 1.11 The PDF of the uniform distributed random variables

$$\mu = \int_{-\infty}^{\infty} xf(x)dx = \int_{a}^{b} \frac{x}{b-a}dx = \frac{a+b}{2} \tag{1.24}$$

$$\sigma = \sqrt{\frac{(b-a)^2}{12}} \tag{1.25}$$

The law of large numbers posits that when a random variable's sample size becomes sufficiently large, its expectation value approaches that of its mean. Consequently, in practical applications, the mathematical expectation value is often taken as an approximation of the mean. The same applies to the later description.

(2) *Normal Distribution*

The normal distribution is also referred to as the Gaussian distribution and is a particularly prevalent continuous probability distribution. The normal distribution is a fundamental concept in statistics. It is frequently assumed to be applicable in situations where the underlying distribution is unclear. When the random variable X follows a normal distribution with the mean of μ and variance of σ^2, it can be denoted as $X \sim N(\mu, \sigma^2)$. Its PDF is,

$$f(x) = \frac{1}{\sigma\sqrt{2\pi}}e^{-\frac{(x-\mu)^2}{2\sigma^2}} \tag{1.26}$$

The CDF of the normal variable is,

$$F(x)\frac{1}{\sigma\sqrt{2\pi}}\int_{-\infty}^{x} \exp\left(-\frac{(t-\mu)^2}{2\sigma^2}\right)dt \tag{1.27}$$

A random variable is said to follow a standard normal distribution when it has mean $\mu = 0$ and standard deviation $\sigma = 1$. The PDF and CDF of the standard normal distribution are respectively given by,

$$f(x) = \frac{1}{\sqrt{2\pi}}e^{-\frac{x^2}{2}} \tag{1.28}$$

$$F(x) = \frac{1}{\sqrt{2\pi}}\int_{-\infty}^{x} e^{-\frac{t^2}{2}}dt \tag{1.29}$$

The transformation of a normally distributed variable, x, into a standard normal variable, u, can be accomplished by

$$u = \frac{x-\mu}{\sigma} \tag{1.30}$$

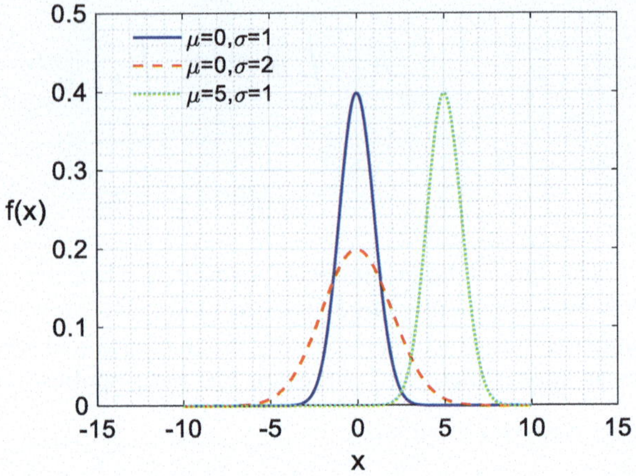

Fig. 1.12 The PDF of the normal random variables

The PDF of the normal random variables can be seen schematically in Fig. 1.12.

(3) *Log-Normal Distribution*

The definition of the lognormal distribution is that when the random variable $\exp(X)$ obeys the normal distribution, then X obeys the lognormal distribution. Similarly, if X is a random variable that obeys a normal distribution, then the $\ln X$ is also normally distributed. In many applications, especially in structural reliability, random variables may not obey the normal distribution. However, the logarithm of the random variable may follow a normal distribution, a situation known as the Log-normal distribution. If a random variable X obeys the log-normal distribution with mean μ and variance σ^2, it can be denoted as $\ln X \sim N\left(\mu, \sigma^2\right)$, and its PDF is,

$$f(x) = \begin{cases} \frac{1}{x\sqrt{2\pi}\sigma} exp\left[-\frac{1}{2\sigma^2}(\ln x - \mu)^2\right] & x > 0 \\ 0 & x \leq 0 \end{cases} \tag{1.31}$$

By given expectation E(X) with variance var(X), the mean μ and standard deviation σ are given by,

$$\mu = \ln(E(X)) - \frac{1}{2}\ln\left(1 + \frac{var(X)}{E(X)^2}\right) \tag{1.32}$$

$$\sigma = \sqrt{\ln\left(1 + \frac{var(X)}{E(X)^2}\right)} \tag{1.33}$$

The PDF of the lognormal random variables is shown in Fig. 1.13.

(4) *Weibull Distribution*

The Weibull distribution plays a critical role in reliability analysis, life testing, and other areas of engineering and science. A continuous probability distribution, the

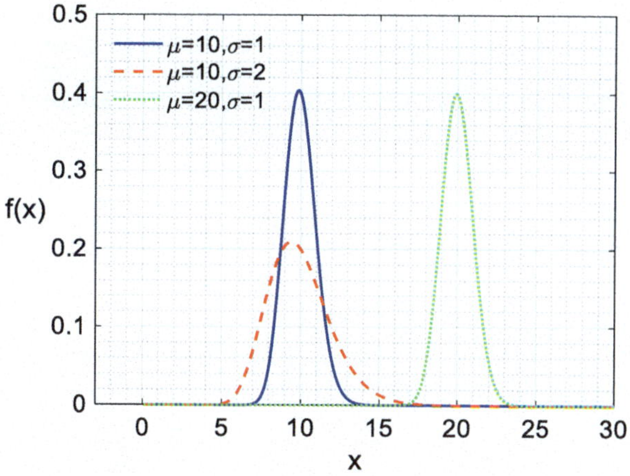

Fig. 1.13 The PDF of the lognormal random variables

Weibull distribution describes the behavior of non-negative random variables. It is particularly useful for modeling systems in which the failure of a single component may lead to the failure of the entire system, a phenomenon known as the weakest link model. It also finds application in the analysis of time-to-failure data. Its PDF is, and the PDF of Weibull distributed random variables are plotted in Fig. 1.14, where $\lambda > 0$ is the scale parameter and $k > 0$ is the shape parameter. Since the CDF is an integral of the PDF, it is clear that it is an extended exponential distribution function. Furthermore, the Weibull distribution is associated with the existence of special cases such as the exponential and Rayleigh distributions. For instance, it is exponentially distributed for $k = 1$ and it is a Rayleigh distribution for $k = 2$. The mean μ and standard deviation σ are respectively given by,

$$f(x; \lambda, k) = \begin{cases} \frac{k}{\lambda}\left(\frac{x}{\lambda}\right)^{k-1} e^{-(x/\lambda)^k} & x \geq 0 \\ 0 & x < 0 \end{cases} \tag{1.34}$$

$$\mu = \lambda\Gamma\left(1 + \frac{1}{k}\right) \tag{1.35}$$

$$\sigma = \sqrt{\lambda^2\left[\Gamma\left(1 + \frac{2}{k}\right) - \Gamma\left(1 + \frac{1}{k}\right)^2\right]} \tag{1.36}$$

where Γ denotes the gamma function.

(5) *Gumbel Distribution*

The Gumbel distribution is employed for the modeling of the distribution of maximum (or minimum) values for multiple samples drawn from a variety of distributions. The distributions in question are derived from the Extreme Value Theorem,

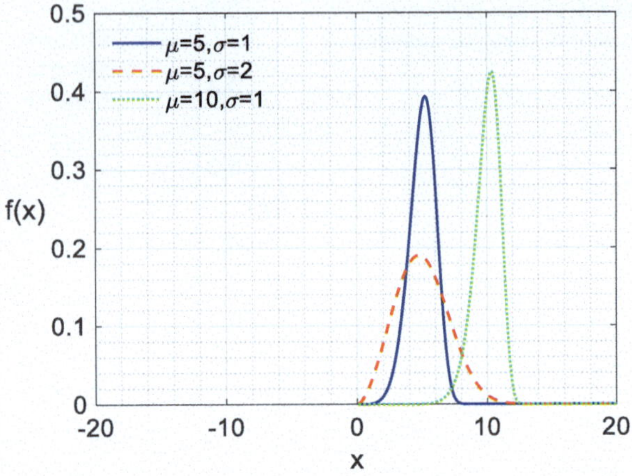

Fig. 1.14 The PDF of the Weibull distributed random variables

the tenets of which dictate that as the number of samples approaches infinity, the maximum or minimum values of said samples will tend toward a particular form of limiting distribution. One of such distributions is the Gumbel distribution. Furthermore, the Gumbel distribution represents a particular instance of what is known as the generalized extreme value distribution, also termed the Fisher-Tippet distribution. The PDF and CDF of the distribution are respectively expressed by,

$$f(x) = \frac{1}{\beta}\exp\left\{-\frac{x-\mu}{\beta} - \exp\left(-\frac{x-\mu}{\beta}\right)\right\} \tag{1.37}$$

$$F(x) = \exp\left\{-\exp\left(\frac{x-\mu}{\beta}\right)\right\} \tag{1.38}$$

where $-\infty < \mu < \infty$ is the location parameter and $\beta > 0$ is the scale parameter. When $\mu = 0$ and $\beta = 1$, it is called the standard Gumbel distribution. The PDF of the random variables following the Gamble distribution is illustrated schematically in Fig. 1.15.

(6) *Exponential Distribution*

As a fundamental component of continuous probability distributions, the exponential distribution is a pivotal concept in statistical analysis. It is employed extensively to characterize intervals between occurrences that manifest at a consistent rate, providing invaluable insight into the dynamics of such phenomena. The probability that an event will occur at a future moment is related only to the length of time from that moment, and not to how long it has been waiting before. The exponential distribution has a rate parameter λ, where $\lambda > 0$. The exponentially distributed random variable X with parameter λ can be denoted as $X \sim Exp(\lambda)$. Its PDF is,

$$f(x) = \begin{cases} \lambda e^{-\lambda x} & x \geq 0 \\ 0 & x < 0 \end{cases} \tag{1.39}$$

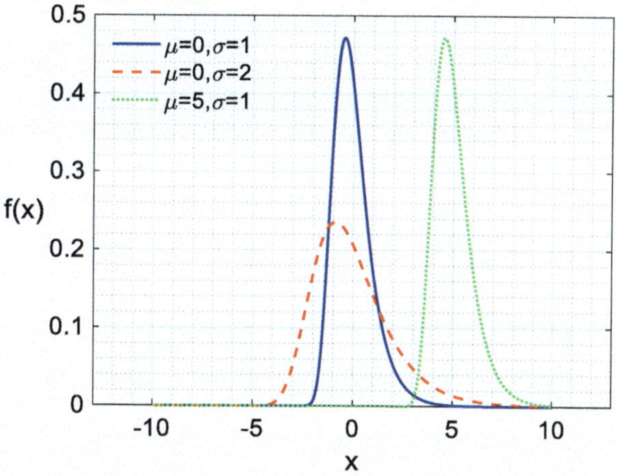

Fig. 1.15 The PDF of the Gumbel distributed random variables

and its CDF is,

$$F(x) = \begin{cases} 1 - e^{-\lambda x} & x \geq 0 \\ 0 & x < 0 \end{cases} \tag{1.40}$$

where λ represents the parameters of exponential distribution that it computed as below,

$$\lambda = 1/\mu \tag{1.41}$$

The PDF of random variables following the exponential distribution is shown schematically in Fig. 1.16.

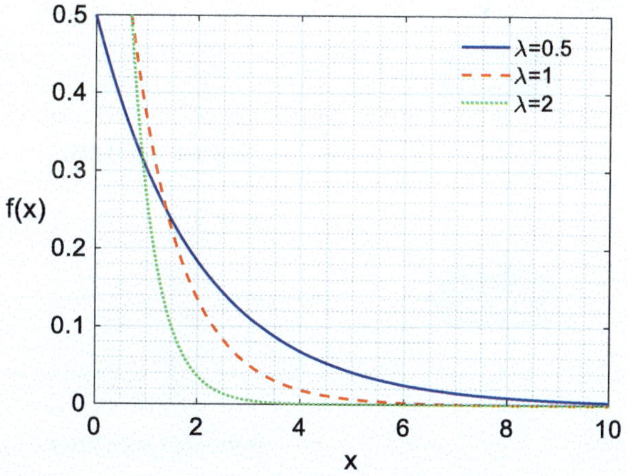

Fig. 1.16 The PDF of the exponential distributed random variables

1.5 Development and Prospects

After a long period of development, and with the increasing complexity of engineering structures, the traditional SRA is no longer meet the increasingly complex reliability problems. Based on this, some typical topics should be briefly mentioned, such as system reliability analysis based on complex structural systems, time-variant reliability analysis considering time-variant effects, and reliability analysis based on artificial intelligence.

(i) *System reliability*

In simple terms, system reliability refers to the reliability analysis of multiple limit state structures. Typical structural systems can be divided into series systems, parallel systems and hybrid systems, as shown in Fig. 1.17. The calculation of the failure probability of a structural system is usually more complicated than the failure probability of a structural element. The analysis of the reliability of a structural system mainly involves the identification of the main failure modes and the estimation of the system failure probability. For structural system reliability analysis, similar to the structural element reliability analysis method, there are also three types of methods: approximate analytical method, simulation method and surrogate model method. The results given by the approximate analytical method are usually in the form of point estimation and interval estimation of system reliability, while the simulation method and surrogate model method usually provide point estimates of reliability.

(ii) *Time-variant reliability*

Conventional reliability analysis and reliability optimization design adopt static model, but in actual engineering, there are a lot of parameters change with time, such as dynamic load, structural size change, material strength decline and so on. Under the influence of these time-variant factors, the performance of the structure

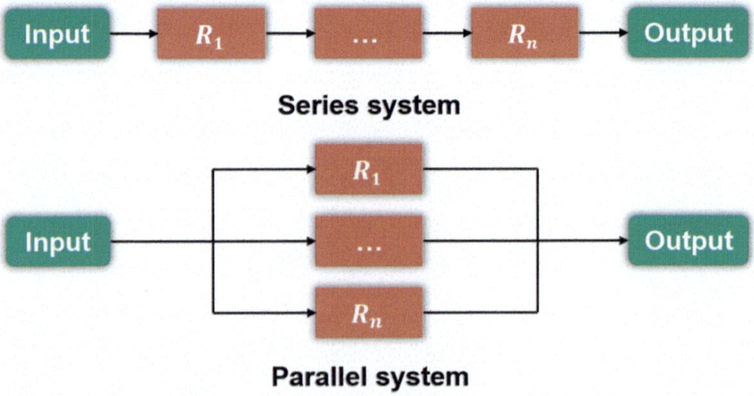

Fig. 1.17 Series and parallel systems

or system will fluctuate with time during long-term use, so its reliability will also show dynamic changes, rather than determining a constant value, that is, there is time-variant reliability. Time-variant reliability is defined as the ability of a system or product to perform a specified function within a given life cycle under the influence of time-variant uncertainty. If the time-variant factors can be fully considered in the design stage to meet the requirements of time-variant reliability, the whole life cycle design of products can be realized, which is of great significance in modern engineering design. Therefore, based on the existing reliability theory, a series of theories and researches on time-variant reliability analysis and time-variant reliability optimization design have appeared. The time-variant reliability theory introduces time into the functional function, and uses stochastic process to describe the time-variant uncertainty existing in practical engineering problems, so as to effectively describe engineering problems mathematically, evaluate the time-variant reliability of structures, and consider the requirements of time-variant reliability in the design.

(iii) *Machine learning*

With the complexity of SRA problems, such as low failure probability, high computational cost, high dimensionality and high nonlinearity, people have gradually realized that traditional SRA methods are increasingly difficult to meet the actual engineering needs [15]. On the other hand, with the rapid development of machine learning and artificial intelligence, the engineering application of machine learning algorithms has injected new vitality into SRA. The combination of machine learning and SRA makes SRA can solve more complex problems. Compared with traditional methods such as response surface models, the main advantage of machine learning algorithms is that they can establish more accurate performance functions. Meanwhile, specific machine learning algorithms can be used to solve different SRA problems.

In the research of SRA methods, commonly used machine learning algorithms include Gaussian process regression, artificial neural network and support vector regression. As a representative algorithm of machine learning and deep learning, neural network is widely used as a modeling method for proxy models because of its high fitting accuracy. However, the performance of the neural network algorithm will be affected by the initial random parameters, and the robustness and repeatability of the algorithm are low. Because of its excellent generalization ability and robustness, support vector regression is also widely used in the field of structural reliability, and its performance is excellent in high-dimensional problems. However, the algorithm depends on the coverage accuracy of the training sample to the limit state function, which often leads to large errors. It's worth mentioning that hybrid methods for SRA based on active learning developed in a blowout recent year. These hybrid methods are still in a period of rapid development in the past five years.

What's more, since the emergence of active learning SRA method based on machine learning algorithm, simulation method and machine learning algorithm have been well combined, and the surrogate model replaces the system response, which can significantly reduce the computing cost. Now, academia has come up with a general framework, that is, machine learning method (such as Gaussian process regression, support vector regression, artificial neural network, etc.), simulation method (such as Monte Carlo simulation, subset simulation, importance sampling, line sampling, direction sampling), learning function (U function, expectation raising function, etc.), stop criteria (such as reliability index, failure probability, etc.). In this way, the response of the system is actively learned by the machine learning algorithm, and the complex system response solving process is replaced by an efficient machine learning surrogate model, so as to realize the SRA. Finally, the application of machine learning methods based on physical information in structural reliability analysis deserves attention.

1.6 Summary

In this chapter, the basic concepts of SRA are introduced. In general, several main defeminations in the SRA and the role of SRA are also briefly discussed. Uncertainty is fundamental to SRA, which can be divided into natural uncertainties and human uncertainties. Based on this, the SRA framework can be developed including the LSF, the concept of failure and safety, failure probability, reliability index as well as the distribution function. Besides, the development and prospects of SRA are discussed including system reliability, time-variant reliability and machine learning.

References

1. J.C. He, S.P. Zhu, C. Luo, W. Li, Q. Liu, Y. He, Q. Wang, Probabilistic notch fatigue assessment under size effect using micromechanics-based critical distance theory. Int. J. Fatigue **183**, 108280 (2024)
2. X.P. Niu, R.Z. Wang, D. Liao, S.P. Zhu, X.C. Zhang, B. Keshtegar, Probabilistic modeling of uncertainties in fatigue reliability analysis of turbine bladed disks. Int. J. Fatigue **142**, 05912 (2021)
3. B. Keshtegar, M. Bagheri, D. Meng, R. Kolahchi, N.T. Trung, Fuzzy reliability analysis of nanocomposite ZnO beams using hybrid analytical-intelligent method. Eng. Comput. **37**, 2575–2590 (2021)
4. X. Deng, S.P. Zhu, L. Wang, C. Luo, S. Fu, Q. Wang. Probabilistic framework for strain-based fatigue life prediction and uncertainty quantification using interpretable machine learning. Int. J. Fatigue **190**, 108647 (2025)
5. D. Meng, S. Yang, A.M.P. De Jesus, T. Fazeres-Ferradosa, S.P. Zhu. A novel hybrid adaptive Kriging and water cycle algorithm for reliability-based design and optimization strategy: application in offshore wind turbine monopile. Comput. Methods Appl. Mech. Eng. **412**, 116083 (2023)
4. M. Bagheri, S.P. Zhu, M.E.A. Ben Seghier, B. Keshtegar, N.T. Trung, Hybrid intelligent method for fuzzy reliability analysis of corroded X100 steel pipelines. Eng. Comput. **37**, 2559–2573 (2021)
5. X.P. Niu, S.P. Zhu, J.C. He, Y. Ai, K. Shi, L. Zhang, Fatigue reliability design and assessment of reactor pressure vessel structures: concepts and validation. Int. J. Fatigue **153**, 106524 (2021)
8. B. Keshtegar, Chaotic conjugate stability transformation method for structural reliability analysis. Comput. Methods Appl. Mech. Eng. **310**, 866–885 (2016)

9. B. Keshtegar, P. Hao, A hybrid self-adjusted mean value method for reliability-based design optimization using sufficient descent condition. Appl. Math. Model. **41**, 257–270 (2017)
10. B. Keshtegar, M.E.A.B. Seghier, E. Zio, J.A. Correia, S.P. Zhu, N.T. Trung, Novel efficient method for structural reliability analysis using hybrid nonlinear conjugate map-based support vector regression. Comput. Methods Appl. Mech. Eng. **381**, 113818 (2021)
11. S.K. Choi, R.A. Canfield, R.V. Grandhi, Reliability-based structural optimization. Springer (2007)
12. B. Keshtegar, P. Hao, A hybrid loop approach using the sufficient descent condition for accurate, robust, and efficient reliability-based design optimization. J. Mech. Des. **138**(12), 121401 (2016)
13. B. Keshtegar, A modified mean value of performance measure approach for reliability-based design optimization. Arab. J. Sci. Eng. **42**(3), 1093–1101 (2017)
14. S.P. Zhu, X. Niu, B. Keshtegar, C. Luo, M. Bagheri. Machine learning-based probabilistic fatigue assessment of turbine bladed disks under multisource uncertainties. Int. J. Struct. Integr. **14**(6), 1000–1024 (2023)
15. C. Luo, S.P. Zhu, Y. Lv, H. Ma, X. Liu, B. Keshtegar. Uniform importance sampling with rejection control for structural reliability analysis. Comput. Methods Appl. Mech. Eng. **436**, 117707 (2025)

Chapter 2
Brief Introduction on Reliability Methods

Abstract In this chapter, 14 typical structural reliability methods are briefly introduced that they are divided into two main catagopries as (i) 6 analytical methods and (ii) 8 simulation or sampling methods. These methods are named as the first-order reliability method, second-order reliability method, moment method, first-order saddle point approximation method, non-probabilistic reliability model, dimension reduction method, Monte Carlo simulation, importance sampling, uniform importance sampling, enhanced Monte Carlo simulation, line sampling, subset simulation, directional sampling as well as direct probabilistic integral method.

Keywords Structural reliability method · Simulation methods · Analytical methods · First-order reliability method · Sampling methods

2.1 Introduction

Difficulty in computing the multi-dimensional integral applied for determining the failure probability has led to developing the reliability analysis methods such as; (i) *simulation methods* including Monte Carlo Simulation (MCS), importance sampling (IS), enhanced Monte Carlo simulation (EMCS), uniform importance sampling (UIS) method and (ii) *analytical approaches* including first-order reliability method (FORM), moment methods, second-order reliability method (SORM). As seen in Fig. 2.1, this chapter, brief descriptions are provided for analytical and simulation reliability methods.

S.-P. Zhu and B. Keshtegar, *Structural Reliability Analysis*, Structural Integrity 31,
https://doi.org/10.1007/978-3-031-75197-4_2

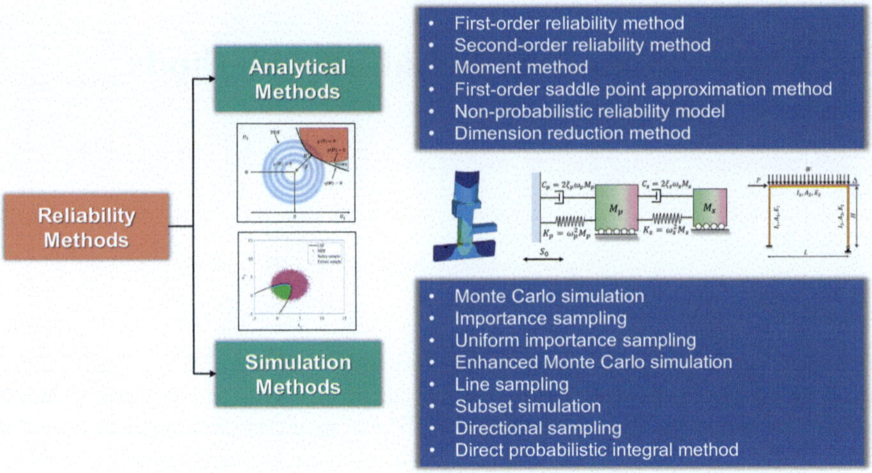

Fig. 2.1 The framework of this chapter

2.2 Analytical Methods

2.2.1 First-Order Reliability Method

In 1974, Hasofer and Lind (HL) developed an efficient method for determining the reliability index, which involved identifying a point on the hyperplane with the highest probability of failure and naming it as the most probable point (MPP). Accordingly, the fundamental process of the approximate reliability method, designated as FORM, is devised to identify the point on the limit state surface that exhibits the minimum Euclidean distance from the origin (i.e., the MPP). Subsequently, the reliability index of the structure is accurately calculated, thereby providing a quantitative foundation for evaluating the safety performance of a structure in the context of uncertainty. Thus, by computing MPP, we have $\beta = \|U^*\|)$ and $P_f \approx \Phi(-\beta)$. The FORM approximates the LSF $g(X)$ with a linear function at the MPP, the plane surface at the MPP divides the design domain into regions of failure $g(\mathrm{X}) < 0$ and safe $g(\mathrm{X}) > 0$ and the basic random variables in LSF i.e. $X = (x_1, x_2, \ldots, x_n)^T$ are described with continuous probability distribution functions. In order to determine the MPP, the performance function needs to be converted into the standard space. Two-dimensional independent random variables of the LSF are plotted in Fig. 2.2 in which, the failure domain, the safe domain and the limit state surface in ordinal ***X***-space and normal standard ***U***-space are shown. As illustrated in this figure, the reliability index is situated in the vicinity of the origin within the standard normal ***U***-space. The integration for computing the failure probability is performed by approximating the procedure with a linear hyper-surface in normal standard space. Thus, it may provide some errors for highly nonlinear limit state surfaces that they have

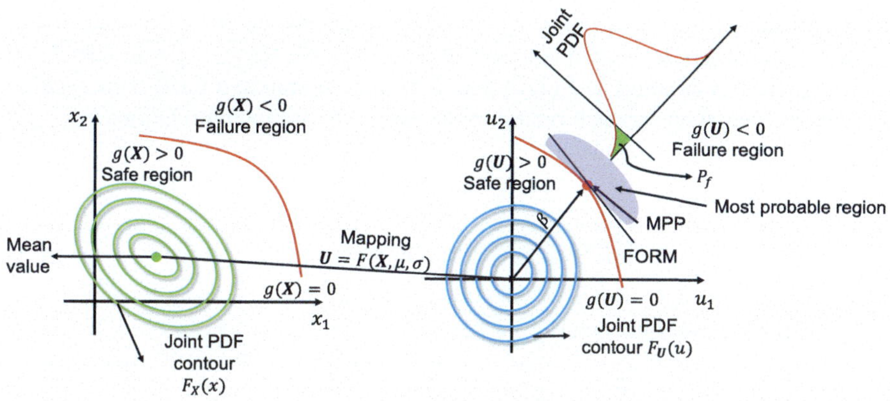

Fig. 2.2 Schematic plot of the failure, safe region and limit state surface in ***X***-space and ***U***-space

high curvature at MPP. However, calculating the integration of failure probability is challenging for engineering problems due to the following reasons:

(i) Applied several basic random variables with non-normal distribution functions, thus we have a multidimensional probability integration that the dimensionality of failure probability integration may be typically high for applicable engineering problems;
(ii) The joint probability density function used in integrating failure probability is typically not normal and is a nonlinear function in ***X***-space;
(iii) In engineering problems, the integration boundary defined by the function LSF $g(X) = 0$ is a complex, multi-dimensional, and nonlinear function. Nevertheless, the LSF for authentic engineering scenarios is a sophisticated function that necessitates computation through diverse methodologies, including finite element analysis, dynamic simulation, and computational fluid dynamics. Consequently, the computation of the multi-dimensional limit state surface is a time-consuming process.

Therefore, the numerical integration to find the solution of the failure probability integration with high-dimension property for engineering applications may be impossible and it is very difficult. Consequently, the FORM approximation method can be applied as a fast and efficient algorithm for the failure probability estimation of a structure. Advantages of the FORM are i) probability integration is computed very easily by simplifying the joint probability function in normal standard space and ii) approximate the integration boundary with a linear hyper-surface in normal standard space. Thus, it is a simple approximated method for failure probability (Fig. 2.2).

The typical FORM process comprises three stages for estimating the probability of failure which are described as below,

(1) The Rosenblatt transformation transforms the initial random variables in the ***X***-space (denoted by ***X***, representing a random variable vector) into ***U***-space

(represented by $\boldsymbol{U}$, denoting a standard normal vector with zero mean, unit variance, and independent components). i.e. $u = \Phi^{-1}[F_X(x^*)]$ where the CDF of $\boldsymbol{X}$ at x^* is denoted as $F_X(x^*)$ and Φ is normal standard CDF. This relation can be computed for each random variable by the following relations [1],

The random variable can be defined at the MPP x^* in the $\boldsymbol{U}$-space using the first-order Taylor series expansion and Rosenblatt transformation as

$$u = \Phi^{-1}[F_X(x*)] + \frac{\partial}{\partial x}\left(\Phi^{-1}[F_X(x*)]\right)(x - x*) \tag{2.1}$$

It can be concluded that,

$$\frac{\partial}{\partial x}\left(\Phi^{-1}[F_X(x*)]\right) = \frac{f_X(x*)}{\varphi\left(\Phi^{-1}[F_X(x*)]\right)} \tag{2.2}$$

By substituting Eq. 2.1 in Eq. 2.2, and subsequently reorganizing the terms according to the appropriate algebraic principles, the following outcome is obtained,

$$u = \frac{x - x * + \left[\Phi^{-1}[F_X(x*)]\right]\varphi\left(\Phi^{-1}[F_X(x*)]\right)/f_X(x*)}{\varphi\left(\Phi^{-1}[F_X(x*)]\right)/f_X(x*)} \tag{2.3}$$

Equations 2.4 and 2.5 demonstrate how to calculate the equivalent mean μ_x^e and standard deviation σ_x^e for non-normal random variables at point x,

$$\sigma_x^e = \frac{1}{f_X(x)}\varphi\left[\Phi^{-1}\{F_X(x)\}\right] \tag{2.4}$$

$$\mu_x^e = x - \sigma_x^e \Phi^{-1}[F_X(x)] \tag{2.5}$$

where $f_X(x)$ and $F_X(x)$ refer to the PDF and CDF of the random variable x, respectively. The function Φ^{-1} is the inverse of the standard normal CDF, while ϕ is the PDF of the standard normal distribution.

(2) Utilize an iterative formulation to identify the MPP. The form is established through the application of Taylor's series expansion at the designated design point, designated as $\boldsymbol{U}_k$. It can be expressed as follows,

$$g(\boldsymbol{U}) \approx g\left(\boldsymbol{U}_k\right) + \nabla^T g\left(\boldsymbol{U}_k\right)\left(\boldsymbol{U}_{k+1} - \boldsymbol{U}_k\right) = 0 \tag{2.6}$$

where, $\nabla g(\boldsymbol{U}_k)$ is the gradient vector of the LSF in the standard normal space at the point $\boldsymbol{U}_k$ as $\nabla g(\boldsymbol{U}) = \left[\partial g/\partial u_1, \partial g/\partial u_2, \ldots, \partial g/\partial u_n\right]^T$. The new design point $\boldsymbol{U}_{k+1}$ is calculated by rearranging the above relation as [2],

$$\boldsymbol{U}_{k+1} = \frac{\nabla^T g(\boldsymbol{U}_k)\boldsymbol{U}_k - g(\boldsymbol{U}_k)}{\nabla^T g(\boldsymbol{U}_k)\nabla g(\boldsymbol{U}_k)}\nabla g(\boldsymbol{U}_k) \tag{2.7}$$

The equation above is an iterative formula used to calculate the MPP in FORM. The LSF minimum distance from the origin in standard normal space, designated by $\boldsymbol{U}$, represents the standard normal variable vector and is defined as such. The MPP $\boldsymbol{U}^* = (u_1^*, u_2^*, ...u_n^*)^T$ search is an iterative process given by Eq. 2.7.

(3) Approximate the probability of failure i.e. $P_f \approx \Phi(-\beta)$. The fundamental objective of FORM is to identify the MPP, which enables the calculation of the reliability index as $\|\beta = \boldsymbol{U}^*\|$. The reliability index β can be computed as [3],

$$\boldsymbol{U}_{k+1} = \beta_{k+1}\alpha_{k+1} \tag{2.8}$$

where

$$\beta_{k+1} = \frac{g(\boldsymbol{U}_k) - \nabla^T g(\boldsymbol{U}_k)\boldsymbol{U}_k}{\|\nabla g(\boldsymbol{U}_k)\|} \tag{2.9}$$

In the context of design point $\boldsymbol{U}_k$, the negative unit normal vector, also known as the negative gradient direction vector, is designated as α,

$$\alpha_{k+1} = -\frac{\nabla g(\boldsymbol{U}_k)}{\|\nabla g(\boldsymbol{U}_k)\|} \tag{2.10}$$

2.2.2 *Second-Order Reliability Method*

Second-order reliability methods (SORM) are commonly applied to calculate failure probability-based MPP search. The principal distinction between this methodology and FORM lies in the second-order approximation of the limit state function in the vicinity of the MPP. The aforementioned methods have been demonstrated to yield highly accurate reliability predictions for problems exhibiting high curvature at MPP, thereby outperforming the FORM approach. As well as FORM, in SORM, also the random variable should be transferred from the original format into a standard normal variable i.e. $\boldsymbol{U}$, which can be accomplished using the Rosenblatt transformation [4] or the Nataf transformation [5].

The SORM is an approximating analytical reliability method that is impressed by second-order approximation. To approximate the performance function at the MPP $\boldsymbol{U}^*$, the Taylor expansion is used as below,

$$q(U) \approx g(\boldsymbol{U}^*) + \nabla^T g(\boldsymbol{U}^*)(U - U^*) + \frac{1}{2}(U - U^*)^T \nabla^2 g(\boldsymbol{U}^*)((U - U^*) = 0 \tag{2.11}$$

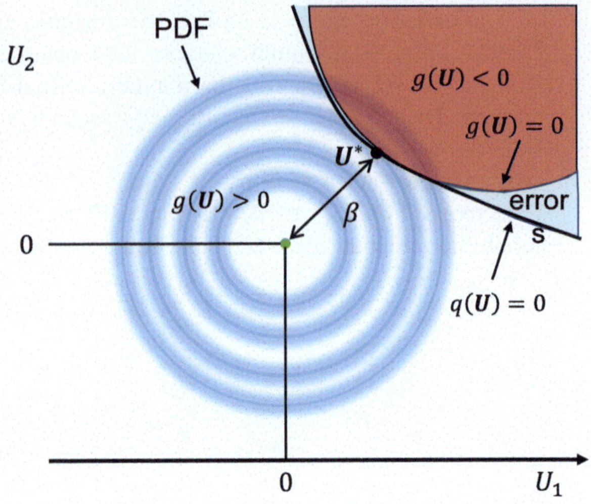

Fig. 2.3 Probability integration in SORM

where $\nabla^2 g(\boldsymbol{U}^*)$ is the Hessian matrix at the MPP, namely, the approximation is given by,

$$\nabla^2 g\left(\boldsymbol{U}^*\right) = \begin{bmatrix} \frac{\partial^2 g}{\partial^2 u_1} & \frac{\partial^2 g}{\partial u_1 \partial u_2} & \cdots & \frac{\partial^2 g}{\partial u_1 \partial u_n} \\ \frac{\partial^2 g}{\partial u_2 \partial u_1} & \frac{\partial^2 g}{\partial^2 u_2} & \cdots & \frac{\partial^2 g}{\partial u_2 \partial u_n} \\ \vdots & \vdots & \ddots & \vdots \\ \frac{\partial^2 g}{\partial u_n \partial u_1} & \frac{\partial^2 g}{\partial u_n \partial u_2} & \cdots & \frac{\partial^2 g}{\partial^2 u_n} \end{bmatrix} \tag{2.12}$$

We have well-known $\nabla^2 g(\boldsymbol{U}^*)$ matrix as the Hessian matrix ($\boldsymbol{H}$) with elements of $\boldsymbol{H}^*_{i,j} = \frac{\partial^2 g}{\partial u_i \partial u_j}$ $i, j = 1, 2, \ldots, n$. The schematic view of the probability integral in SORM is shown in Fig. 2.3. It can be seen that the approximation accuracy of SORM is much higher.

The SORMs can be classified into two main categories: (i) Parabolic SORMs and (ii) General quadratic SORMs. This section will examine some representative SORMs from each category.

2.2.2.1 Parabolic-Based SORM

Due to the second-order term, it is not possible to derive an exact solution for the failure probability, directly. Therefore, the traditional second-order reliability approach requires two transformations of the variables. The first transformation is $\boldsymbol{Z} = \boldsymbol{Pu}$, where $\boldsymbol{P}$ is obtained by a Gram-Schmidt orthogonalization, which is a transformation from $\boldsymbol{U}$-space to $\boldsymbol{Z}$-space. By considering an orthonormal matrix $\boldsymbol{P}$ that leads to a rotation of axes then limit-state function can be expressed as,

$$q(\boldsymbol{U}) \approx \beta - Z_n^* + \frac{1}{2}(\mathbf{Z} - \mathbf{Z}^*)^T \boldsymbol{D}(\mathbf{Z} - \mathbf{Z}^*) \tag{2.13}$$

where $Z_n^* = \boldsymbol{\alpha P}^T \boldsymbol{Z}'$ and $\boldsymbol{D} = \boldsymbol{PHP}^T / \| \nabla G(*)\|$. $\beta = \|\boldsymbol{u}^*\|$, $\boldsymbol{Z}^*$ is the MPP in $\boldsymbol{Z}$-space. $\boldsymbol{D}$ is transformed Hessian matrix, according to,

$$\boldsymbol{D} = \boldsymbol{T}\frac{\nabla^2 g(\boldsymbol{u}^*)}{\|\nabla^2 g(\boldsymbol{u}^*)\|}\boldsymbol{T}^T = \begin{bmatrix} \tilde{\boldsymbol{H}} & \tilde{\boldsymbol{h}} \\ \tilde{\boldsymbol{h}}^T & \lambda \end{bmatrix} \tag{2.14}$$

where $\tilde{\boldsymbol{H}} \in \mathbb{R}^{(n-1)\times(n-1)}$ and $\tilde{\boldsymbol{h}} \in \mathbb{R}^{(n-1)\times 1}$. A second variable transformation is then implemented to diagonalize $\tilde{\boldsymbol{H}}$. The transformation takes the form of $\boldsymbol{Y}^T = [\tilde{\boldsymbol{Z}}^T \boldsymbol{T}', \boldsymbol{Z}_n]$, where $\tilde{\boldsymbol{Z}}^T = [\boldsymbol{Z}_1, \ldots \boldsymbol{Z}_{n-1}]$. The columns of $\boldsymbol{T}'$ are the eigenvectors of $\tilde{\boldsymbol{H}}$, and the corresponding eigenvalue is κ_i. Thus, LSF can be rewritten as,

$$q(\boldsymbol{Y}) = -Y_n + \beta + \frac{1}{2}\sum_{i=1}^{n-1}\kappa_i Y_i^2 + (Y_n - \beta)\sum_{i=1}^{n-1}\gamma_i Y_i + \frac{1}{2}\lambda(Y_n - \beta^2) \tag{2.15}$$

where $\gamma_i = \frac{\partial^2 g_y}{\partial Y_n Y_i}$, $i = 1, \ldots, \mathrm{n}-1$, and $\lambda = \frac{\partial^2 g_y}{\partial Y_n^2}$ is the components of the last column of the symmetric Hessian matrix computed at the MPP. The conventional parabolic SORMs introduce an additional parabola, designated as $p(\boldsymbol{Y})$, which is employed to approximate the function $q(\boldsymbol{Y})$. It can be given, for a two-dimensional reliability problem, only one principal curvature is needed to determine a parabola, so for an n-dimensional reliability problem, $n-1$ principal curvatures are needed. Thus, the approximate parabola based on the MPP can be expressed as,

$$q(\boldsymbol{Y}) \approx p(\boldsymbol{Y}) = -Y_n + \beta + \frac{1}{2}\sum_{i=1}^{n-1}\kappa_i Y_i^2 \tag{2.16}$$

It should be noted that the derivation of Eq. 2.16 involves the neglect of the final row and column of the Hessian matrix, which is expressed as $\gamma_i = 0, \forall i$ and $\lambda = 0$. The function $p(\boldsymbol{Y})$ is defined by a reliability index in $\boldsymbol{U}$-space, the main curvatures κ_i at the MPP, and the number of random variables n. While equation is more straightforward than the general quadratic approximation, it does not yield an explicit solution for approximating the failure probability.

(i) *Breitung's method*

In Breitung's method [6], the failure probability is calculated by integral boundaries of $p(\boldsymbol{Y})$ in Eq. 2.17 as below,

$$P_f = (2\pi)^{-\frac{n}{2}} \int\limits_{-y_n+\beta+\frac{1}{2}\sum_{i=1}^{n-1}\kappa_i y_i^2<0} \exp\left(-\frac{\|\mathbf{y}\|^2}{2}\right) dy \tag{2.17}$$

After a transformation of variable as $\overline{\mathbf{Y}} = \beta^{-1}\mathbf{Y}$, resulting in,

$$P_f = (2\pi)^{-\frac{n}{2}}\beta \int\limits_{-\overline{y}_n+1+\frac{1}{2}\sum_{i=1}^{n-1}\beta\kappa_i\overline{y}_i^2<0} \exp\left(-\frac{\beta^2\|\overline{\mathbf{y}}\|^2}{2}\right)d\overline{\mathbf{y}} \tag{2.18}$$

The integral form of Eq. 2.18 can be expressed simply in Eq. 2.19.

$$P_f = \int\limits_{\mathcal{D}} \exp\left(\beta^2\mathcal{F}(\overline{\mathbf{y}})\right)d\overline{\mathbf{y}} \tag{2.19}$$

where $\mathcal{F}(\overline{\mathrm{y}})$ is a two times continuously differentiable function whose expression has been derived by Bleistein [7]. The integral boundary $\mathcal{D}$ is defined by the parabola $-\overline{y}_n + 1 + \frac{1}{2}\sum_{i=1}^{n-1}\beta\kappa_i\overline{y}_i^2$. Based on the above derivation, the asymptotic expansion becomes,

$$P_f = (2\pi)^{-\frac{1}{2}}\beta^{-1}\exp\left(-\frac{\beta^2}{2}\right)\prod_{i=1}^{n-1}(1 + k_i\beta)^{-\frac{1}{2}} \tag{2.20}$$

Based on the asymptotic relation $(2\pi)^{-\frac{1}{2}}\beta^{-1}\exp\left(-\frac{\beta^2}{2}\right) \to \Phi(-\beta)$ as $\beta \to \infty$, the failure probability can be expressed as,

$$P_{f,\text{Breitung}} = \Phi(-\beta)\prod_{i=1}^{n-1}(1 + k_i\beta)^{-\frac{1}{2}} \tag{2.21}$$

Then $\prod_{i=1}^{n-1}(1 + k_i\beta)^{-\frac{1}{2}}$ is an adjusting factor for correction of FORM approach. The FORM is needed to find the MPP for computing the parameters of SORM. It is known that Breitung's method applies to the case where beta is large. Furthermore, Breitung' method only applies to $k_i > -1/\beta$, by assumption in Eq. 2.21.

(ii) *Tvedt's method*

In the Tvedt's method [8, 9], the performance function is approximated using the paraboloid in Eq. 2.20 with a safety domain is $p(\mathbf{Y}) \geq 0$, i.e., $Y_n \leq \beta + \frac{1}{2}\sum_{i=1}^{n-1}k_iY_i^2$. Therefore, the asymptotic expansion of the reliability integral can be derived, and the reliability is calculated by,

$$R = \int\limits_{-\infty}^{\infty}\cdots\int\limits_{-\infty}^{\infty}\phi(y_1)\ldots\phi(y_{n-1})\int\limits_{-\infty}^{\beta+\frac{1}{2}\sum_{i=1}^{n-1}k_iY_i^2}\phi(y_n)dy_n dy_{n-1}\ldots dy_1 \tag{2.22}$$

Tvedt obtained a three-term approximation to Eq. 2.22 by means of a power series expansion of $\frac{1}{2}\sum_{i=1}^{n-1}k_iY_i^2$, wherein terms exceeding the second order were disregarded. The closed form of the failure probability P_f can be approximated by a trinomial polynomial

$$P_{f,\mathrm{Tvedt}} = T_1 + T_2 + T_3 \tag{2.23}$$

where

$$\begin{cases} T_1 = \Phi(-\beta) \prod_{i=1}^{n-1} (1 + k_i\beta)^{-1/2} \\ T_2 = [\beta\Phi(-\beta) - \phi(\beta)] \left[\prod_{i=1}^{n-1} (1 + k_i\beta)^{-1/2} \right. \\ \qquad \left. - \prod_{i=1}^{n-1} (1 + k_i(\beta + 1))^{-1/2} \right] \\ T_3 = (\beta + 1)[\beta\Phi(-\beta) - \phi(\beta)] \left[\prod_{i=1}^{n-1} (1 + k_i\beta)^{-1/2} \right. \\ \qquad \left. -Re\left[\prod_{i=1}^{n-1} (1 + k_i(\beta + 1))^{-1/2} \right] \right] \end{cases} \tag{2.24}$$

$Re(\cdot)$ is the real part of the complex number. According to the expression of the first term of Eq. 2.24, the value is equal to the failure probability calculated by the method of Breitung's method, i.e., $T_1 = P_{f,\mathrm{Breitung}}$. Then the second and third terms in Eq. 2.24 can be viewed as correctors for Breitung's method for improving the accuracy of the moderate values of β.

(iii) *Hohenbichler and Rackwitz's method*

Similar to the work of Tvedt et al. [10], Hohenbichler and Rackwitz used a Taylor series-based method to expand $\frac{1}{2}\sum_{i=1}^{n-1} k_i Y_i^2$, obtaining an expression in closed form. This closed-form expression provides an exact or approximate mathematical description of the original function or model in the vicinity of a specific point for subsequent numerical analysis or engineering design. The expression for the probability integral P_f can initially be expressed by Eq. 2.23. In this integral, the part of the inner integral is replaced by,

$$\int_{-\infty}^{-\beta - \frac{1}{2}\sum_{i=1}^{n-1} k_i Y_i^2} \phi(y_n)\, dy_n = \Phi\left(-\beta - \frac{1}{2}\sum_{i=1}^{n-1} k_i Y_i^2\right) \tag{2.25}$$

where the integral boundary is $p(\boldsymbol{Y}) < 0$, i.e., $Y_n < -\beta - \frac{1}{2}\sum_{i=1}^{n-1} k_i Y_i^2$. Taylor expansion of the $\frac{1}{2}\sum_{i=1}^{n-1} k_i Y_i^2$ on the right-hand side of Eq. 2.23 results in,

$$\Phi\left(-\beta - \frac{1}{2}\sum_{i=1}^{n-1} k_i Y_i^2\right) = \Phi(-\beta)\exp\left[-\frac{1}{2}\sum_{i=1}^{n-1} k_i Y_i^2 \frac{\phi(-\beta)}{\Phi(-\beta)}\right] \tag{2.26}$$

to simplify the computation of the probability integral, the inner integral in Eq. 2.23 is replaced by the first-order Taylor approximation given in Eq. 2.26, which yields,

$$P_{f,\text{Hohenbichler}} = \Phi(-\beta)\prod_{i=1}^{n-1}(1+k_i)^{-1/2} \tag{2.27}$$

Compared to the Breitung's method, this probabilistic formula is primarily designed to improve the precision and accuracy of reliability estimation over a range of moderate β. In other words, the formula provides more reliable and accurate predictions for reliability estimation under medium β conditions, thus optimizing the performance of the Breitung's method under these specific conditions. Moreover, it can be seen from Eq. 2.26 that $\frac{\phi(\beta)}{\Phi(\beta)} \to \beta$ as $\beta \to \infty$, so for large β, the method is equivalent to the Breitung's method.

(iv) *Koyluoglu and Nielsen's method*

The reliability integral domain is $\left[-\infty, \beta + \frac{1}{2}\sum_{i=1}^{n-1}\kappa_i Y_i^2\right]$, Koyluoglu and Nielsen's method [11] divides the integral domain into two intervals $[-\infty, \beta]$ and $\left[\beta, \beta + \frac{1}{2}\sum_{i=1}^{n-1}\kappa_i Y_i^2\right]$, the reliability integral becomes,

$$R = \Phi(\beta) + \int_{-\infty}^{\infty}\ldots\int_{-\infty}^{\infty}\phi(y_1)\ldots\phi(y_{n-1})\left[\left(\beta + \frac{1}{2}\sum_{i=1}^{n-1}\kappa_i Y_i^2\right)\right. \\ \left. +\Phi(\beta)\right]dy_{n-1}\ldots dy_1 \tag{2.28}$$

To further simplify the integral, the expression $\Phi\left(\beta + \frac{1}{2}\sum_{i=1}^{n-1}k_i Y_i^2\right) - \Phi(\beta)$ is expanded by McLaurin series such that Eq. 2.28 can be integrated analytically. Based on the different signs of curvatures, the following three different formulas can be derived,

1. The failure probability for the case where all curvatures are positive, i.e., $k_i > 0$,

$$P_{f,Koyluoglu} = \Phi(-\beta)\prod_{i=1}^{n-1}\left(1 + k_i\frac{\phi(\beta)}{\Phi(\beta)}\right)^{-\frac{1}{2}} \tag{2.29}$$

2. The failure probability for the case where all curvatures are negative, i.e., $k_i < 0$,

$$P_{f,Koyluoglu} = 1 - \Phi(-\beta)\prod_{i=1}^{n-1}\left(1 - k_i\frac{\phi(\beta)}{\Phi(\beta)}\right)^{-\frac{1}{2}} 1 \tag{2.30}$$

3. For all cases of positive and negative curvatures, i.e., $k_i > 0$, and $k_j < 0$, the Failure Probability is,

$$
\begin{aligned}
P_{f,Koyluoglu} &= 1 - \Phi(-\beta) \\
&-\Phi(-\beta)\left\{\prod_{j=m}^{n-1}\left(1 - k_j\frac{\phi(\beta)}{2\Phi(\beta)}\right)^{-1/2}\left[1 - \prod_{i=1}^{m-1}\left(1 + k_i\frac{\phi(\beta)}{\Phi(\beta)}\right)^{-1/2}\right]\right\} \\
&+\Phi(\beta)\left\{\sum_{i=1}^{m-1}\left(1 + k_i\frac{\phi(\beta)}{2\Phi(-\beta)}\right)^{-1/2}\left[1 - \sum_{i=m}^{n-1}\left(1 - k_i\frac{\phi(\beta)}{\Phi(\beta)}\right)^{-1/2}\right]\right\}
\end{aligned}
\tag{2.31}
$$

(v) *Cai and Elishakoff's method*

In the method of Cai and Elishakoff [12], which also solves for the integral. In order to facilitate the integration process, the inner integral is divided into two parts, according to,

$$
\int_{-\infty}^{\beta+\frac{1}{2}\sum_{i=1}^{n-1}k_iY_i^2} \phi(y_n)dy_n = \Phi(\beta) + \frac{1}{\sqrt{2\pi}}\exp\left(-\frac{\beta^2}{2}\right)\int_0^{\sum_{i=1}^{n-1}k_iY_i^2} \exp\left(-\beta v - \frac{v^2}{2}\right)dv
\tag{2.32}
$$

where $v = y_n - \beta$. Expanding $\exp\left(-\beta v - \frac{v^2}{2}\right)$ in a Taylor series at $v = 0$ and associating it in the last integral in Eq. 2.33, three approximations of the failure probability can be obtained according to,

$$
P_{f,\mathrm{Cai}} = \Phi(-\beta) - \phi(\beta)(D_1 + D_2 + D_3) \tag{2.33}
$$

where

$$
\begin{cases}
D_1 = \sum_{i=1}^{n-1}\frac{1}{2}k_i \\
D_2 = -\frac{1}{2}\beta\left(3\sum_{i=1}^{n-1}\left(\frac{1}{2}k_i\right)^2 + \sum_{i\neq j}^{n-1}\left(\frac{1}{2}k_i\right)\left(\frac{1}{2}k_j\right)\right) \\
D_3 = \frac{1}{6}(\beta^2 - 1)\left(15\sum_{i}^{n-1}\left(\frac{1}{2}k_i\right)^3 + 9\sum_{i\neq j}^{n-1}\left(\frac{1}{2}k_i\right)^2\left(\frac{1}{2}k_j\right) + \sum_{i\neq j\neq l}^{n-1}\left(\frac{1}{2}k_i\right)\left(\frac{1}{2}k_j\right)\left(\frac{1}{2}k_l\right)\right)
\end{cases}
\tag{2.34}
$$

(f) *Zhao and Ono's method*

The method proposed by Zhao and Ono [13] approximates the LSF not by using the main curvatures at the MPP, but by employing a rotating parabola with curvatures equivalent to the mean of the mean curvatures, and the approximated limit function can be expressed as,

$$
P(\mathbf{Y}) = -Y_n + \beta + \frac{1}{2}\frac{K_s}{n-1}\sum_{i=1}^{n-1}Y_i^2 \tag{2.35}
$$

where

$$K_s = \sum_{i=1}^{n-1} k_i = \frac{\text{tr}\left(\nabla^2 g(\boldsymbol{u}^*)\right) - \boldsymbol{\alpha}^T \nabla^2 g(\boldsymbol{u}^*)\boldsymbol{\alpha}}{|\nabla g(\boldsymbol{u}^*)|} \tag{2.36}$$

in which K_s is the sum of the main curvatures and $\boldsymbol{\alpha} = -\frac{\nabla g(\boldsymbol{u}^*)}{\nabla g(\boldsymbol{u}^*)}$. The failure probability can be computed by,

$$P_f = \int_0^\infty \Phi\left(\frac{t}{2R} - \beta\right) f_{\chi^2_{n-1}}(t)dt \tag{2.37}$$

where $f_{\chi^2_{n-1}}(t)$ is the PDF of the central chi-square distribution and $R = \frac{n-1}{K_s}$. the failure probability can be given by,

$$P_{f,\text{Zhao}} = \begin{cases} \Phi(-\beta)\left(1 + \frac{\phi(\beta)}{R\Phi(-\beta)}\right)^{-\left(\frac{n-1}{2}\right)\left[1+\frac{2K_s}{10(1+2\beta)}\right]}, K_s \geq 0 \\ \Phi\left(-\beta\left(1 + \frac{2.5K_s}{2n-5R+\frac{25(23-5\beta)}{R^2}}\right) - \frac{1}{2}K_s\left(1 + \frac{K_s}{40}\right)\right), K_s < 0 \end{cases} \tag{2.38}$$

The schematic of all the parabolic-based SORMs presented above is illustrated in Fig. 2.4. These methods necessitate the resolution of second-order derivatives of the limit state function, a process that is particularly costly in terms of computational resources, particularly when addressing reliability problems with implicit performance functions. Furthermore, the parabola fitted based on the MPP is contingent upon the location of the MPP, which is typically sought through iterative optimization methods. However, this approach may yield non-convergent solutions when confronted with high-dimensional nonlinearities in the performance function.

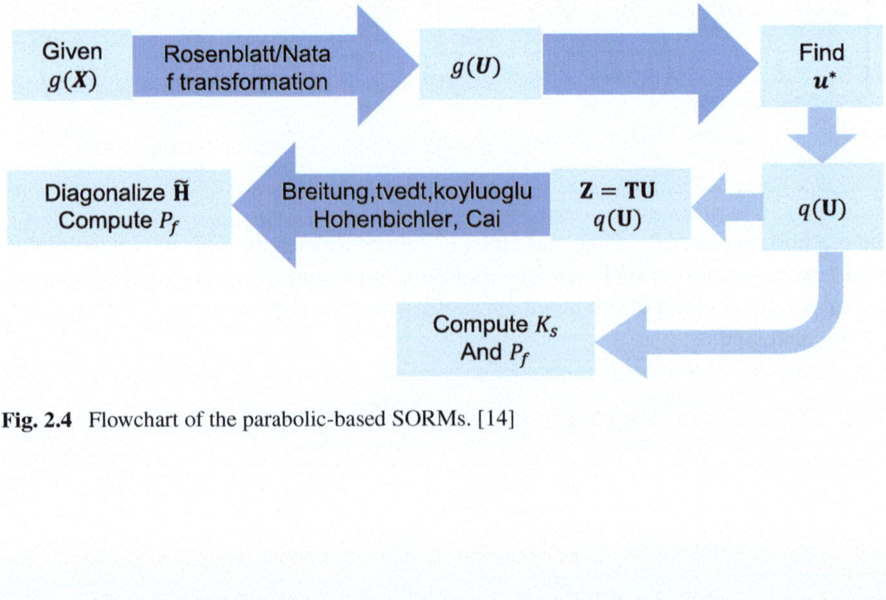

Fig. 2.4 Flowchart of the parabolic-based SORMs. [14]

2.2.2.2 General Quadratic Second-Order Reliability Methods

In recent years, research on SORM has been conducted with the main objective of proposing a closed form of the general quadratic approximation function $q(\boldsymbol{U})$ with the standardized form,

$$q(\boldsymbol{U}) = a + \boldsymbol{b}^T \boldsymbol{U} + \boldsymbol{U}^T \boldsymbol{C}\boldsymbol{U} \tag{2.39}$$

where

$$\begin{cases} a = \frac{1}{2}\boldsymbol{u}^{*T}\nabla^2 g(\boldsymbol{u}^*)\boldsymbol{u}^* - \nabla g(\boldsymbol{u}^*)^T \boldsymbol{u}^* \\ b = \nabla g(\boldsymbol{u}^*) - \nabla^2 g(\boldsymbol{u}^*)\boldsymbol{u}^* \\ C = \frac{1}{2}\nabla^2 g(\boldsymbol{u}^*) \end{cases} \tag{2.40}$$

Three novel methodologies have been devised that, in contrast to parabolic-based second-order reliability techniques, employ the full range of components within the Hessian matrix. The MPP $\boldsymbol{u}^*$ in Eq. 2.36 can be given by,

$$\boldsymbol{u}^* = \beta\boldsymbol{\alpha} \tag{2.41}$$

$$\boldsymbol{\alpha} = -\frac{\nabla g(\boldsymbol{u}^*)}{\nabla g(\boldsymbol{u}^*)} \tag{2.42}$$

Thus, the LSF can be expressed as,

$$\begin{aligned} \frac{q(\boldsymbol{U})}{\nabla g(\boldsymbol{u}^*)} &= \beta + \frac{\beta^2}{2}\boldsymbol{\alpha}^T \frac{\nabla^2 g(\boldsymbol{u}^*)}{\nabla g(\boldsymbol{u}^*)}\boldsymbol{\alpha} \\ -\boldsymbol{\alpha}^T &\left(1 + \beta\frac{\nabla^2 g(\boldsymbol{u}^*)}{\nabla g(\boldsymbol{u}^*)}\right)\boldsymbol{U} + \frac{1}{2}\boldsymbol{U}^T \frac{\nabla^2 g(\boldsymbol{u}^*)}{\nabla g(\boldsymbol{u}^*)}\boldsymbol{U} \end{aligned} \tag{2.43}$$

The method eliminates the second-order cross terms. Achieved by applying the variable transformation $\tilde{\boldsymbol{U}} = \boldsymbol{D}^{-1}\boldsymbol{U}$, $\boldsymbol{D}$ is an orthogonal matrix with column vectors that are eigenvectors of $\boldsymbol{C}$, $\tilde{\boldsymbol{U}} = \left(\breve{U}_1, \breve{U}_2, \ldots, \breve{U}_n\right)$ is a n-dimensional vector containing independent standard normal variables, and the LSF is expressed as,

$$\begin{cases} q(\tilde{\boldsymbol{U}}) = a + \tilde{\boldsymbol{b}}^T \tilde{\boldsymbol{U}} + \tilde{\boldsymbol{U}}^T \tilde{\boldsymbol{C}}\tilde{\boldsymbol{U}} \\ \tilde{\boldsymbol{b}} = \tilde{\boldsymbol{D}}^T \tilde{\boldsymbol{b}} \\ \tilde{\boldsymbol{C}} = \tilde{\boldsymbol{D}}^T \tilde{\boldsymbol{C}}\tilde{\boldsymbol{D}} \end{cases} \tag{2.44}$$

thus $\tilde{\boldsymbol{C}}$ is an orthogonal matrix, then we have,

$$q(\tilde{\mathbf{U}}) = \sum_{i=1}^{n} \tilde{C}_{ii}\left(\tilde{U}_i + \frac{\tilde{b}_i}{2\tilde{C}_{ii}}\right)^2 + a - \sum_{i=1}^{n} \frac{\tilde{b}_i^2}{4\tilde{C}_{ii}} \tag{2.45}$$

The failure probability is

$$P_f = P\left[q(\tilde{U}) < 0\right] = P\left\{h(\tilde{U}) = \sum_{i=1}^{n} \tilde{C}_{ii}\left(\tilde{U}_i + \frac{\tilde{b}_i}{2\tilde{C}_{ii}}\right)^2 < \sum_{i=1}^{n} \frac{\tilde{b}_i^2}{4\tilde{C}_{ii}} - a\right\} \tag{2.46}$$

In the study proposed by Konishi et al. [15], the function $h(\cdot)$ is defined as a linear combination of non-centered cardinal variables based on which two formulas for calculating the probability of failure are proposed,

$$P_{f,\text{Mansour}} = \Phi(\kappa) - \phi(\kappa)\left[\frac{\sqrt{2}H_2(\kappa)}{3}\frac{m_3}{m_2\sqrt{m_2}} + \frac{H_5(\kappa)}{9}\frac{m_3^2}{m_2^3} + \frac{H_3(\kappa)}{2}\frac{m_4}{m_2^2}\right] \tag{2.47}$$

$$\begin{aligned} P = \Phi(\kappa') - \phi(\kappa')\Big[&H_3(\kappa')\left(\frac{m_4}{2m_2^2} - \frac{20m_3^2}{27m_2^3} + \frac{2m_3}{9m_1m_2}\right) \\ &+H_1(\kappa')\left(-\frac{2m_3^2}{3m_2^3} + \frac{2m_3}{3m_1m_2}\right)\Big] \end{aligned} \tag{2.48}$$

where $H_i(\cdot)$ is the Hermite polynomials, for the i-th, can be expressed as,

$$\begin{cases} \kappa = -\frac{a+\sum_{i=1}^{n}\tilde{C}_{ii}}{\sqrt{\sum_{i=1}^{n}(2\tilde{C}_{ii}^2+\tilde{b}_i^2)}} \\ m_r = \sum_{i=1}^{n}\left(\tilde{C}_{ii}^r + \frac{r}{4}\tilde{C}_{ii}^{r-2}\tilde{b}_i^2\right) \end{cases} \tag{2.49}$$

For the case where the LSF is elliptic, i.e., when the sign of all $\tilde{C}_{ii}$ are the same, the probability of failure can be calculated using Eq. 2.47, and the probability of failure is $P_f = P$ for $\text{sign}(\tilde{C}_{ii}) \cdot s < 0$ and $P_f = 1 - P$ for $\text{sign}(\tilde{C}_{ii}) \cdot s < 0$. $\text{sign}(\cdot)$ is a symbol function, $\text{sign}(\cdot) = +1, -1$ or 0, depending on whether the independent variable is positive, negative or zero.

In their seminal work, Hu and Du [16] developed the second-order saddle-point approximation, which provides the Cumulant Generating Function (CGF) of the limit state function in standard normal space. This approach enables the use of saddle points to approximate the probability of failure. The limit state function can be expressed as follows,

$$q(\tilde{U}) = \sum_{i=1}^{n} q_i(\tilde{U}) = \sum_{i=1}^{n}\left(\tilde{a}_i + \tilde{a}_i\tilde{U}_i + \tilde{C}_{ii}\tilde{U}_i^2\right) \tag{2.50}$$

where $\tilde{a}_i = \frac{a}{n}$, Eq. 2.50 can be further formulated as,

$$q_i(\tilde{\mathbf{U}}) = \begin{cases} \left(\sqrt{\tilde{C}_{ii}}\tilde{U}_i + \frac{\tilde{b}_i}{2\sqrt{\tilde{C}_{ii}}}\right)^2 + \tilde{a}_i - \frac{\tilde{b}_i^2}{4\tilde{C}_{ii}}, \tilde{C}_{ii} > 0 \\ -\left(\sqrt{-\tilde{C}_{ii}}\tilde{U}_i - \frac{\tilde{b}_i}{2\sqrt{-\tilde{C}_{ii}}}\right)^2 + \tilde{a}_i - \frac{\tilde{b}_i^2}{4\tilde{C}_{ii}}, \tilde{C}_{ii} < 0 \\ \tilde{a}_i + \tilde{b}_i\tilde{U}_i, \tilde{C}_{ii} = 0 \end{cases} \tag{2.51}$$

The CGF of $q(\tilde{\mathbf{U}})$ can be obtained by $K_q(t) = \sum_{i=1}^{n} K_{qi}(t)$, and the $K_{qi}(t)$ is computed from,

$$K_{qi}(t) = \begin{cases} \frac{\lambda_i\sigma_i^2 t}{1-2\sigma_i^2 t} - \frac{1}{2}\log\left(1 - 2\sigma_i^2 t\right) + d_i t, \tilde{C}_{ii} > 0 \\ -\frac{\lambda_i\sigma_i^2 t}{1+2\sigma_i^2 t} - \frac{1}{2}\log\left(1 + 2\sigma_i^2 t\right) + d_i t, \tilde{C}_{ii} < 0 \\ \tilde{a}_i t + \frac{1}{2}b_i^2 t^2, \tilde{C}_{ii} = 0 \end{cases} \tag{2.52}$$

where $\tilde{d}_i = \tilde{a}_i - \frac{\tilde{b}_i^2}{4\tilde{C}_{ii}}$, $\lambda_i = \frac{\tilde{b}_i^2}{4\tilde{C}_{ii}}$ and $\sigma_i = \sqrt{\text{sign}\left(\tilde{C}_{ii}\right)\tilde{C}_{ii}}$, and the CGF can be rewritten as,

$$K_{qi}(t) = \tilde{a}_i t + \frac{1}{2}\frac{b_i^2 t^2}{1 - 2t\tilde{C}_{ii}} + \frac{1}{2}\log\left(\frac{1}{1 - 2t\tilde{C}_{ii}}\right) \tag{2.53}$$

The saddle point t_s can be obtained by solving for the first-order derivative $K_q'(t) = 0$ of $K_q(t)$, failure probability is computed by,

$$P_{f,SOSPA} = P\left[q\left(\tilde{U}\right) < 0\right] = \Phi(w) + \phi(w)\left(\frac{1}{w} - \frac{1}{v}\right) \tag{2.54}$$

where

$$\begin{cases} w = \text{sign}(t_s)\left\{2\left[K_q(t_s)\right]\right\}^{1/2} \\ v = t_s\left[K_q''(t_s)\right]^{1/2} \end{cases} \tag{2.55}$$

A SORM based on convolutional integration, which is similar to the SOSPA, with a limit state function of,

$$\frac{q\left(\tilde{\boldsymbol{U}}\right)}{\left\|\nabla g(\boldsymbol{u}^*)\right\|} = a_0 - \boldsymbol{a}_1^T\tilde{\boldsymbol{U}} + \tilde{\boldsymbol{U}}^T\tilde{\boldsymbol{C}}\tilde{\boldsymbol{U}} \tag{2.56}$$

where,

$$\begin{cases} a_0 = \beta + \frac{\beta^2}{2}\boldsymbol{\alpha}^T \frac{\nabla^2 g(\boldsymbol{u}^*)}{\|\nabla g(\boldsymbol{u}^*)\|}\boldsymbol{\alpha} \\ \boldsymbol{a}_1^T = \boldsymbol{\alpha}^T\left(1 + \beta\frac{\nabla^2 g(\boldsymbol{u}^*)}{\|\nabla g(\boldsymbol{u}^*)\|}\right)\boldsymbol{D} \\ \tilde{\boldsymbol{C}} = \frac{1}{2\|\nabla g(\boldsymbol{u}^*)\|} diag|\lambda_1, \lambda_1, \ldots \lambda_n| \end{cases} \tag{2.57}$$

and λ_i is the i-th eigenvalue of the Hessian matrix $\nabla^2 g(\boldsymbol{u}^*)$, Eq. 2.52 becomes,

$$\begin{aligned} q(\tilde{\mathbf{U}}) = \|\nabla g(\mathbf{u}^*)\| a_0 \\ + \frac{1}{2}\Bigg[\sum_{i\in\{\mathbb{N}|\lambda_i=0\}} \left(-2\|\nabla g(\mathbf{u}^*)\| a_{1i}\tilde{U}_i + \lambda_i\tilde{U}_i^2\right) \\ + \sum_{i\in\{\mathbb{N}|\lambda_i=0\}} \left(-2\|\nabla g(\mathbf{u}^*)\| a_{1i}\tilde{U}_i\right)\Bigg] \end{aligned} \tag{2.58}$$

The limit state function can be further written as,

$$q(\tilde{U}) = \frac{1}{2}\sum_{i\in\{\mathbb{N}|\lambda_i=0\}} \lambda_i(\tilde{U}_i + \delta_i)^2 + \|\nabla g(\boldsymbol{u}^*)\| \sum_{i\in\{\mathbb{N}|\lambda_i=0\}} a_{1i}\tilde{U}_i - \frac{1}{2}r \tag{2.59}$$

where $\delta_i = -\|\nabla g(\boldsymbol{u}^*)\| a_{1i}/\lambda_i$ and $r = \sum_{i\in\{\mathbb{N}|\lambda_i=0\}} \lambda_i\delta_i^2 - 2\|\nabla g(\boldsymbol{u}^*)\| a_0$, and the failure probability can be given by,

$$P_{f,Park} = P\left[Q(\mathbf{W}) = \sum_{i=1}^{N} Q_i = \sum_{i=1}^{N} d_i W_i < r\right] = \int_{-\infty}^{r} f_Q(q)dq \tag{2.60}$$

where $d_i = \lambda_i$ and $W_i \sim \chi\left(1, \delta_i^2\right)$ when $\lambda_i \neq 0$ while $d_i = 2\|\nabla g(\boldsymbol{u}^*)\| a_{1i}$ and $W_i \sim \chi(0, 1)$, and $\lambda_i = 0$, the PDF of Q can be computed by convolution, according to,

$$f_Q(q) = \left(f_{Q_1} * f_{Q_2} * \cdots * f_{Q_n}\right) \tag{2.61}$$

where the PDF of Q_i can be derived from the PDF of W_i as $f_{Q_i}(q_i) = f_{W_i}(q_i/d_i)/|d_i|$. The selection of integration boundaries and the implementation of discretization steps in numerical convolution have a direct impact on the efficacy and precision of the method. The flow of three general quadratic SORMs presented above is schematized in Fig. 2.5.

The employment of quadratic polynomials for the approximation of the limit state function yields enhanced precision for SORM in comparison to FORM. However, this increased accuracy is accompanied by a greater computational complexity. In the case of highly nonlinear reliability problems, the approximation accuracy of

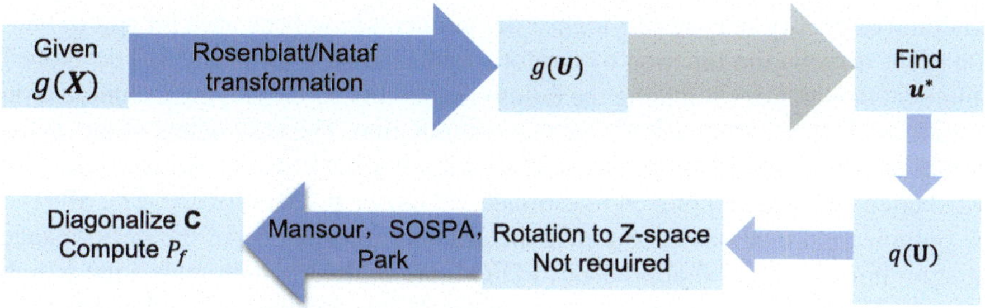

Fig. 2.5 Flowchart of the general quadratic SORMs [14]

second-order reliability methods may also prove inadequate. Among these SORMs mentioned in this section, the SOSPA method is the most advantageous in terms of accuracy, capability, and robustness because the performance of the method does not depend on the sign and magnitude of the curvatures of the MPP, the dimensionality of the performance function, and the magnitude of the first-order reliability indexes. The initial generation of SORMs, based on parabolic approximations, was primarily focused on the development of a limit state function that closely approximated a parabola at the MPP. In contrast, current methods tend to construct fully quadratic approximations to the LSF, which can be employed to address highly nonlinear reliability problems.

Given that the approximation of the performance function in SORM is associated with the second-order format, it can be concluded that this method is superior in terms of accuracy when compared to FORM, particularly for concave LSF. However, for convex reliability problems with negative principal curvatures, the results have not been adjusted with a suitable factor. Moreover, SORM necessitates the computation of the second-order derivative, rendering it less efficient than FORM when the derivatives are evaluated numerically. Therefore, it can be concluded that SORM requires a greater number of function evaluations when compared to FORM.

2.2.3 Moment Method

The approximation of a random variable's distribution using its moments represents a fascinating avenue of inquiry in the computation of failure probability. The distribution function can be approximated using the well-known statistical techniques, including the normal, Pearson, Johnson, and Burr methods. It was with the objective of approximating the moment of a probability density function that the Edgeworth and Cornish-Fisher expansions were developed [17]. The Hermite polynomial transformation has been raised as a potential methodology for the application of moment methods in the approximation of the probability density function within the context of structural reliability analyses. Its high-order moments have been used for approximation of the distribution function by Grigoriu and Lind [18]. Grigoriu [19] developed a moment method for estimating the failure probability by employing a combination of the Lamda distribution and Monte Carlo simulation. The parameters of the distributions are calculated using point estimates for a PDF of LSF. Subsequently, a fitted distribution of LSF is employed to estimate the failure probability [20, 21]. Most of the optimization methods need to use the gradient information of the performance function in the analytical approaches FORM and SORM, which makes the search difficult. This section presents moment methods for structural reliability analysis, which obviate the necessity for an iterative search for the MPP. Instead, in order to obtain the moments of the performance function, point estimation is conducted in a standard normal space. Subsequently, the reliability index is computed by applying a normalized function.

(i) *Second moment method*

In the case of an LSF utilizing two moments, the distribution function of the LSF is assumed to be normal. Consequently, the reliability index and failure probability are determined through the application of the second-moment method, as outlined below,

$$\beta = \frac{\mu_g}{\sigma_g} \tag{2.62}$$

Attempts have been made to enhance the precision of FOSM through the utilization of high-order moments. This has led to the development of a first-order third-moment reliability method and a higher-order moments standardization technique, which have been employed to improve the accuracy of FOSM. In contrast to conventional FORM, which employs only the first two moments of random variables to compute the second moment of the performance function by simulating the performance function e.g., $g = R - S$, μ_g and σ_g in Eq. 2.62 are obtained by estimating the random variables in the performance function's probability distribution. However, for practical structural reliability problems, the performance function may be complex, and the second-moment information of random variables cannot be computed.

(ii) *Third-moment method*

In the case of a performance function $Z = g(\boldsymbol{X})$, once the first three moments of the performance function have been obtained, it is possible to assume that the standardized performance function $Z_u = \frac{Z-\mu_g}{\sigma_g}$ follows a lognormal distribution comprising three parameters. This allows the standard normal variable u to be expressed by the following equation,

$$\frac{sign(\alpha_{3g})}{\sqrt{\ln(A)}} \ln\left[\sqrt{A}\left(1 - \frac{Z_u}{u_b}\right)\right] \tag{2.63}$$

where

$$A = 1 + \frac{1}{u_b^2} \tag{2.64}$$

$$u_b = (a + b)^{\frac{1}{3}} + (a - b)^{\frac{1}{3}} - \frac{1}{\alpha_{3g}} \tag{2.65}$$

$$a = \frac{1}{\alpha_{3g}}\left(\frac{1}{\alpha_{3g}^2} + \frac{1}{2}\right) b = -\frac{1}{2\alpha_{3g}^2}\sqrt{\alpha_{3g}^2 + 4} \tag{2.66}$$

where α_{3G} is the third central moment of the performance function, i.e., the skewness. Since,

$$P[Z < 0] = P\left[Z_u < -\frac{\mu_g}{\sigma_g}\right] = P[Z_u < -\beta_{SM}] \tag{2.67}$$

Subsequently, the calculated reliability index and failure probability by the third-moment method are as follows,

$$\beta_{TM} = \frac{-sign(\sigma_{3g})}{\sqrt{ln(A)}} ln\left[\sqrt{A}\left(1 + \frac{\beta_{SM}}{u_b}\right)\right] \tag{2.68}$$

$$P_{fTM} = \Phi(-\beta_{TM}) \tag{2.69}$$

Since $u_b < 0$ for $\sigma_{3G} < 0$ and $u_b < 0$ for $\sigma_{3G} > 0$, Eq. 2.67 is always increasing for $\beta_{\mathrm{TM}} > 0$. It should be noted that when $\sigma_{3G} = 0$, it is impossible to derive the proper results, and at that time the second-moment method should be used for that case, i.e.,

$$\beta_{\mathrm{TM}} = \beta_{\mathrm{SM}} \tag{2.70}$$

(iii) *Fourth-moment method*

The performance function of the structure is $Z = g(\boldsymbol{X})$. Assuming that the first four moments of performance function have been obtained, the principle can be used to relate the standardized variable Z_u to a standard normal variable, according to,

$$u = \frac{\alpha_{3g} + 3(\alpha_{4g} - 1)Z_u - \alpha_{3g}Z_u^2}{\sqrt{\left(5\alpha_{3g}^2 - 9\alpha_{4g} + 9\right)(1 - \alpha_{4g})}} \tag{2.71}$$

where, α_{4G} is the fourth central moment of the performance function, i.e., the kurtosis. Using Eq. 2.71, the reliability index and failure probability can be calculated based on the first fourth moment according to,

$$\beta_{FM} = \frac{3(\alpha_{4g} - 1)\beta_{SM} + \alpha_{3g}\left(\beta_{SM}^2 - 1\right)}{\sqrt{\left(9\alpha_{4g} - 5\alpha_{3g}^2 - 9\right)(\alpha_{4G} - 1)}} \tag{2.72}$$

$$P_{fFM} = \Phi(-\beta_{FM}) \tag{2.73}$$

since $\beta_{\text{FM}} = \beta_{\text{TM}}$ for $\alpha_{3g} = 0$, Eq. 2.72 is always monotonically increased for $\beta_{\text{SM}} > 0$. In addition, the fourth-moment method can be implemented by Edgeworth expansion, the probability distribution function of expanded standardized variables Z_u can be expressed by the fourth moment's information, according to,

$$F(Z_u) = \Phi(Z_u) - \phi(Z_u)\left[\frac{1}{6}\alpha_{3g}H_2(Z_u) + \frac{1}{24}\left(\alpha_{4g} - 3\right)H_3(Z_u) + \frac{1}{72}\alpha_{3g}^2 H_5(Z_u)\right] \tag{2.74}$$

where the second-order Hermite polynomial is represented by the function $H_2(x) = x^2 - 1$, the third-order polynomial is represented by $H_3(x) = x^3 - 3x$, and the fifth-order polynomial is represented by $H_5(x) = x^5 - 10x^3 + 15x$. The failure probability and reliability index can be expressed as,

$$\begin{gathered} P_{fFM} = \Phi(-\beta_{SM}) - \phi(\beta_{SM}) \\ \left[\tfrac{1}{6}\alpha_{3g}H_2(-\beta_{SM}) + \tfrac{1}{24}\left(\alpha_{4g} - 3\right)H_3(-\beta_{SM}) + \tfrac{1}{72}\alpha_{3g}^2 H_5(-\beta_{SM})\right] \end{gathered} \tag{2.75}$$

$$\beta_{\text{FM}} = -\Phi^{-1}\left(P_{f\text{FM}}\right) \tag{2.76}$$

In the same way, $\beta_{\text{FM}} = \beta_{\text{TM}}$ for $\alpha_{3g} = 0$.

In addition, the fourth-moment method can be applied to existing frequency response systems, such as the Pearson system. In the Pearson system, the normalized variable Z_u satisfies the following differential equation,

$$\frac{1}{f}\frac{df}{dZ_u} = -\frac{aZ_u + b}{c + bZ_u + dZ_u^2} \tag{2.77}$$

$$a = 10\alpha_4 - 12\alpha_3^2 - 18 \tag{2.78}$$

$$b = \alpha_3(\alpha_4 + 3) \tag{2.79}$$

$$c = 4\alpha_4 - 3\alpha_3^2 \tag{2.80}$$

$$d = 2\alpha_4 = 3\alpha_3^2 - 6 \tag{2.81}$$

where f is the PDF of the Z_u, see Table 2.1. The reliability index is calculated from the fourth moment of,

Table 2.1 Probability density functions for the Pearson system

Parameter	PDF $f(Z_u)$	Range of Z_u
$b \neq 0\ \Delta > 0 d < 0$	$K(Z_u - r_2)^{\frac{-1}{\sqrt{\Delta}}(ar_2+b)}(r_1 - Z_u)^{\frac{-1}{\sqrt{\Delta}}(ar_2+b)}$	(r_2, r_1)
$d = 0$	$K(c + bZ_u)^{(ac+b^2)/b^2} exp\left[-\frac{aZ_u}{b}\right]$	$\left(-\infty, -\frac{c}{b}\right)$ for $b > 0$
		$\left(-\frac{c}{b}, +\infty\right)$ for $b > 0$
$d > 0$	$K\lvert Z_u - r_1\rvert^{\frac{-1}{\sqrt{\Delta}}(ar_1+b)}\lvert Z_u - r_2\rvert^{\frac{-1}{\sqrt{\Delta}}(ar_2+b)}$	$(-\infty, r_1)$ for $b < 0$
		$(r_2, +\infty)$ for $b > 0$
$\Delta = 0$	$K\lvert Z_u - r_0\rvert^{\frac{-a}{d}} \exp\left[\frac{ar_0+b}{d(Z_u-r_0)}\right]$	$(-\infty, r_0)$ for $b < 0$
		$(r_0, +\infty)$ for $b > 0$
$\Delta < 0$	$K\left(c + bZ_u + dZ_u^2\right)^{\frac{-a}{2d}} exp\left[\frac{ab-2bd}{d\sqrt{-\Delta}} \tan^{-1}\left(\frac{b+2dZ_u}{\sqrt{-\Delta}}\right)\right]$	$(-\infty, +\infty)$
$b = 0,\ \Delta > 0,\ d < 0$	$K\left[-\frac{c}{d} - Z_u^2\right]^{\frac{-a}{2d}}$	$\left(\frac{\sqrt{\Delta}}{2d}, \frac{-\sqrt{\Delta}}{2d}\right)$
$\Delta = 0,\ d = 0$	$exp\left[-\frac{Z_u^2}{2}\right]$	$(-\infty, +\infty)$
$\Delta\langle 0, d\rangle 0$	$K\left[c + +dZ_u^2\right]^{\frac{-a}{2d}}$	$(-\infty, +\infty)$

$^*\Delta = b^2 - 4cd, r_1 \frac{-b-\sqrt{\Delta}}{2d}, r_2 \frac{-b+\sqrt{\Delta}}{2d}, r_0 = \frac{-b}{2d}, K$ is determined by $F_x(+\infty) = 1$

$$\beta_{\mathrm{FM}} = -\Phi^{-1}\left[\int_{-\infty}^{-\beta_{\mathrm{SM}}} f(Z_u)dZ_u\right] \tag{2.82}$$

The moments of the performance function are derived through point estimation, employing the structural reliability moment method previously delineated. For a performance function $Z = g(x)$ containing only one random variable, its moments can be obtained by point estimates, according to,

$$\mu_g = \sum_{i=1}^{m} P_i y\left[T^{-1}(u_i)\right] \tag{2.83}$$

$$\sigma_g^2 = \sum_{i=1}^{m} P_i\left\{y\left[T^{-1}(u_i)\right] - \mu_g\right\}^2 \tag{2.84}$$

$$\alpha_{rg}\sigma_g^r = \sum_{i=1}^{m} P_i\{y[T^{-1}(u_i)] - \mu_g\}^r \tag{2.85}$$

where μ_g, σ_g and α_{rg} are the first order moment, second moment and the r-th dimensionless moment of the $g(x)$, respectively; T^{-1} is the inverse Rosenblatt transformation. u_i is the estimating point and P_i is the corresponding weight factor, and they can be represented as,

$$u_i = \sqrt{2x_i},\ P_i = \frac{w_i}{\sqrt{\pi}} \tag{2.86}$$

where x_i and w_i are the abscissa and weight of the Hermite integral with weight function $exp\left(-x^2\right)$. For a performance function containing multiple random variables $Z = g(\mathbf{X}), \boldsymbol{X} = [X_1, X_2, \cdots X_n]$, the performance function $g(\boldsymbol{X})$ can be approximated as,

$$g'(\boldsymbol{X}) = \sum_{i=1}^{n} (g_i - g_\mu) + g_\mu \tag{2.87}$$

where,

$$g_\mu = g(\boldsymbol{\mu}) \tag{2.88}$$

$$g_i = g\left[T^{-1}(\boldsymbol{U}_i)\right] \tag{2.89}$$

where $\boldsymbol{\mu}$ is the vector consisting of the means of all random variables, and $\boldsymbol{U}_i = \left[u_{\mu 1}, u_{\mu 2}, \ldots, u_i, u_{\mu i+1}, \ldots, u_{\mu m}\right]^T$, which is composed of the means of random variables in standard normal space. G_μ is a constant, G_i is a function related only to u_i. The first four central moments of $g'(\boldsymbol{X})$ are obtained by,

$$\mu_g = \sum_{i=1}^{n} (\mu_i - g_\mu) + g_\mu \tag{2.90}$$

$$\sigma_g^2 = \sum_{i=1}^{n} \sigma_i^2 \tag{2.91}$$

$$\alpha_{3g}\sigma_g^3 = \sum_{i=1}^{n} \alpha_{3i}\sigma_i^3 \tag{2.92}$$

$$\alpha_{4g}\sigma_g^4 = \sum_{i=1}^{n} \alpha_{4i}\sigma_i^4 + 6\sum_{i=1}^{n-1}\sum_{j>i}^{n} \sigma_i^2\sigma_j^2 \tag{2.93}$$

where μ_i, σ_i, α_{3i} and α_{4i} are the first to fourth moments of g_i, respectively. Because g_i is a function containing only one random variable, its first four moments of order can be obtained using the Eqs. 2.88 to 2.91.

The moment method of structural reliability is implemented in the following steps:

(1) Determine g_μ and g_i using the formulas Eqs. 2.89 and 2.88;
(2) Computing the first four order moments of g_i;
(3) Using Eqs. 2.88–2.91, calculate the first four moments of the performance function.

The first four moments of the performance function, as delineated in the aforementioned method of moments, are employed to ascertain the reliability index.

2.2.4 *First-Order Saddle Point Approximation Method*

In approximate analytical methods, it is typically essential to transform the random variable $\boldsymbol{X}$ from its original $\boldsymbol{X}$-space to the standard normal $\boldsymbol{U}$-space. This transformation is, in the majority of cases, a nonlinear transformation, which increases the nonlinearity of the LSF and thus results in a reduction in approximation accuracy. This shortcoming can be effectively addressed through the use of the first-order saddle-point approximation method (FOSAM), which serves to mitigate the escalation of nonlinearity observed in the limit state function through the approximation of the function at the most likelihood point (MPL) within the $\boldsymbol{X}$-space domain, complemented by the utilization of an accurate saddle-point approximation technique.

2.2.4.1 Saddle Point Approximation

A saddle point is defined as a singularity that is stable in one direction but unstable in the other and is a stationary point of nonlocal extrema. The saddle-point approximation method can efficiently and accurately compute the PDF and CDF of the structural performance function. Let the performance function be $Z = g(\boldsymbol{X})$ and the PDF of the random variable be $f(x)$. Subsequently, the moment generating function (MGF) and cumulant generating function (CGF) of x can be expressed as,

$$M(t) = \int_{-\infty}^{+\infty} \exp(tx) f(x) dx \tag{2.94}$$

$$K(t) = ln[M(t)] \tag{2.95}$$

Table 2.2 shows the PDF and CGF of commonly used probability distributions for random variables.

Table 2.2 PDF and CGF of some common distributions

PDF	CGF	Distribution
$f(x) = \frac{1}{\sqrt{2\pi}\sigma} e^{\frac{(x-\mu)^2}{\sigma^2}}$	$K(t) = \mu t + \frac{1}{2}\sigma^2 t^2$	Normal
$f(x) = \frac{1}{b-a}$	$K(t) = \ln\left(\frac{e^{bt} - e^{at}}{t(b-a)}\right)$	Uniform
$f(x) = \frac{e^{\frac{x-\mu}{\sigma}} e^{-e^{\frac{x-\mu}{\sigma}}}}{\sigma}$	$K(t) = \mu t + \ln\Gamma(1 - \sigma t)$	Gumbel
$f(x) = \alpha e^{-\alpha x}$	$K(t) = -\ln\left(\frac{1-t}{\alpha}\right)$	Exponential
$f(x) = \frac{\beta^\alpha}{\Gamma(\alpha)x^{\alpha-1}e^{-\beta x}}$	$K(t) = \alpha[\ln\beta - n(\beta - t)]$	Gamma
$f(x) = \frac{x^{\frac{n}{2}-1} e\left(-\frac{1}{2}x\right)}{\Gamma(\frac{n}{2})2^{\frac{n}{2}}}$	$K(t) = -\frac{1}{2}n\ln(1-2t)$	χ^2

The inverse Fourier formula can be implemented to invert the solution of $f(x)$ from $K(t)$.

$$f(x) = \frac{1}{2\pi}\int_{-\infty}^{+\infty} M(it)e^{-itx}dt = \frac{1}{2\pi}\int_{-i\infty}^{+i\infty} \exp\{K(t) - tx\}dt \tag{2.96}$$

The saddle point approximation of the PDF of the random variable x is,

$$f(x) = \left\{\frac{1}{2\pi K''(t)}\right\}^{\frac{1}{2}} \exp\{K(t_s) - t_s x\} \tag{2.97}$$

Equation 2.97 is obtained by expanding the integral in Eq. 2.96 by an exponential power series and computing its Hermite polynomial approximation, where t_s is the saddle point computed from $K'(t_s) = x$. $K'(\cdot)$ is the first order derivative of CGF and $K''(\cdot)$ is the second order derivative of CGF. The first and second order derivatives of the CGF for commonly used variable distributions are shown in Table 2.3.

The probability of failure is shown below,

$$P_f = \Phi\left\{w + \frac{1}{w}\log\left(\frac{v}{w}\right)\right\} \tag{2.98}$$

It can be found that the accuracy of the fault probability based on the saddle point approximation is high, especially in the tail region of the distribution, and it is computationally efficient because the implementation of the saddle point approximation only requires searching for the saddle point locations without the need for integration operations.

Table 2.3 Derivative of CGF of some common distributions

K'	K''	Distribution
$K' = \mu + \sigma^2$	$K'' = 2\sigma$	Normal
$K' = \frac{\beta}{(1-\beta t)}$	$K'' = \frac{\beta^2}{(1-\beta t)^2}$	Exponential
$K' = a - \frac{\Psi(1-bt)}{b}$	$K'' = \Psi(1, 1-bt)b^2$	Extreme value
$K' = \frac{n}{1-2t}$	$K'' = \frac{2n}{(1-2t)^2}$	χ^2
$K' = \frac{[b\exp(bt) - a\exp(at)]}{[\exp(bt) - \exp(at)]} - \frac{1}{t}$	$K'' = \frac{[b^2\exp(bt) - a^2\exp(at)]}{[\exp(bt) - \exp(at)]} + \frac{1}{s^2} - \frac{[b\exp(bt) - a\exp(at)]^2}{[\exp(bt) - \exp(at)]^2}$	Uniform

2.2.4.2 Principles and Steps of FOSAM

The FORM and the SORM are commonly used as approximate analytical methods for structural reliability analysis, and a common disadvantage of these two methods is the need to transform the random variables, which can be achieved by Rosenblatt transformation, and it is important to note that this process tends to be nonlinear, and therefore increases the degree of nonlinearity of the limit state function, which results in the lack of approximation accuracy and may also produce chaotic non-convergent solutions during the MPP point search process. For improving the approximate accuracy, consideration should be given to avoiding the risk of increasing the nonlinearity of the limit state function by the transformation of the random variables. In other words, try to approximate the limit state function in the original space. This disadvantage can be better improved by the proposal of the FOSAM, CGF which is a linear approximation in the original space at the MPP of the limit state function. But not all random variables have the closed form of CGF, so they need to be converted to random variables with CGF and the FOSAM is discussed next in three cases.

(1) For the case where all random variables have the closed form of the CGF. A linear approximation is made at the MPP of the limit state function, where the MPP is defined in the same way as the MPP, as,

$$\begin{cases} \max\limits_{X} \prod\limits_{i=1}^{n} f_i(x_i) \\ \textit{subject to } g(\boldsymbol{X}) = 0 \end{cases} \tag{2.99}$$

The performance function for linearization is,

$$g(\boldsymbol{X}) \approx L(\boldsymbol{X}) = \nabla g(\boldsymbol{x}^*)(\boldsymbol{X} - \boldsymbol{x}^*) \tag{2.100}$$

The CGF of the performance function can be computed by,

$$K(t)=\sum_{i=1}^{n}K_i(t) \tag{2.101}$$

where $K_i(t)$ is the CGF of $\nabla g_i(\boldsymbol{x}^*)\left(\boldsymbol{X}_i-\boldsymbol{x}_i^*\right)$. The first two derivatives of $K(t)$ are,

$$K'(t)=\sum_{i=1}^{n}K_i'(t) \tag{2.102}$$

$$K''(t)=\sum_{i=1}^{n}K_i''(t) \tag{2.103}$$

The saddle point t_s is determined by,

$$K'(t)-x=\sum_{i=1}^{n}K_i'(t)-x=0 \tag{2.104}$$

The failure probability can be computed using Eq. 2.98, where,

$$\begin{cases} w=sign(t_s)\{2[t_s x-K(t_s)]\}^{1/2}=sign(t_s)\left\{2\left[t_s x-\sum_{i=1}^{n}K_i(t_s)\right]\right\}^{1/2} \\ v=t_s\{K''(t_s)\}^{1/2}=t_s\left\{\sum_{i=1}^{n}K_i''(t_s)\right\}^{1/2} \end{cases} \tag{2.105}$$

(2) For the case where some of the random variables have CGF closed form. Random variables that do not have CGF could be transformed into standard normal variables using the same variable transformation methods as those used in FORM and SORM. Denote the random variable with CGF closed form as $\boldsymbol{X}^t=\left\{X_i^t;i=1,2,\cdots,n_i\right\}$ and the random variable without CGF closed form as $\boldsymbol{X}^{\sim t}=\left\{X_j^{\sim t};j=1,2,\cdots,n_{\sim i}\right\}$, and after variable transformation, $\boldsymbol{X}^{\sim t}$ becomes $\boldsymbol{U}^t=\left\{U_j;j=1,2,\cdots,n_{\sim i}\right\}$. The MPP is defined as,

$$\begin{cases} \max\limits_{\boldsymbol{x}',\boldsymbol{u}}\prod_{i=1}^{n_i}f_i\left(x_i^t\right)\prod_{j=1}^{n_{\sim I}}\phi\left(u_j\right) \\ subject\ to\ g\left(\boldsymbol{x}',\ \boldsymbol{u}\right)=0 \end{cases} \tag{2.106}$$

The performance function after linearization at the MPP $\left\{\boldsymbol{x}^{t*},\boldsymbol{u}^*\right\}$ is given by,

$$g(\boldsymbol{X}) = L(\boldsymbol{X}', \boldsymbol{U}) = \sum_{i=1}^{n_i} \frac{\partial g}{\partial x_i^t}\bigg|_{\boldsymbol{x}^{t*}, \boldsymbol{u}^*} (\boldsymbol{X}_i^t - x_i^{t*}) + \sum_{j=1}^{n_{\sim i}} \frac{\partial g}{\partial u_j}\bigg|_{\boldsymbol{x}^{t*}, \boldsymbol{u}^*} \left(\boldsymbol{U}_j - u_j^*\right) \tag{2.107}$$

Equation 2.106 can be combined with Eqs. 2.101 to 2.105 to solve for the failure probability.

(3) For the case where all random variables do not have CGF. All random variables need to be transformed, i.e., from the original variable to the standard normal variable, at which point the MPP are defined as,

$$\begin{cases} \max\limits_{\boldsymbol{u}} \prod\limits_{i=1}^{n} \phi(u_i) \\ subject\ to\ g(\boldsymbol{u}) = 0 \end{cases} \tag{2.108}$$

The definition of MPP in Eq. 2.108 is the same as the definition of MPP. Thus, the linearized performance function is shown below,

$$g(\boldsymbol{X}) = \sum_{i=1}^{n} \frac{\partial g}{\partial u_i}\bigg|_{\boldsymbol{u}^*} (\boldsymbol{U}_i - u_i^*) \tag{2.109}$$

The failure probability obtained by using the FOSAM in this case is the same as the result of FORM, therefore FORM is a special case of FOSAM.

The general procedure of the first-order saddle-point approximation is summarized as below,

(1) Distinguish which random variables have the closed form of CGF and which do not, and denote them respectively as $\boldsymbol{X}^t$ and $\boldsymbol{X}^{\sim t}$. Transform $\boldsymbol{X}^{\sim t}$ into standard normal variable $\boldsymbol{U}^t$;
(2) Use Eq. 2.106 to define MPP;
(3) Linearize the performance function according to Eq. 2.107;
(4) Solve for the saddle point using Eqs. 2.101 to 2.105;
(5) Obtain the failure probability according to Eq. 2.98.

The difference between the FOSAM and the FORM is that it does not necessarily need to transform each random variable, so it declines the possibility of increasing the degree of nonlinearity of the limit state function, and therefore the accuracy of the saddle-point approximation of the limit state function is higher. The FORM is a special case of FOSAM, and the two are equivalent in these cases: (1) All random variables have the closed form of CGF, i.e., no variable transformation is required; (2) Random variables with a CGF closed form all obey a normal distribution, and some random variables that do not have a CGF closed form are transformed into standard normal variables; (3) All random variables follow a normal distribution.

In conclusion, the first-order saddle-point approximation method is an efficient and reliable method that balances accuracy and computational efficiency. The FOSAM reduces the likelihood of the degree of nonlinearity of the performance function as compared to the FORM and SORM so that the FOSAM can be considered when the degree of nonlinearity of the limiting state function in the primal space is less than the degree of nonlinearity in its standard normal space.

2.2.5 *Non-probabilistic Reliability Model*

The probabilistic model can solve the reliability index and failure probability of the structure by determining the uncertainty parameters of the random variables and then combining them with the existing probabilistic methods, such as MCS [22], FORM, etc. However, the use of probabilistic models requires the provision of precise uncertainty parameters and a large number of sample points, otherwise, there is a risk of large errors in the assessment of structural reliability due to small deviations in the estimation of random variables. Uncertainty parameters for random variables are nevertheless easy to obtain, hence non-probabilistic reliability (NPR) models. The difference between NPR and SRA is that the former solves for the range of structurally acceptable behavior, while the latter emphasizes the probability of structurally acceptable behavior. This section describes the NPR models, such as the interval set model and the convex model.

2.2.5.1 Convex NPR Model

Random variables are symmetrically and uniformly distributed in the NPR. The centroids of the intervals of random variables are within the uncertainty interval, and the corresponding interval centroids $\boldsymbol{x}^C$ (nominal values) and radii $\boldsymbol{x}^w$ (deviations) are as follows

$$\boldsymbol{x}^c = \frac{x^U + x^L}{2}, \boldsymbol{x}^c = \frac{x^U - x^L}{2} \tag{2.110}$$

The random variable is standardized as,

$$q = \boldsymbol{x}^c + \boldsymbol{x}^c \delta \tag{2.111}$$

where $\delta \in [-1, +1]$ represents a standardized interval variable. Assuming a linear approximation function of,

$$y = \sum_{i=1}^{n} a_i q_i \tag{2.112}$$

where $a_i \in R(i = 1, 2, \cdots, n)$ is an arbitrary real number, the mean and STD of y can be respectively expressed as,

$$y^c = \sum_{i=1}^{n} a_i q_i^c,\ y^r = \sum_{i=1}^{n} |a_i|\, q_i^r \tag{2.113}$$

The performance functions are shown below,

$$Z = g(\boldsymbol{X}) = g(x_1, x_2, \ldots, x_n) \tag{2.114}$$

According to the definition of the limit state function $g(\boldsymbol{X})$, which divides the design space of the structure into two parts, $g(\boldsymbol{X}) > 0$ for the safety domain and $g(\boldsymbol{X}) < 0$ for the failure domain. When $g(\boldsymbol{X})$ is a continuous function, Z is also an interval variable face with a maximum value of Z^U and a minimum value of Z^L, then its mean Z^c and deviation Z^r are expressed respectively as,

$$Z^c = \frac{Z^U + Z^L}{2}, Z^r = \frac{Z^U - Z^L}{2} \tag{2.115}$$

Combined with the knowledge related to non-probabilistic reliability theory, the non-probabilistic reliability index of the structure is shown as follows,

$$\eta = \frac{Z^c}{Z^r} = \frac{Z^U + Z^L}{Z^U - Z^L} \tag{2.116}$$

when $g(\boldsymbol{X}) > 0$ for $\eta > 1$, the structure is safe; $g(\boldsymbol{X}) < 0$ for $\eta < -1$, the structure is failure; when $-1 \leq \eta \leq 1$, both $g(\boldsymbol{X}) > 0$ and $g(\boldsymbol{X}) < 0$ are possible, it means that the structure may be safe or failing. In summary, the larger the value of η, the more reliable the structure is. The η_j can be also given by,

$$\begin{cases} \mathit{find}\ q \\ \max\limits_{\boldsymbol{d}} \eta = sign(g(0)) \cdot \min\left\{\max\left(\sqrt{\boldsymbol{q}_1^T \boldsymbol{q}_1}, \sqrt{\boldsymbol{q}_2^T \boldsymbol{q}_1}, \ldots, \sqrt{\boldsymbol{q}_k^T \boldsymbol{q}_k}\right)\right\} \\ \mathit{subjec\ to}\ g_i(\boldsymbol{q}) = 0\ i = 1, 2, \ldots, n_g \end{cases} \tag{2.117}$$

In which $g_i(\cdot)$ is the ith performance function, the number n_g of performance functions and the number $g_i(\boldsymbol{q})$ of grouped ellipsoids, respectively. The schematic of the problem with 3 uncertain parameters is plotted in Fig. 2.6.

2.2.5.2 Interval Set NPR Model

The implementation of the interval set model, one of the most commonly used of the non-probabilistic models, requires the standardization of uncertain design variables $\boldsymbol{X}$ and uncertain variables $\boldsymbol{q}$. The expression for the non-probabilistic model based on the set of intervals is given by,

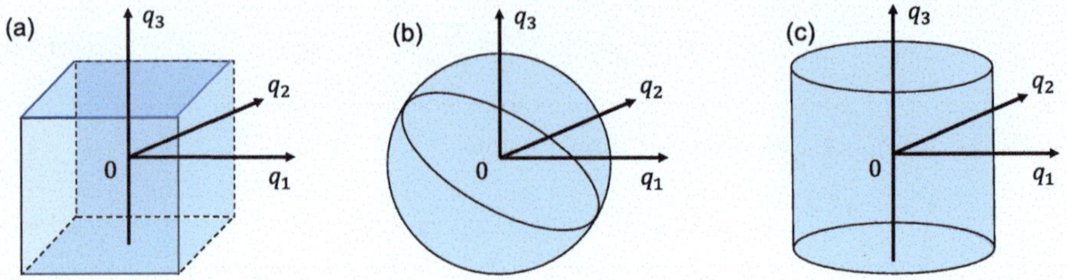

Fig. 2.6 Convex models: **a** interval set model $\{\|q_i\| \leq 1,\ i = 1, 2, 3\}$, **b** single-ellipsoid, model,$\{\|q_1, q_2, q_3\| \leq 1\}$, **c** multi-ellipsoid model $\{\|q_1, q_2\| \leq 1;\ q_3 \leq 1\}$. Reprinted from [23], with permission from Elsevier

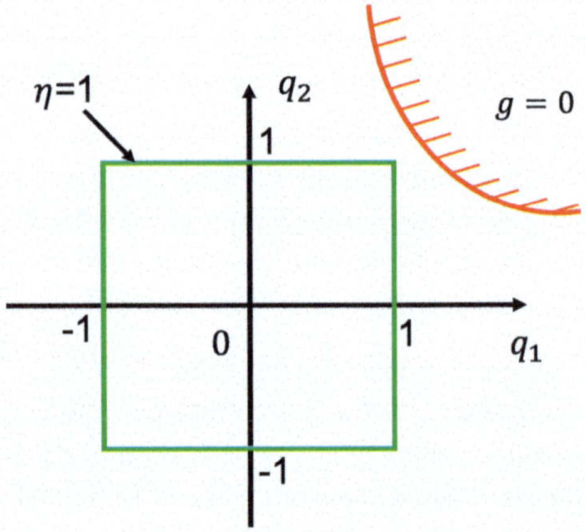

Fig. 2.7 Interval set model. Reprinted from [24], with permission from Elsevier

$$\begin{cases} \mathit{find}\ \boldsymbol{q} \\ \min\limits_{d} \eta = sign(g(0)) \cdot \min\{\max(|q_1|, \ldots, |q_i|, \ldots, |q_n|)\} \\ subjec\ to\ g(\boldsymbol{q}) = 0 i = 1, 2, \ldots, n_g \end{cases} \tag{2.118}$$

where $g(\cdot)$ indicate the performance function; $\boldsymbol{q}$ is the normalized uncertain variable and $sign(\cdot)$ is the sign function. When $\eta \geq 1$, it indicates that all possible values of uncertainty lie in the security domain, otherwise, the other cases are strictly considered as structural failures. Figure 2.7 shows a two-dimensional interval set model.

2.2.5.3 Ellipsoid Model

The interval set model, although it shows good performance in dealing with uncertainties but bounded variables, it may encounter the problem of over-conservatism. Therefore, the introduction of ellipsoid modeling effectively overcomes this problem to some extent. The mathematical expression for the ellipsoid model is,

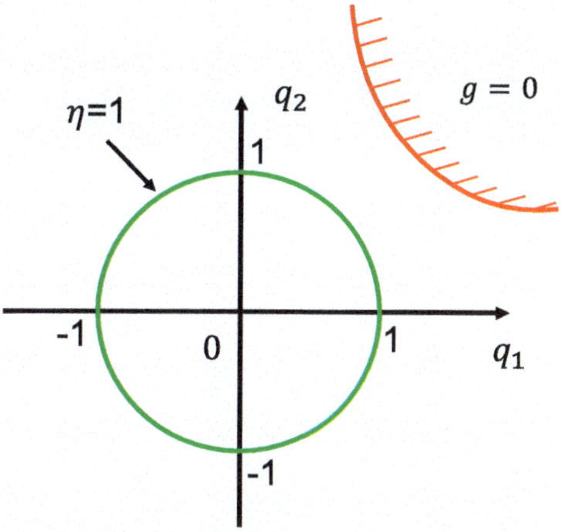

Fig. 2.8 Ellipsoid model. Reprinted from [24], with permission from Elsevier

$$\begin{cases} \textit{find } \boldsymbol{q} \\ \min\limits_{d} \eta = sign(g(0)) \cdot \|\boldsymbol{q}\| \\ \textit{subjec to } g(\boldsymbol{q}) = 0 \end{cases} \tag{2.119}$$

Figure 2.8 shows an example of a two-dimensional ellipsoid model that can be represented by a circle with center (0, 0) and radius 1. The FORM-based on CFSL is to take the structural reliability index as the closest distance from the origin to the limit state function $g(\boldsymbol{U})$ in standard normal space as $\beta = \|\boldsymbol{U}\|$. Therefore, ignoring the role of the sign function in the ellipsoid model, which is essentially the same as the FORM, the well-established optimization methods of FORMcan be used, but it is worth noting that these optimization methods may not be used for other non-probabilistic models.

2.2.5.4 Super Ellipsoid Model

The super ellipsoid can be introduced to non-probabilistic Reliability-Based Design Optimization (NRBDO) due to the fact that it provides more options for geometry which has been verified in analytic geometry theory, and the super ellipsoid model can be given by,

$$\left|\frac{q_1}{a_1}\right|^n + \left|\frac{q_2}{a_2}\right|^n + \cdots + \left|\frac{q_n}{a_n}\right|^n = 1 \tag{2.120}$$

Equation 2.120 defines a multidimensional rectangle with closed curves contained in the rectangle $-a_i \leq q_i \leq a_i, i = 1, 2, \ldots, n$, where a_i is the semidiameters of the curve. There is a two-dimensional super ellipsoid model with $a_1 = a_2 = 1$ is shown in Fig. 2.9. If $n = 2$, the curve is a normal circle, and if $n > 2$, the curve is a rounded rectangle.

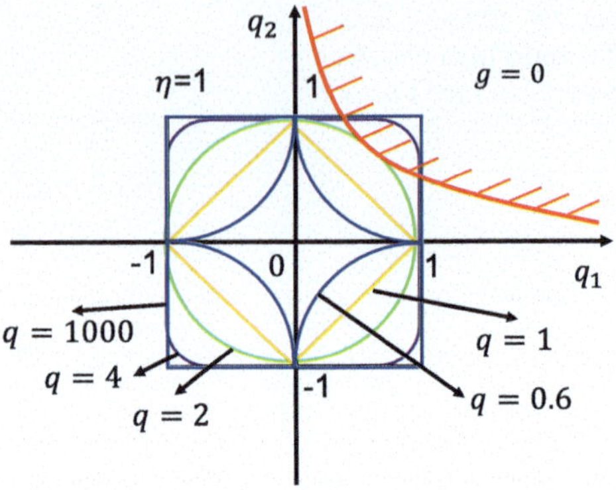

Fig. 2.9 Super ellipsoid model. Reprinted from [24], with permission from Elsevier

2.2.6 Dimension Reduction Method

The dimensionality reduction method [25] performs structural reliability analysis by calculating the statistical moments of the performance function. Depending on the degree of dimensionality reduction, i.e., the difference in the dimensionality of the function into which the performance function is decomposed, it can be classified into three different methods: (1) univariate dimensionality reduction; (2) binary dimensionality reduction; and (3) multivariate dimensionality reduction. In this section, we only introduce univariate dimensionality reduction methods.

2.2.6.1 Univariate Dimension Reduction Method

The performance function $h(x)$ is shown below,

$$h(\boldsymbol{X}) \cong \hat{h}(\boldsymbol{X}) \equiv \sum_{i=1}^{N} h(\mu_1, \ldots, \mu_{i-1}, x_i, \mu_{i+1}, \ldots, \mu_N) - (N-1)h(\mu_1, \ldots, \mu_N) \tag{2.121}$$

In which u_i is the average value of the random variable. N is the number of design variables.

The univariate additive decomposition of a two-dimensional performance function is,

$$h(\boldsymbol{X}) \cong \hat{h}(\boldsymbol{X}) \equiv h(x_1, \mu_2) + h(\mu_1, x_2) - h(\mu_1, \mu_2) \tag{2.122}$$

The use of a univariate DRM, which turns an N-dimensional integral into N one-dimensional integrals, which greatly reduces the number of function evaluations. According to MBIR, the kth statistical moment of a one-dimensional function is,

$$E\left(\{h(\boldsymbol{X})\}^k\right) = \sum_{i=1}^{n} w_i h^k(x_i) \tag{2.123}$$

where w_i indicates weight sand x_i indicates quadrature points. n is the number of weights and orthogonal points.

Combining Eqs. 2.121 and 2.122, the mean and variance of the performance function are,

$$\begin{aligned}\mu_H \equiv E[h(X)] &\cong E\left\{\sum_{i=1}^{N} h(\mu_1, \ldots, \mu_{i-1}, X_i, \mu_{i+1}, \ldots, \mu_N) - (N-1)h(\mu_1, \ldots, \mu_N)\right\} \\ &\cong \sum_{j=1}^{n}\sum_{i=1}^{N} w_i^j h\left(\mu_1, \ldots, \mu_{i-1}, x_i^j, \mu_{i+1}, \ldots, \mu_N\right) - (N-1)h(\mu_1, \ldots, \mu_N)\end{aligned} \tag{2.124}$$

$$\begin{aligned}\sigma_H^2 &\equiv E\left[(h(\boldsymbol{X}) - \mu_H)^2\right] = E\left[h^2(\boldsymbol{X})\right] - \mu_H^2 \\ &\cong E\left\{\sum_{i=1}^{N} h^2(\mu_1, \ldots, X_i, \ldots, \mu_N) - (N-1)h^2(\mu_1, \ldots, \mu_N)\right\} - \mu_H^2 \\ &\cong \sum_{i=1}^{n}\sum_{i=1}^{N} w_i^j h^2\left(\mu_1, \ldots, x_i^j, \ldots, \mu_N\right) - (N-1)h^2(\mu_1, \ldots, \mu_N) - \mu_H^2\end{aligned} \tag{2.125}$$

As shown in Eqs. 2.123 and 2.124, the univariate DRM uses a sum of one-dimensional functions to approximate the performance function $h(x)$. The approximation is exact if $h(\boldsymbol{X}) = \sum_{i=1}^{N} h_i\ (x_i)$ in which $h_i(x_i)$ is an arbitrary function of x_i. However, the use of a one-dimensional function and approximation of a non-diagonal term will produce some error if a non-diagonal or mixed term exists. To minimize this error, a binary DRM can be used. The second approximation involves the use of weights and the numerical integration of orthogonal points. Based on Gaussian quadrature theory, n quadrature points and weights have an accuracy of $2n-1$. Therefore, the error in the numerical integration of the quadratic performance function will disappear.

The efficiency of the method for estimating the central moments is described next, and the number of function evaluations required to estimate the output moments using the univariate DRM and MBIR is,

$$FE = n \times N + 1 \tag{2.126}$$

In which n and N are the number of orthogonal points and the number of variables, respectively. When the distribution of the input variables is symmetric and the number is odd, the number of function evaluations is,

$$FE = (n-1) \times N + 1 \tag{2.127}$$

In summary, the number of function evaluations decreases significantly when the number of design variables is large, but it is still proportional to the number of design variables.

The number of function estimates using the bivariate DRM is,

$$FE = \frac{N(N-1)}{2} \times n^2 + n \times N + 1 \tag{2.128}$$

The model for solving the MPP point is shown below,

$$\begin{aligned} &\text{minimize} \quad \|\boldsymbol{u}\| \\ &\text{subject to } h(x(\boldsymbol{u})) = 0 \end{aligned} \tag{2.129}$$

The model is to find the point $\boldsymbol{u}^* = \{u_1^*, u_2^*, \ldots, u_N^*\}^{\mathrm{T}}$ on $h(x)$ that is closest to the origin of standard normal $\boldsymbol{U}$-space.

$\boldsymbol{U}$-space can generally be found using the Rosenblatt transform and $x(\boldsymbol{u})$,

$$F_{X_i}(x_i) = \Phi(u_i) \tag{2.130}$$

Solve for the probability of failure using the MPP point as a reference point,

$$P_{\mathrm{F}} \equiv P[h(\boldsymbol{X}) > 0] = \int\limits_{G(\boldsymbol{X})>0} f_X(\boldsymbol{x})\,\mathrm{d}\boldsymbol{x} \tag{2.131}$$

in which $h(x) > 0$ is defined here as the structure failure. Decomposing the performance function into a one-dimensional function at the MPP yields,

$$h(\boldsymbol{X}) \cong \hat{h}(\boldsymbol{X}) \equiv \sum_{i=1}^{N} h\left(x_1^*, \ldots, x_{i-1}^*, X_i, x_{i+1}^*, \ldots, x_N^*\right) - (N-1)h\left(\boldsymbol{x}^*\right) \tag{2.132}$$

where $\boldsymbol{x}^* = \{x_1^*, x_2^*, \ldots, x_N^*\}^T$ indicates the MPP point according to Eqs. 2.125 and 2.126.

Next, convert $\boldsymbol{U}$-space to rotated standard normal $\boldsymbol{V}$-space. Construct an $N \times N$ orthogonal matrix R.

$$\alpha^* \equiv \frac{\mathrm{u}^*}{\beta_{\mathrm{HL}}} \tag{2.133}$$

i.e., $R = [R_1|\alpha^*]$, where $N \times (N-1)$ matrix R_1 satisfies $(\alpha*)TR1 = 0$ and N is the number of random variables. Using the orthonormal transformation $u = Rv$, v represents the rotated standard normal $\boldsymbol{V}$-space with the MPP $\boldsymbol{v}^* = \{0, \ldots, 0, \beta_{\mathrm{HL}}\}^{\mathrm{T}}$.

In $\boldsymbol{V}$-space, the failure probability can be expressed as,

$$P_{\mathrm{F}} \equiv P\big[\tilde{h}(\boldsymbol{V}) > 0\big] = \int\limits_{\tilde{h}(\boldsymbol{V})>0} f_{\boldsymbol{V}}(\boldsymbol{v})\,\mathrm{d}\boldsymbol{v}, \tag{2.134}$$

where $h(\boldsymbol{V}) = h(\boldsymbol{X}(\boldsymbol{V}))$. Next, the probability of failure is calculated in conjunction with the constraint transfer.

$$\widetilde{h^s}(\boldsymbol{v}) \equiv \tilde{h}(\boldsymbol{v}) - \tilde{h}(\boldsymbol{v}^*) \tag{2.135}$$

where $\widetilde{h^s}(\boldsymbol{v})$ is the displacement performance function. $\boldsymbol{v}^* = \{0, \ldots, 0, \beta\}^{\mathrm{T}}$ is the FORM-based MPP in $\boldsymbol{V}$-space given the reliability index β, so $\tilde{h}^s(\boldsymbol{v}^*) = \tilde{h}(\boldsymbol{v}^*) - \tilde{h}(\boldsymbol{v}^*) = 0$, Assuming that the performance function has three random variables,$\widetilde{h^s}(\boldsymbol{v})$ at the MPP is denoted as,

$$\begin{aligned}\tilde{h}^s(v) &\cong \hat{\tilde{h}}^s(v) \equiv \sum_{i=1}^{3} \tilde{h}_i^s(v_i) - (3-1)\tilde{h}^s(\boldsymbol{v}^*)\\ &= \tilde{h}^s\big(v_1, v_2^*, v_3^*\big) + \tilde{h}^s\big(v_1^*, v_2, v_3^*\big) + \widetilde{h}^s\big(v_1^*, v_2^*, v_3\big)\\ &= \tilde{h}^s(v_1, 0, \beta) + \tilde{h}^s(0, v_2, \beta) + \tilde{h}^s(0, 0, v_3)\end{aligned} \tag{2.136}$$

where $\boldsymbol{v^*} = \left\{v_1^*, v_2^*, v_3^*\right\}^{\mathrm{T}} = \{0, 0, \beta\}^{\mathrm{T}}$. Due to the rotational transformation of the coordinates, the last univariate component $\tilde{h}^s(0, 0, v_3)$ can be approximated as,

$$h^s(0, 0, v_3) \cong \left.\frac{\partial h(\boldsymbol{v})}{\partial v_3}\right|_{v=v^*}(v_3 - \beta) = b_1(v_3 - \beta) \tag{2.137}$$

By using the linear assumption, Eqs. 2.136 and 2.134 are denoted as,

$$\begin{aligned}P_{\mathrm{F}} \equiv P[h^s(V) > 0] &\cong P[b_1(V_3 - \beta) + h^s(V_1, 0, \beta) + h^s(0, V_2, \beta) > 0]\\ &= P[b_1 V_3 > b_1\beta - h^s(V_1, 0, \beta) - h^s(0, V_2, \beta)]\end{aligned} \tag{2.138}$$

Equation 2.139 can be rewritten as,

$$\begin{aligned}P_F &\cong P\big[b_1 V_3 > b_1\beta - \widetilde{h^s}(V_1, 0, \beta) - \widetilde{h^s}(0, V_2, \beta)\big]\\ &= P\Big[V_3 < -\beta + \tfrac{1}{b_1}\big\{\widetilde{h^s}(V_1, 0, \beta) + \widetilde{h^s}(0, V_2, \beta)\big\}\Big]\\ &= E\Big[\Phi\Big(-\beta + \tfrac{1}{b_1}\big\{\tilde{h}^s(V_1, 0, \beta) + \tilde{h}^s(0, V_2, \beta)\big\}\Big)\Big]\end{aligned} \tag{2.139}$$

Introduction of Eq. 2.138

$$\ln\Big[\widetilde{h}^s(V_1, V_2)\Big] = \ln\Big[\Phi\Big(-\beta + \frac{1}{b_1}\{h^s(V_1, 0, \beta) + h^s(0, V_2, \beta)\}\Big)\Big] \tag{2.140}$$

Combining Eqs. 2.137 and 2.121, it follows that,

$$\ln\left[\overline{h}^{s}(V_1, V_2)\right] \cong \ln\left[\Phi\left(-\beta + \frac{1}{b_1}h^{s}(V_1, 0, \beta)\right)\right] + \ln\left[\Phi\left(-\beta + \frac{1}{b_1}h^{s}(0, V_2, \beta)\right)\right] - \ln[\Phi(-\beta)] \tag{2.141}$$

$$\overline{h}^{s}(V_1, V_2) \cong \frac{\Phi\left(-\beta + \frac{1}{b_1}h^{s}(V_1, 0, \beta)\right)\Phi\left(-\beta + \frac{1}{b_1}h^{s}(0, V_2, \beta)\right)}{\Phi(-\beta)} \tag{2.142}$$

According to Eqs. 2.141 and 2.142.

$$P_{\mathrm{F}} \cong E\left[\overline{\tilde{h}}^{s}(V_1, V_2)\right] = \frac{E\left[\Phi\left(-\beta + \frac{1}{b_1}\tilde{h}^{s}(V_1, 0, \beta)\right)\right]E\left[\Phi\left(-\beta + \frac{1}{b_1}\tilde{h}^{s}(0, V_2, \beta)\right)\right]}{\Phi(-\beta)} \tag{2.143}$$

A further generalization of Eq. 2.140, for n-dimensional functions, Eq. 2.140 can be transformed as,

$$P_{\mathrm{F}}^{\mathrm{DRM}} \cong \frac{\prod_{i=1}^{N-1}\int_{-\infty}^{\infty}\Phi\left(-\beta + \frac{\tilde{h}_i^{s}(v_i)}{b_1}\right)\phi(v_i)\,\mathrm{d}v_i\left(-\beta + \frac{\tilde{h}_i^{s}(v_i)}{b_1}\right)\phi(v_i)\,\mathrm{d}v_i}{\Phi(-\beta)^{N-2}} \tag{2.144}$$

where $\tilde{h}_i^{s}(v_i) \equiv \tilde{h}^{s}\left(v_1^{*}, \ldots, v_{i-1}^{*}, v_i, v_{i+1}^{*}, \ldots, v_N^{*}\right) = \tilde{h}^{s}(0, \ldots, 0, v_i, 0, \ldots, \beta)$ is a function of v_i. Using the Moment-Based Integration Rule (MBIR), Eq. 2.145 can be transformed into,

$$P_{\mathrm{F}}^{\mathrm{DRM}} \cong \frac{\prod_{i=1}^{N-1}\sum_{j-1}^{n} w_j\Phi\left(-\beta + \frac{\tilde{h}_i^{s}\left(v_i^{j}\right)}{b_1}\right)}{\Phi(-\beta)^{N-2}} \tag{2.145}$$

where v_i^{j} and w_j are the orthogonal points and weights, respectively, and n is the number of theirs.

Unlike the conventional DRM, the MPP-based DRM tends to move the samples and the quadrature points to the vicinity of the failure region, respectively, whereas the former concentrates the samples and the quadrature points near the estimated value of the parameter, respectively, which is the reason why the method of the MPP-based DRM is more accurate.

2.3 Simulation Methods

2.3.1 Monte Carlo Simulation

The SRA integral can be transformed as follows,

$$P_f = \int_{g(X)\leq 0} \int I_{g(X)\leq 0}(X) f_X(x_1, \ldots, x_n) dx_1 \ldots dx_n \tag{2.146}$$

In practice, the MCS is generally used as a reference for other methods, and its estimates are as follows:

$$\hat{P}_f^{MC} = E\left[I_{g(X)\leq 0}(X)\right] = \frac{\sum_{i=1}^{NS} I_{g(X)\leq 0}(X)}{NS} \tag{2.147}$$

in which, $E[\cdot]$ and NS are the expectation operator and the number of sample sizes. $\sum_{i=1}^{NS} I_{g(X)\leq 0}(X)$ the number of samples giving negative or zero values of LSF ($g(X)$). Sample size is extremely important for accurate estimation of failure probability. A large number of random samples is required to obtain a very low-probability of failure. The accuracy of the MCS method is generally tested by the variance and coefficient of variation of the prediction results, estimated as follows,

$$Var\left(\hat{P}_f^{MC}\right) \approx \frac{\hat{P}_f^{MC} - \left(\hat{P}_f^{MC}\right)^2}{NS - 1} \tag{2.148}$$

The coefficients of variation are as follows,

$$CoV_{P_f}^{MC} = \frac{\sqrt{Var\left(\hat{P}_f^{MC}\right)}}{E\left(\hat{P}_f^{MC}\right)} = \sqrt{\frac{1 - \hat{P}_f^{MC}}{(NS - 1)\,\hat{P}_f^{MC}}} \tag{2.149}$$

If the coefficient of variation of $\hat{P}_f^{MC}$ is low enough (i.e.$CoV_{P_f}^{MC} < 0.05$), MCS approach to solving the failure probability is highly accurate when it is applied a suitable sample size.

The sampling error is inversely proportional to the number of samples. MCS is a general method that obtains basic sample points by directly extracting the distribution function of random variables. Therefore, as shown in Fig. 2.10, for the case where the performance function has a large degree of nonlinearity, although the MCS simulation can obtain more failure samples, there are very few samples near the MPP point.

Due the engineering problem is time-consuming, MCS requires a huge number of samples to evaluate the probability of failure, so it is generally only used as a test method.

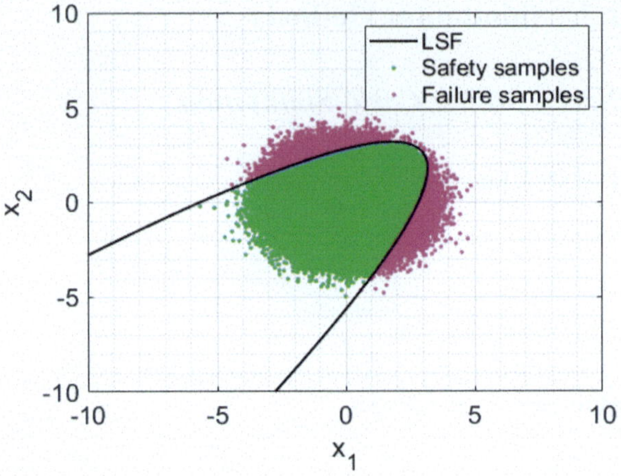

Fig. 2.10 MCS simulation data points in safe and failure domain for a two-dimension reliability problem

2.3.2 *Importance Sampling*

The importance of sampling can improve the computational burden of MCS when it may be impossible for applying in the engineering reliability analysis problems. Instead of using samples, the IS technique introduces an alternative sampling method that it can reduce the variance of simulated data for LSF. In IS the MCS integration is expressed as follows,

$$P_f = \int_{g(X)\leq 0} \int I_{g(X)\leq 0}(X)\frac{f_X(x)}{h_X(x)}h_X(x)dx_1 \ldots dx_n \tag{2.150}$$

where, $h_X(x)$ is an empirical density function and it should be satisfied $f_X(x) > 0$ and $h_X(x) > 0$. In the IS, the integration is computed based on another joint distribution function of $h_X(x)$ where $\int_{all} \int h_X(x)dx_1 \ldots dx_n = 1$ and factor $\frac{f_X(x)}{h_X(x)}$ indicated to sampling ratio (weight). Therefore, the failure probability is determined as below,

$$\hat{P}_f^{IS} = E\left[I_{g(X)\leq 0}(X)\right] = \frac{\sum_{i=1}^{NS} I_{g(X)\leq 0}(X)\frac{f_X(x)}{h_X(x)}}{NS} \tag{2.151}$$

An appropriate choice for empirical density function in the IS i.e. $h_X(x)$ can provide the smallest variance than Monte Carlo estimate. The data points in the safe and failure domain is presented in Fig. 2.11 for a nonlinear LSF. As seen, IS generates normally distributed samples in important areas based on MPP information and calculates the failure probability in a weighted manner. Therefore, for single MPP point problems, its results are usually more accurate, but for other situations, IS may be less effective. In addition, the cost of obtaining MPP is also worth considering.

IS samples are usually more concentrated, so the sampling efficiency is also higher. It may be provided a variance for IS estimate as zero if it is chosen perfect.

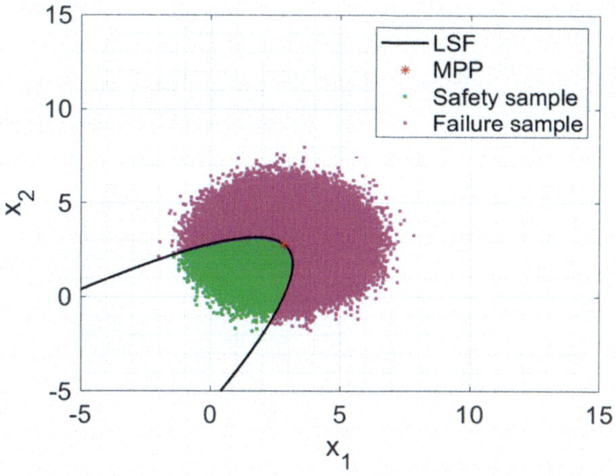

Fig. 2.11 Simulated data points in IS for illustrating the safety and failure domain of a two-dimension reliability problem

Commonly, its choices are as unimodal or multimodal density functions which are defined based on initial sampling.

2.3.3 Uniform Importance Sampling

Rashki et al. [26] proposed a uniform sampling method. The method simulates the variables of the limit state function $g(X_1, X_2, \ldots X_n)$

$$\begin{cases} L_i = F_i^{-1}(P_A) \\ U_i = F_i^{-1}(1 - P_A) \end{cases} \tag{2.152}$$

where i is the i-th random variable. The upper and lower limits of the uniform distribution, U_i and L_i, respectively, are defined. The assumed failure probability, P_A, is also specified. Finally, the inverse function of the cumulative distribution function of the random variable X_i, F_i^{-1}, is given. Subsequently, the failure probability can be estimated as follows [27],

$$\hat{P}_f = \frac{\sum_{j=1}^{NS} I_{g(X)\leq 0} \cdot w_j}{\sum_{j=1}^{NS} w_j} \tag{2.153}$$

In this instance, $\hat{P}_f$ represents the estimated failure probability utilizing the UIS methodology. In this context, N represents the sample size. The term $w_j = \prod_{i=1}^{n} f_i(j)$ denotes the weight attributed to the j-th sample. This is calculated as the probability density function of the j-th sample of variable X_i, represented by $f_i(j)$. The indicator function, $I(j)$, is defined as follows: when $g \leq 0$, $I_{g(X)\leq 0} = 1$ and when $g > 0$, $I = 0$ and when g > 0, $I = 0$. The simulated data for a two-dimension problem

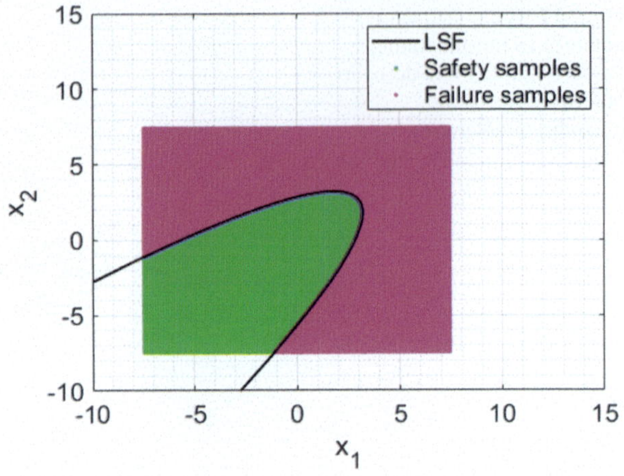

Fig. 2.12 Simulated safe and failure domains using UIS for a two-dimension reliability problem

is schematically shown in Fig. 2.12. As seen, UIS generates uniformly distributed samples based on the information of the random distribution function. It can better cover all areas of the performance function. Therefore, the sampling efficiency is higher than MCS [28, 29]. However, due to this feature, the samples are not concentrated enough, so the efficiency is lower than IS in most cases. However, it does not require MPP information and is a general and simple method.

The method requires reasonable assumptions about the probability of failure. The recommended intervals for sampling are shown in Table 2.4 [26].

Importance sampling is a common method proposed by researchers that can increase the efficiency and reduce the error of multiple indicator cluster surveys (MICS). When the sampling density is appropriate, the variance of the method is smaller. Standardized Importance Sampling (SIS) [30] has also been developed to address the problem of sampling density selection for IS,

$$\hat{p}_{f(SIS)} = \frac{\sum_{j=1}^{N} w_j I(\boldsymbol{X}_j)}{\sum_{j=1}^{N} w_j} \tag{2.154}$$

where w_j is the weight, defined as,

Table 2.4 The least required interval for UIS [26]

Interval (β)	3	4	5	6
L	$F^{-1}(5 \times 10^{-7})$	$F^{-1}(5 \times 10^{-9})$	$F^{-1}(5 \times 10^{-12})$	$F^{-1}(5 \times 10^{-16})$
U	$F^{-1}(1 - 5 \times 10^{-7})$	$F^{-1}(1 - 5 \times 10^{-9})$	$F^{-1}(1 - 5 \times 10^{-12})$	$F^{-1}(1 - 5 \times 10^{-16})$

$$w_j = \frac{f(\boldsymbol{X}_j)}{h(\boldsymbol{X}_j)} \tag{2.155}$$

It should be noted that for a uniform distribution $X \sim [a, b]$, its probability density function is given by,

$$h(\boldsymbol{X}) = \frac{1}{b-a} \tag{2.156}$$

In the event that the distribution of $h(\boldsymbol{X})$ is a uniform distribution, then,

$$\hat{P}_{f(UIS)} = \frac{\sum_{j=1}^{N} w_j I(\boldsymbol{X}_j)}{\sum_{j=1}^{N} w_j} = \frac{\sum_{j=1}^{N} f(\boldsymbol{X}_j) I(\boldsymbol{X}_j)}{\sum_{j=1}^{N} f(\boldsymbol{X}_j)} \tag{2.157}$$

As illustrated in Fig. 2.13, this is the UIS presented by Rashki et al., wherein distinct design points yield disparate weightings. It should be noted that $\hat{P}_{f(IS)}$ is an unbiased and strongly consistent estimate of the failure probability P_f. Nevertheless, $\hat{P}_{f(SIS)}$ and $\hat{P}_{f(UIS)}$ are strongly consistent estimates of P_f, yet not unbiased. Moreover, the mean squared error of $\hat{P}_{f(UIS)}$ is typically smaller than that of $\hat{P}_{f(IS)}$.

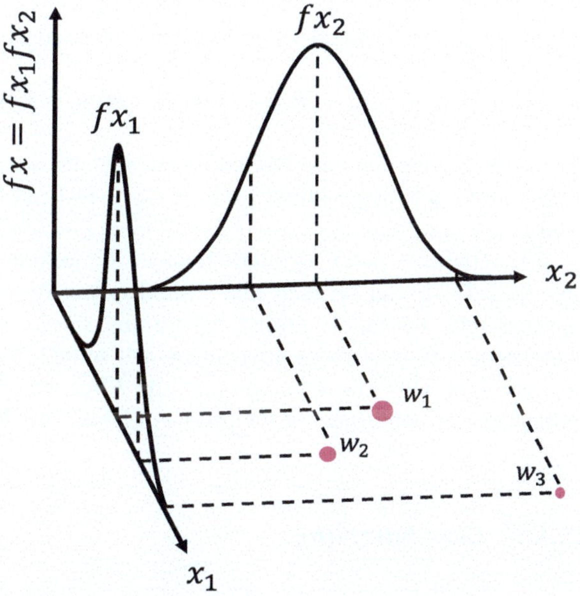

Fig. 2.13 Weighting of samples in the design space. Reprinted from [27], with permission from Elsevier

2.3.4 Enhanced Monte Carlo Simulation

The EMCS proposed by Naess et al. [31] is formulated as follows

$$g = R - Q \tag{2.158}$$

where R represents the structural resistance, while Q denotes the load effect. If we assume that R and Q are mutually independent normal distributions and consider the introduction of a scaling factor, we can rewrite Eq. 2.145 as follows [32],

$$g(\lambda) = R - Q - \mu_G(1 - \lambda) \tag{2.159}$$

where μ_G is the mean value of the performance function (g). Obviously, when $\lambda = 1$, there is $g = g(\lambda)$. Given that the sum of two independent normal distributions remains a normal distribution, and in accordance with the approximate formula for the cumulative distribution function of the normal distribution presented, the probability of failure can be evaluated through the following methodology,

$$P_f(\lambda) \approx \frac{1}{\sqrt{2\pi}}\left[\frac{1}{\lambda\beta} - \frac{1}{(\lambda\beta)^3} + \frac{1}{(\lambda\beta)^5} - \cdots\right] \cdot e^{-\frac{\beta^2\lambda^2}{2}} \frac{-b \pm \sqrt{b^2 - 4ac}}{2a} \tag{2.160}$$

where $0 \leq \lambda \leq 1$. The above equation is expressed as an exponential function as,

$$P_f(\lambda) = m_1 \exp\left(m_2(\lambda - m_3)^{m_4}\right) \tag{2.161}$$

where m_1, m_2, m_3, m_4 are the parameters of the estimated exponential function, respectively. The logarithmic form of Eq. 2.160 can be solved by the Levenberg–Marquardt algorithm to estimate the ultimate probability of failure.

The reliability index for predictions of the failure curve is presented for EMCS in Fig. 2.14. It can be seen that EMCS is based on the basic MCS sample and approximates the failure probability through the Levenberg–Marquardt algorithm. This method is dimension-independent and suitable for high-dimensional problems and problems with low failure probability. When the sample size is low, the sampling efficiency and robustness of the results of EMCS are higher than those of MCS.

2.3.5 Line Sampling

Line sampling is an effective method to solve the reliability problem of high-dimensional small failure probability, for the reliability analysis in the case of n-dimensional random variables, the line sampling method efficiently calculates the failure probability by the difference in the important direction and random sampling in the $(n-1)$ dimension. In the independent standard space, the vector direction from

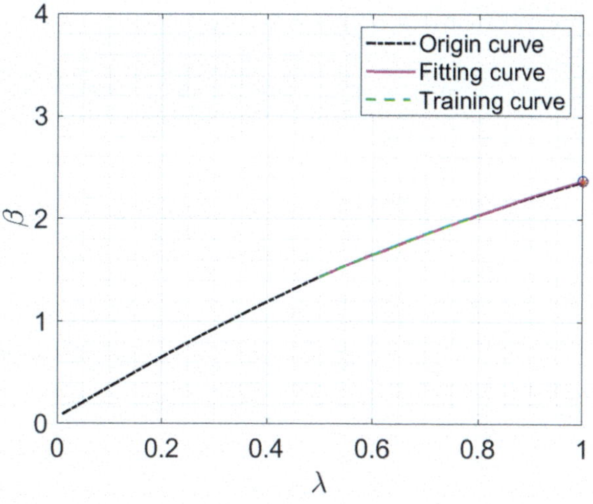

Fig. 2.14 Reliability index corresponding to various λ in EMCS

the coordinate origin to the real design point is the optimal and important direction. By regularizing the optimal important direction, the optimal important direction of the unit can be obtained.

There are two key steps to solve the reliability of line sampling: (i) obtain points on a functional function so that they can be expanded linearly; (ii) The other is to obtain an important direction in order to determine the direction of linear expansion. Typically, step (i) requires the use of a zero-point solution method, and step (ii) requires the use of a design point search technique. The core of the line sampling principle is to obtain the optimal important direction to improve accuracy. It's the optimal and important direction obtained through the design point search technology. When the vector direction of the sample point is different from the optimal important direction, the sample point can be decomposed along the important direction and perpendicular to the important direction.

$$y^j = \mathrm{c}_i \boldsymbol{e}_\alpha + \boldsymbol{y}_j^\perp \tag{2.162}$$

where c_i is a coefficient. In this way, a line $l(j)$ parallel to $\boldsymbol{e}_\alpha$ is obtained, and the probability of failure corresponding to this sample point can be expressed as,

$$\hat{p}_{f_{LS}} = \frac{1}{N}\sum_{j=1}^{N}\Phi(-\tilde{\mathrm{c}}_j) \tag{2.163}$$

where φ is a function of normally distributed probability density; β is a fixed value, the physical meaning of which is the distance from the intersection of $l(j)$ and the functional function in the standard normal space to the vertical plane in the important direction, which can be understood as the sample point reliability index. The schematic view for line samples is presented in Fig. 2.15 for a nonlinear problem. As

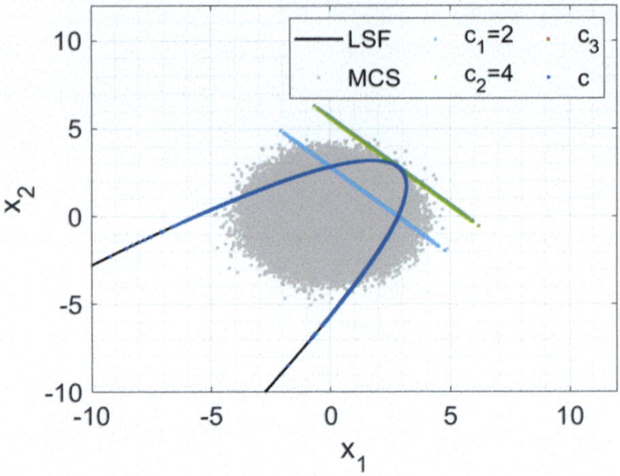

Fig. 2.15 Schematic view of line samples for a nonlinear problem

seen, the idea of LS combines the ideas of sampling method and analytical method. That is, the sampling result is corrected by the maximum probability point so that the sample can cover LSF well. As shown in this figure, this chapter obtains the final sample of LS (dark blue) through three-point quadratic interpolation, but due to the high nonlinearity of LSF, the result error of LS is still large.

In the traditional line sampling method, the important direction needs to be selected before the failure probability is solved, thus the design point must be solved first. The improved first-order second-order moment method is a classical method for solving design points, but the low efficiency of iterative solution is its own problem. Subsequently, the line sampling method was continuously improved and refined. Among them, the introduction of Markov Chain Monte Carlo (MCMC) simulation, that is, MCMC simulation technology, has provided a new impetus for the development of line sampling, and a line sampling method for Markov chain sample simulation to seek design points and important directions has been developed. The traditional line sampling method needs to find the important directions of the functional functions in the standard normal space in advance, which will increase the computational workload on the basis of the stochastic simulation.

2.3.6 *Subset Simulation*

The subset simulation method is a reliability analysis method that solves for small failure probabilities [33]. Compared with the traditional MCS, the SS converts the failure probability into the product of a series of conditional failure probabilities by introducing intermediate states, and the Markov chain simulation can be used to efficiently estimate the conditional sample points for these large failure probabilities, so as to reduce the sample requirement and improve the computational efficiency.

The basic idea of SS is to transform the problem of a small probability event into the product of a series of events with a high probability of failure, so as to avoid directly sampling the small probability events. This section begins with an overview of the basic concepts of SS and a description of several representative approaches to SS.

Subset simulation addresses structural reliability analysis with the following two key issues,

(1) how to select intermediate failure events reasonably;
(2) how to accurately estimate the conditional probabilities.

Usually, the first problem concerning the selection of intermediate failure regions is solved by specifying a generic parameter to define the failure regions and determining the number of intermediate failure regions as well as their conditional failure probability.

Figure 2.16 shows a schematic representation of the intermediate failure probability. Let $g(x)$ denote the limit state function, and let $F = \{x : g(x) < 0\}$ be its failure domain. Introduces a series of critical values:$b_1 > b_2 > \cdots > b_m = 0$. Introducing a series of critical values: to form a series of intermediate failure domains:$F_i(i = 1, 2, \ldots, m)$ with inclusion relationships. The intermediate failure domain is defined as,

$$F_i = \{x : g(x) < b_k\}, i = 1, 2, \ldots, m \tag{2.164}$$

According to the knowledge of probability statistics and theory, the probability of failure is defined as,

$$P_f(F) = P(F_m) = \prod_{i=1}^{m} P(F_i|F_{i-1}) \tag{2.165}$$

Let $P_1 = P(F_1), P_i = P(F_i|F_{i-1})(i = 2, \ldots, m)$, Eq. 2.165 can be rewritten as,

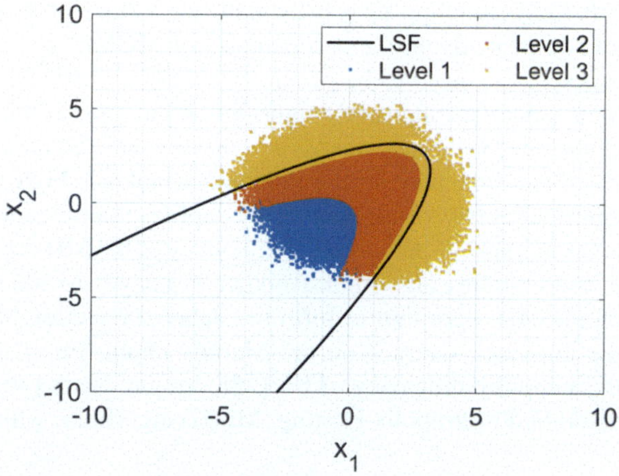

Fig. 2.16 Schematic view of SS domains for a nonlinear problem

$$P_f = \sum_{i=1}^{m} P_i \tag{2.166}$$

The failure probability P_1 of the first state F can be solved by Monte Carlo methods,

$$P_1 = \frac{1}{N_1} \sum_{k=1}^{N_1} I_{F_1}\left[x_k^{(1)}\right] \tag{2.167}$$

where $x_k^{(1)}$ are random samples obeying the joint probability density function and superscript (1) denotes its corresponding 1st state. N is the number of samples. $I_{F_1}[]$ is the schematic function corresponding to the first state,

$$I_{F_i}[\cdot] = \begin{cases} 1 g\left(x^i\right) \leq b_i \\ 0 g\left(x^i\right) > b_i \end{cases} \tag{2.168}$$

Let P_1 be a fixed value, according to Eqs. 2.167 and 2.168, we can calculate b_1, taking the $P_1 \bullet N_1$ samples that satisfy $g\left(x^i\right) \leq b_i$ as a subset to simulate the next state, the conditional failure probability P_i of the i-th state F_i is,

$$\hat{P}_i = \frac{1}{N_i} \sum_{k=1}^{N_i} I_{F_i}\left[x_k^{(i)}\right] \tag{2.169}$$

$$x^{(i)} \sim \pi_i(x) = \frac{f(x) I_{F_i}[g(x)]}{P(F_i)} \tag{2.170}$$

The SS simulation for a nonlinear function is presented in Fig. 2.16 for a nonlinear problem. It is shown that the SS domain that SS obtains simulation samples through the concept of intermediate failure events and Markov Monte Carlo (MCMC) sampling to approximate the failure probability. However, this method greatly depends on the performance of MCMC. Random samples of intermediate states for SS need to be generated using the MCMC sampling algorithm, and the Metropolis-Hasting (MH) algorithm is the most commonly used MCMC sampling algorithm [34]. Since the MH algorithm samples the random variables as a whole, when the dimension of the random variables is high, its alternative sample acceptance rate is low, resulting in many duplicate samples and low sampling efficiency. In order to improve the sampling efficiency of the MH algorithm from the point of view of the sampling algorithm. Another strategy is to add one step of delayed rejection on the MH algorithm to form the delayed rejection MH algorithm, which reduces the duplicate samples on the Markov chain. From the perspective of improving the sampling efficiency of SS [35], Au and Beck [36] proposed an improved MH (Modified Metropolis-Hasting, MMH) algorithm, which solves the problem of low sampling efficiency of MH algorithm for high-dimensional random variables by sampling the components of the random variables.

In addition, the MCMC method based on the diffusion equation further uses the information of the target distribution to construct the proposed distribution so as to improve the acceptance rate of the alternative samples. pCN (preconditioned Crank-Nicolson) algorithm uses the information of the gradient of the probability density function of the target distribution to achieve the MCMC sampling with 0-mean normal distribution so that the generated alternative samples are accepted as the next state samples; Papaioannou et al. [37] developed a conditional sampling (CS) algorithm in standard normal space, which introduces the number of interrelationships between the current sample and the corresponding components of the alternative samples, so that the generated alternative samples are all accepted as the next state samples and reduces the duplicated samples of the Markov Chain, and Au [36] further proved that the CS algorithm satisfies the Markov Chain requirements. Currently, for complex probability density function sample generation and SS calculation computation, the selection of the proposed distributions and parameters in different MCMC sampling algorithms and their effects on the sampling efficiency and accuracy of failure probability calculation need to be further studied.

2.3.7 Directional Sampling

Directional sampling (DS) is a means of estimating failure probabilities in the context of structural reliability [38]. DS first introduces a hypersphere in an independent standard normal space, then generates a set of directions and performs a closed-form or multiple closed-form joint integration in the directions to estimate the conditional failure probability of the direction as a weighted average of the direction integrals, and then repeats the above operation with multiple choices of directions to obtain the failure probability of the whole model. Compared with MCS, the DS method can reduce the number of dimensions in the limit state probability integral, thus reducing the computational cost and improving the efficiency to a certain extent. In this section, the basic procedure of direction sampling is outlined, and then the direction sampling method is analyzed through a case study.

The DS method first requires sampling in an independent standard normal space (hypersphere Z-space), thus the DS first needs to construct an independent standard normal space, where the variables need to be transformed into the original non-normal variables, which can be transformed into independent standard normalized variables by the Rosenblatt transform or Nataf transform [39]. For the independent variable X_i in the performance function $g(x)$, the value of the cumulative distribution density function can be transformed into the value of a single variable Z_i in the standard normal function.

For DS, how to select the direction vectors uniformly concerns the accuracy and computational efficiency of the direction sampling method. In general, the method of generating these direction vectors is to generate the number of N_D points

$[z_i i = 1, 2, \ldots, N_D]$ in the hypersphere according to the distribution density function, and then the vectors formed by these points are unitized to obtain the uniformly distributed directions $[a_i i = 1, 2, \ldots, N_D]$, and then the failure probabilities in these directions are solved.

In general, solving for the failure probability P_i in a single direction is the same as solving for the length R_i of the direction of the unit vector of the independent variable in the limit state function, which can be expressed by Eq. 2.171.

$$g(R_i \cdot a_i) = 0 \tag{2.171}$$

Solving the equation to derive the diameter length R_i then the probability of failure in that direction P_i can be found, which can be expressed by Eq. 2.172.

$$P_i = P\left[g(R_i \cdot a_i) \leq 0\right] \tag{2.172}$$

Exceptionally, for the case where only one R_i can satisfy the limit state equation, then P_i can be simplified by the following equation,

$$P_i = 1 - \chi^2(\mathrm{R}_i^2) \tag{2.173}$$

After solving the failure probabilities P_i in the N_D directions, the total structural failure probability P_f can be expressed as follows,

$$P_f = \frac{1}{N_D} \sum_{i=1}^{N_D} P_i \tag{2.174}$$

As presented in Fig. 2.17, the idea of DS is similar to that of LS. They both use three-point quadratic interpolation to obtain the final samples. However, DS samples the chi-square distribution to approximate the failure probability. For performance functions that are approximately spherical, DS usually works better. It should be noted that how to generate uniform hypersphere vectors is the key to the DS method.

In summary, the DS method is to transform the structural reliability problem into a single-direction rooting problem, which also makes DS more general and to some extent more effective for single-design point, multi-design point and other problems.

2.3.8 Direct Probabilistic Integral Method

The direct probability integration method (DPIM) is a new method proposed by Chen and Yang [40] on the basis of probability density integral equation (PDIE), which can efficiently calculate the probability density function (PDF) of the stochastic response as well as the failure probability of linear and nonlinear stochastic structures under static and dynamic loads.

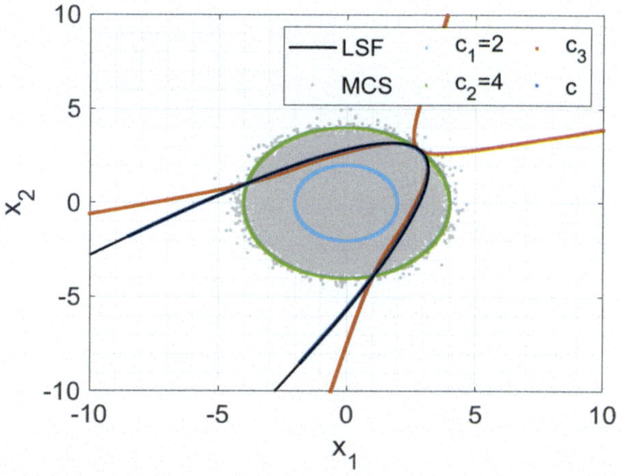

Fig. 2.17 Schematic view of DS line search for a nonlinear problem

Assuming that the limit state function is $g(\boldsymbol{X}) = R - Q$, where Q is the structural response, R is the structural threshold, and $\boldsymbol{X}$ is an n-dimensional random vector. If the actual function value R of the structure is greater than or equal to the expected function value S of the structure, the structure is reliable and the structural reliability is denoted by P_f. If the actual function value R of the structure is less than the expected function value S of the structure, the structure fails, and the probability of the failure of the structure is P_f.

If the limit state function is uncorrelated with time, whether it is explicit or implicit, it can be defined by the following mapping,

$$\xi : Y = g(\Theta) \tag{2.175}$$

where Y is the associated random response of the structure, e.g., displacements, stresses, etc., and $Y = [\Theta_1, \Theta_2, \ldots, \Theta_S]$ denotes the s-dimensional input random variable vector.

The probability of a random event is conserved from input to output in an uncertain engineering system, i.e., the principle of conservation of probability,

$$\int_{\Omega_Y} p_Y(y)dy = \int_{\Omega_\Theta} p_\Theta(\theta)d\theta \tag{2.176}$$

where $p_\Theta(\theta)$ and $p_Y(y)$ are the PDF of the input random variable Θ and the output response Θ, respectively. Ω_Θ is the sample space of input random vectors and Ω_Y is the corresponding sample spaces of output random vectors.

By introducing the Dirac delta function and based on its screening properties, the input PDF can be derived to be related to the output PDF as follows,

$$p_Y(Y) = \int_{\Omega_\Theta} p_\Theta(\theta)\delta\left[y - g(\theta)\right]d\theta \tag{2.177}$$

where $\delta(\bullet)$ denotes the Dirac delta function.

Analogous to the stochastic static system, the PDIE for the response function of the dynamic system is,

$$p_Y(y, t) = \int_{\Omega_\Theta} p_\Theta(\theta)\delta\left[y - g(\theta, t)\right]d\theta \tag{2.178}$$

Note that PDIE in Eq. 2.178, stochastic dynamic systems imply that the total response probability is constant at each moment. That is, the probability that the system response evolves over time has time translation invariance or symmetry.

In engineering, the limit state functions are often multidimensional and highly complex, which leads to the complexity of the mappings, so it is difficult to obtain the mappings by using PDIE directly. Generally, for direct probabilistic integration methods, with the help of numerical integration and smoothing techniques for Dirac δ-functions, the PDF of the response is obtained by solving the partial differential equation.

In this section, quasi-Monte Carlo methods [41] and sparse grid numerical integration methods [42] will be used to calculate the PDIE with high dimensions and non-smooth integrands.

Before examining numerical integration algorithms, we first consider multidimensional integrals of the form $I(f)$.

$$\mathrm{I}(f) = \int_{\mathrm{R}^s} f(\boldsymbol{x})p(\boldsymbol{x})\mathrm{d}\boldsymbol{x} \tag{2.179}$$

where R^s is the s-dimensional integration space of $x, f(x)$ is a function of x, and $p(x)$ is PDF of x.

The integral formula for numerical integration is,

$$I(f). = \sum_{q=1}^{N} w_q f\left(x_q\right) \tag{2.180}$$

in which $x_q \in R^s$ is the qth integration point and w_q is the weight of the integration point pair used. N is the number of points.

Firstly, this section describes the quasi-Monte Carlo method. The quasi-Monte Carlo method uses a high uniformity sequence instead of pseudo-random points which is different from Monte Carlo method. Here, the Sobol' sequence $u_q = \left(u_{1,q}, u_{2,q}, \ldots u_{s,q}\right), q = 1, 2, \ldots, N$ is the basic point set. The N points $\phi_q = \left(\phi_{1,q}, \phi_{2,q}, \ldots \phi_{s,q}\right), q = 1, 2, \ldots, N$ in the distribution domain of interest can be solved by the following CDF inverse transformation,

$$\phi_{i,q} = F_i^{-1}(u_{i,q}) \tag{2.181}$$

where $F_i^{-1}(\cdot)$ is the inverse CDF in the ith dimension. Taking the set of points as an integration point, it is weighted as $w_q = 1/N$. Then the PDIEs in Eq. 2.180 and 2.181 can be approximated as,

$$p_Y(y) = \int_{\Omega_\Theta} p_\Theta(\boldsymbol{\theta})\delta[y - g(\boldsymbol{\theta})]\mathrm{d}\boldsymbol{\theta} = \sum_{q=1}^{N} \{\delta[y - g(\phi_q)]w_q\} \tag{2.182}$$

$$p_Y(y,t) = \int_{\Omega_\Theta} p_\Theta(\boldsymbol{\theta})\delta[y - g(\boldsymbol{\theta},t)]\mathrm{d}\boldsymbol{\theta} = \sum_{q=1}^{N} \{\delta[y - g(\phi_q,t)]w_q\} \tag{2.183}$$

Sparse grid numerical integration approach (SGNI) combines one-dimensional orthogonal nodes into multidimensional nodes using a special tensor multiplication operation that lets $\boldsymbol{U}_1^i$ be an orthogonal point in one-dimensional space and w_1^i be a weight. They can be obtained by Gaussian orthogonality. SGNI generates the set of nodes $\boldsymbol{U}_s^k$ in s-dimensional space with k-level ($k \geq 0$) accuracy in the form of a tensor product, which is defined as,

$$\boldsymbol{U}_s^k = \bigcup_{r-s+1 \leq \sum_i (\cdot) \leq r} \left(U_1^{i_1} \otimes \cdots \otimes U_1^{i_s}\right) \tag{2.184}$$

where $r = k + s$ and $\sum_i = i_1 + i_2 + \cdots + i_s$ denote the summation of the multi-indices $\{i_1, i_2, \ldots i_s\}$. It is automatically bounded. On the other hand, the number of points is related to the dimension s and the precision level k. Therefore, the number of points cannot be set arbitrarily. The weight w_q with respect to the q-th multidimensional point $\xi_q = \left[\xi_{j_{i_1},q}^{i_1}, \ldots \xi_{j_{i_s},q}^{i_s}\right] \in U_s^k$ is formulated by,

$$w_q = (-1)^{s+k-1}\binom{s-1}{s+k-|i|} \cdot \left(w_{j_{i_1}}^{i_1} \times \cdots \times w_{j_{i_s}}^{i_s}\right) \tag{2.185}$$

where ξ_j^i and w_j^i are the j-th element in the node U_1^i and its weight, respectively.

Next, the N points $\xi_q, q = 1, 2, \ldots, N$, are transformed into points $\eta_q, q = 1, 2, \ldots, N$, in a specific distribution domain. The corresponding PDIEs in Eqs. 2.177 and 2.178 are then as follows,

$$p_Y(y) = \int_{\Omega_\Theta} p_\Theta(\boldsymbol{\theta})\delta[y - g(\boldsymbol{\theta})]\mathrm{d}\boldsymbol{\theta} = \sum_{q=1}^{N} \{\delta[y - g(\boldsymbol{\eta}_q)]w_q\} \tag{2.186}$$

$$p_Y(y,t) = \int_{\Omega_\Theta} p_\Theta(\boldsymbol{\theta})\delta[y - g(\boldsymbol{\theta},t)]\mathrm{d}\boldsymbol{\theta} = \sum_{q=1}^{N} \{\delta[y - g(\boldsymbol{\eta}_q,t)]w_q\} \tag{2.187}$$

A measure of the difference in representative point sets for GF difference-based point selection methods is GF Difference,

$$D_{\mathrm{GF}} = \max_{1 \le i \le s} \left\{ sup \left| \hat{F}_i(\boldsymbol{\theta}) - F_i(\boldsymbol{\theta}) \right| \right\} \tag{2.188}$$

where $F_i(\theta)$ denotes the CDF of the ith random variable. $\hat{F}_i(\boldsymbol{\theta})$ is the i-th component of the following empirical CDF,

$$\hat{F}(\boldsymbol{\theta}) = \sum_{q=1}^{N} P_q \cdot I\{\boldsymbol{\theta}_q < \boldsymbol{\theta}\} \tag{2.189}$$

where I(·) is an indicator function.

Compared to Generalized F-discrepancy-based point selection approach (GFD), Quasi-Monte Carlo approach (QMC) is more effective as it reduces the variance by rearranging the set of points $\phi_q = (\phi_{1,q}, \phi_{2,q}, \ldots \phi_{s,q}), q = 1, 2, \ldots, N$. The exact process of the method is as follows:

Firstly, the following transformations are performed in each dimension using the set of points $\phi_q = (\phi_{1,q}, \phi_{2,q}, \ldots \phi_{s,q}), q = 1, 2, \ldots, N$ based on the Sobel sequence in Eq. 2.181,

$$\theta_{i,m} = F_i^{-1}\left(\sum_{q=1}^{N} \frac{1}{N} \cdot I\{\phi_{i,q} < \phi_{i,m}\} + \frac{1}{2} \cdot \frac{1}{N} \right) \tag{2.190}$$

where $i = 1, 2, \ldots s$; is the dimension of the input variable and $m = 1, 2, \ldots, N$ is the number of representative points.

The next step is to obtain a representative region of N points $\theta_q = (\theta_{1,q}, \theta_{2,q}, \ldots \theta_{s,q}), q = 1, 2, \ldots, N$ using the segmentation technique of probability space:

$$\Omega_{\boldsymbol{\Theta},q} = \left\{ \mathbf{x} \in \mathbb{R}^s : \left\| \mathbf{x} - \boldsymbol{\theta}_q \right\| \le \left\| \mathbf{x} - \boldsymbol{\theta}_j \right\| \text{ for } all\, j \right\},\ q = 1, 2, \ldots, N \tag{2.191}$$

where θ_q denotes the representative points in subdomain $\Omega_{\boldsymbol{\Theta},q}$ and θ_j denotes the representative points in subdomain $\Omega_{\boldsymbol{\Theta},q}$.

The probability of Voronoi cell $\Omega_{\boldsymbol{\Theta},q}$ representing the point θ_q can be expressed as,

$$P_q = \Pr\{\Theta \in \Omega_{\Theta,q}\} = \int_{\Omega_{\Theta,q}} p_{\Theta}(\boldsymbol{\theta}) d\boldsymbol{\theta} \tag{2.192}$$

where P_q is the allocation probability of the q-th point.

Next, the point set is arranged according to the transformation expressed in Eq. 2.193.

$$\theta'_{i,m} = F_i^{-1}\left(\sum_{q=1}^{N} P_q \cdot I\{\theta_{i,q} < \theta_{i,m}\} + \frac{1}{2} \cdot P_m\right) \tag{2.193}$$

Use $\theta_q = (\theta_{1,q}, \theta_{2,q}, \ldots \theta_{s,q}), q = 1, 2, \ldots, N$ to denote the final set of representative points $\theta'_q = \left(\theta'_{1,q}, \theta'_{2,q}, \ldots \theta'_{s,q}\right), q = 1, 2, \ldots, N,$

Combining Eqs. 2.177 and 2.178, the response PDF of the static and dynamic system is shown below,

$$p_Y(y) = \sum_{q=1}^{N} \{\delta[y - g(\boldsymbol{\theta}_q)]P_q\} \tag{2.194}$$

$$p_Y(y, t) = \sum_{q=1}^{N} \{\delta[y - g(\boldsymbol{\theta}_q, t)]P_q\} \tag{2.195}$$

Then, Eqs. 2.183–2.187 are expressed uniformly in the form of Eqs. 2.194 and 2.195. As shown in Fig. 2.18, the probability that the random response is at the point is equal to the probability carried by the point in the initial random source. It reflects the principle of conservation of probability.

The smoothing technique of the Dirac δ-function can also be used to solve the partial differential harmonizer. Chen and Yang found that it is effective to smooth the Dirac delta function using a Gaussian function, i.e.

$$\delta(y) = \lim_{\sigma \to 0} \hat{\delta}(y; \sigma) = \lim_{\sigma \to 0} \frac{1}{\sqrt{2\pi}\sigma} e^{-y^2/2\sigma^2} \tag{2.196}$$

The response PDF for static and dynamic systems is shown below,

$$p_Y(y) = \sum_{q=1}^{N} \left\{\frac{1}{\sqrt{2\pi}\sigma} e^{-[y-g(\theta_q)]^2/2\sigma^2} P_q\right\} \tag{2.197}$$

$$p_Y(y, t) = \sum_{q=1}^{N} \left\{\frac{1}{\sqrt{2\pi}\sigma} e^{-[y-g(\theta_q,t)]^2/2\sigma^2(t)} P_q\right\} \tag{2.198}$$

where σ represents the smoothing parameter. If the smoothing parameter is too small, the response PDF will have numerical fluctuations; if it is too large, the PDF will be too smooth. The parameters σ of the static and dynamic systems are expressed,

$$\sigma = \frac{0.9}{N^{1/5}} \min_{q=1,2,\ldots,N} \left\{\text{std}[g(\boldsymbol{\theta}_q)], \frac{\text{iqr}[g(\boldsymbol{\theta}_q)]}{1.34}\right\} \tag{2.199}$$

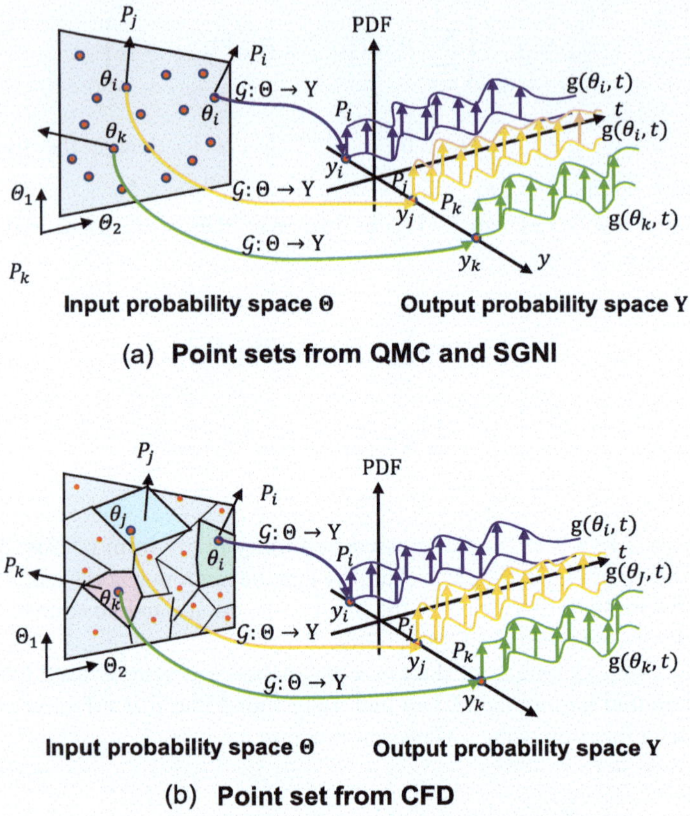

Fig. 2.18 Transformation of the input probability space to the output probability space **a** the point sets of QMC and SGNI; **b** the point set of GFD. Reprinted from [43], with permission from Elsevier

$$\sigma(t) = \frac{0.9}{N^{1/5}} \min_{q=1,2,\ldots,N} \left\{ \mathrm{std}\big[g(\boldsymbol{\theta}_q, t)\big], \frac{\mathrm{iqr}\big[g(\boldsymbol{\theta}_q, t)\big]}{1.34} \right\} \tag{2.200}$$

where 0.9 represents the smoothing coefficient and $iqr[\cdot]$ denotes the interquartile spacing. Based on numerical experience, the coefficients of 0.9 were reduced to 0.5 with appropriate modifications,

$$\sigma = \frac{0.5}{N^{1/5}} \min_{q=1,2,\ldots,N} \left\{ \mathrm{std}\big[g(\boldsymbol{\theta}_q)\big], \frac{\mathrm{iqr}\big[g(\boldsymbol{\theta}_q)\big]}{1.34} \right\} \tag{2.201}$$

$$\sigma(t) = \frac{0.5}{N^{1/5}} \min_{q=1,2,\ldots,N} \left\{ \mathrm{std}\big[g(\boldsymbol{\theta}_q, t)\big], \frac{\mathrm{iqr}\big[g(\boldsymbol{\theta}_q, t)\big]}{1.34} \right\} \tag{2.202}$$

From the above description of DPIM, it can be seen that the equations of control and the calculation of PDIE are decoupled. Hence, DPIM is commonly used to solve the reliability analysis of multidimensional dynamic structures.

The PDF for calculating the structural response of a stochastic static system using DPIM is as follows,

$$P_f = \int_{\Omega_{f,S}} p(s)da \tag{2.203}$$

where $\Omega_f = \{S|R < S\}$ is the failure domain. $p_s(s)$ is the PDF of the structural response calculated by DPIM.

Combining failure probability of a stochastic static system evaluated by using DPIM is shown as below,

$$\begin{aligned} P_f^{\text{DPIM}} &= \int_{\Omega_{f,s}} \sum_{q=1}^{N} \left\{ \frac{1}{\sqrt{2\pi}\sigma} e^{-[s-g(\theta_q)]^2/2\sigma^2} P_q \right\} ds \\ &= \sum_{q=1}^{N} \left\{ \int_{\Omega_{f,s}} \frac{1}{\sqrt{2\pi}\sigma} e^{-[s-g(\theta_q)]^2/2\sigma^2} ds \right\} \cdot P_q \end{aligned} \tag{2.204}$$

The dynamic reliability of a general stochastic dynamic system based on the one-pass criterion is expressed as,

$$R(t) = \Pr\{Y(\Theta, \tau) \in \Omega_s, \tau \in (0, t]\} \tag{2.205}$$

where $Y(\Theta, \tau)$ is the stochastic response of interest for the dynamic system and Ω_s represents the safety domain. Conversely, the dynamic reliability cannot be obtained directly through the instantaneous integration PDF.

According to Eq. 2.205, a failure condition occurs if the dynamic response extremes in the time interval (0, t) exceed a critical value. Therefore, the first-time dynamic reliability base can be transformed into an extreme value distribution problem incorporating DPIM as follows.

Firstly, a stochastic dynamic response with polar mapping is constructed in the time interval,

$$Y_{ext}(\Theta, t) = \underset{0<\tau\leq t}{ext} \{Y(\Theta, \tau)\} \tag{2.206}$$

where the extremal is the maximum or minimum value. For bilateral symmetric boundary problems, the extremal can be the maximum absolute value of the associated response. Thus, in Eq. 2.205, the reliability of the structural dynamic at instantaneous t can be expressed as,

$$R(t) = \Pr\{Y_{ext}(\Theta, t) \in \Omega_s\} \tag{2.207}$$

Then, DPIM is used to solve the dynamic response of the pole-mapped PDF $Y_{ext}(\Theta, t)$,

$$\begin{aligned} p_{Y_{\text{ext}}}(y_{\text{ext}}, t) &= \int_{\Omega_\Theta} \delta\left[y_{\text{ext}} - Y_{\text{ext}}(\theta, t)\right] p_\Theta(\theta) d\theta \\ &= \sum_{q=1}^{N} \left\{ \frac{1}{\sqrt{2\pi}\sigma} e^{-\left[y_{\text{ext}} - Y_{\text{ext}}(\theta_q, t)\right]^2 / 2\sigma^2} P_q \right\} \end{aligned} \tag{2.208}$$

where y_{ext} are the variables of the Dirac delta function. Dynamic reliability within (0, t) is shown below,

$$R(t) = \int_{\Omega} p_{Y_{\text{ext}}}(y_{ext}, t) dy_{ext} \tag{2.209}$$

In summary, the probability of failure can be calculated by the following equation,

$$P_f(t) = 1 - \int_{\Omega_s} p_{Y_{\text{ext}}}(y_{\text{ext}}, t) dy_{\text{ext}} \tag{2.210}$$

2.4 Summary

In this chapter, the typical SRA methods are briefly introduced which can be mainly divided into analytical methods and simulation methods. Specifically, the first-order reliability method, second-order reliability method, moment method, first-order saddle point approximation method, non-probabilistic reliability model and dimension reduction method are the six classical analytical methods. Similarly, the Monte Carlo simulation, importance sampling, uniform importance sampling, enhanced Monte Carlo simulation, line sampling, subset simulation, directional sampling as well as direct probabilistic integral method are the classical simulation methods introduced in this chapter.

References

1. B. Keshtegar, Enriched FR conjugate search directions for robust and efficient structural reliability analysis. Eng. Comput. **34**, 117–128 (2018)
2. B. Keshtegar, M. Bagheri, Fuzzy relaxed-finite step size method to enhance the instability of the fuzzy first-order reliability method using conjugate discrete map. Nonlinear Dyn. **91**, 1443–1459 (2018)
3. S.P. Zhu, B. Keshtegar, N.T. Trung, Z.M. Yaseen, D.T. Bui, Reliability-based structural design optimization: hybridized conjugate mean value approach. Eng. Comput. **37**, 381–394 (2021)

4. M. Rosenblatt, Remarks on a multivariate transformation. Ann. Math. Stat. **23**(3), 470–472 (1952)
5. A. Nataf, Determination des distribution don't les marges sont donnees. Comptes rendus de l'Académie des Sciences **225**, 42–43 (1962)
6. K. Breitung, Asymptotic approximations for multinormal integrals. J. Eng. Mech. **110**(3), 357–366 (1984)
7. N. Bleistein, R.A. Handelsman, *Asymptotic Expansions of Integrals* (Courier Corporation, 1986)
8. L. Tvedt, Two second-order approximations to the failure probability. Veritas report RDIV/ 20-004083 (1983)
9. L. Tvedt, Distribution of quadratic forms in normal space—application to structural reliability. J. Eng. Mech. **116**(6), 1183–1197 (1990)
10. M. Hohenbichler, R. Rackwitz, Improvement of second-order reliability estimates by importance sampling. J. Eng. Mech. **114**(12), 2195–2199 (1988)
11. H.U. Köylüoğlu, S.R. Nielsen, New approximations for SORM integrals. Struct. Saf. **13**(4), 235–246 (1994)
12. G. Cai, I. Elishakoff, Refined second-order reliability analysis. Struct. Saf. **14**(4), 267–276 (1994)
13. Y.G. Zhao, T. Ono, New approximations for SORM: part 1. J. Eng. Mech. **125**(1), 79–85 (1999)
14. Z. Hu, R. Mansour, M. Olsson, X. Du, Second-order reliability methods: a review and comparative study. Struct. Multidiscip. Optim. **64**, 1–31 (2021)
15. S. Konishi, N. Niki, A.K. Gupta, Asymptotic expansions for the distribution of quadratic forms in normal variables. Ann. Inst. Stat. Math. **40**, 279–296 (1988)
16. Z. Hu, X. Du, Saddlepoint approximation reliability method for quadratic functions in normal variables. Struct. Saf. **71**, 24–32 (2018)
17. A. Stuart, K. Ord, *Kendall's Advanced Theory of Statistics, Distribution Theory* (John Wiley & Sons, 2010), p. 1
18. M. Grigoriu, N.C. Lind, Optimal estimation of convolution integrals. J .Eng. Mech. Div. **106**(6), 1349–1364 (1980)
19. M. Grigoriu, Approximate analysis of complex reliability problems. Struct. Saf. **1**(4), 277–288 (1982)
20. H. Hong, Point-estimate moment-based reliability analysis. Civ. Eng. Syst. **13**(4), 281–294 (1996)
21. E.S. Pearson, N.L. Johnson, I. Burr, Comparisons of the percentage points of distributions with the same first four moments, chosen from eight different systems of frequency curves. Commun. Stat.-Simul. Comput. **8**(3), 191–229 (1979)
22. B. Echard, N. Gayton, M. Lemaire, AK-MCS: an active learning reliability method combining Kriging and Monte Carlo simulation. Struct. Saf. **33**(2), 145–154 (2011)
23. Z. Kang, Y. Luo, A. Li, On non-probabilistic reliability-based design optimization of structures with uncertain-but-bounded parameters. Struct. Saf. **33**(3), 196–205 (2011)
24. Z. Meng, H. Hu, H. Zhou, Super parametric convex model and its application for non-probabilistic reliability-based design optimization. Appl. Math. Model. **55**, 354–370 (2018)
25. H. Xu, S. Rahman, A generalized dimension-reduction method for multidimensional integration in stochastic mechanics. Int. J. Numer. Meth. Eng. **61**(12), 1992–2019 (2004)
26. M. Rashki, M. Miri, M.A. Moghaddam, A new efficient simulation method to approximate the probability of failure and most probable point. Struct. Saf. **39**, 22–29 (2012)
27. C. Luo, S.P. Zhu, B. Keshtegar, X. Niu, O. Taylan, An enhanced uniform simulation approach coupled with SVR for efficient structural reliability analysis. Reliab. Eng. Syst. Saf. **237**, 109377 (2023)
28. C. Luo, S.P. Zhu, Y. Lv, H. Ma, X. Liu, B. Keshtegar. Uniform importance sampling with rejection control for structural reliability analysis. Comput. Methods Appl. Mech. Eng. **436**, 117707 (2025)
29. S.P. Zhu, X. Niu, B. Keshtegar, C. Luo, M. Bagheri. Machine learning-based probabilistic fatigue assessment of turbine bladed disks under multisource uncertainties. Int. J. Struct. Integr. **14**(6), 1000-1024 (2023)

30. J.E. Gentle, W.K. Härdle, Y. Mori, *Springer Handbooks of Computational Statistics* (Springer, 2010)
31. A. Naess, B. Leira, O. Batsevych, System reliability analysis by enhanced Monte Carlo simulation. Struct. Saf. **31**(5), 349–355 (2009)
32. C. Luo, B. Keshtegar, S.P. Zhu, O. Taylan, X.P. Niu, Hybrid enhanced Monte Carlo simulation coupled with advanced machine learning approach for accurate and efficient structural reliability analysis. Comput. Methods Appl. Mech. Eng. **388**, 114218 (2022)
33. O. Ditlevsen, H.O. Madsen, *Structural Reliability Methods* (Wiley, New York, 1996), p.178
34. N. Metropolis, A.W. Rosenbluth, M.N. Rosenbluth, A.H. Teller, E. Teller, Equation of state calculations by fast computing machines. J. Chem. Phys. **21**(6), 1087–1092 (1953)
35. A. Mira, On Metropolis-Hastings algorithms with delayed rejection. Metron **59**(3–4), 231–241 (2001)
36. S.K. Au, J.L. Beck, Estimation of small failure probabilities in high dimensions by subset simulation. Probab. Eng. Mech. **16**(4), 263–277 (2001)
37. I. Papaioannou, W. Betz, K. Zwirglmaier, D. Straub, MCMC algorithms for subset simulation. Probab. Eng. Mech. **41**, 89–103 (2015)
38. J. Nie, B.R. Ellingwood, Directional methods for structural reliability analysis. Struct. Saf. **22**(3), 233–249 (2000)
39. P.L. Liu, A. Der Kiureghian, Multivariate distribution models with prescribed marginals and covariances. Probab. Eng. Mech. **1**(2), 105–112 (1986)
40. G. Chen, D. Yang, Direct probability integral method for stochastic response analysis of static and dynamic structural systems. Comput. Methods Appl. Mech. Eng. **357**, 112612 (2019)
41. C. Lemieux, *Monte carlo and quasi-monte carlo sampling* (Springer, 2009), p. 20
42. H.J. Bungartz, M. Griebel, Sparse grids. Acta numerica **13**, 147–269 (2004)
43. L. Li, G. Chen, M. Fang, D. Yang, Reliability analysis of structures with multimodal distributions based on direct probability integral method. Reliab. Eng. Syst. Saf. **215**, 107885 (2021)

Chapter 3
Steepest Methods for FORM

Abstract In this chapter, based on the steepest descent sensitivity vector, the formulation of first-order reliability method is disscused, the improved vesions of the FORM formulated by steepest descent iterative snesetivety vector are collected that these improvd version of steepest decent formulation are enhanced for robustanss using a step size strategy. Four improved HL-RF methods as well-known steepest descent method are developed including (i) improve HL-RF, (ii) chaos control method, (iii) relaxed HL-RF, (iv) Non-negative constraint method that they are formulated for improving gradient line search direction based on a step size. Also, several cast studies are provided for comparing the efficiency and robustanss of these methods.

Keywords Improve steepest descent method · Chaos control method · Relaxed HL-RF · Non-negative constraint method · Steepest descent sensitivity vector

3.1 Introduction

Analytical reliability methods are suitable approaches to consider different uncertainties applied using a probabilistic model. The first-order second moment [1], advanced mean value (AMV) [2], first-order reliability method (FORM) [3], dimension reduction [4], and second-order reliability method (SORM) [5] are a kind of analytical approaches. In structural reliability analysis (SRA), approximating the failure probability (i.e. P_f) is the main effort. Commonly, the FORM can be utilized by considering the balance between efficiency and accuracy for applicable SRA. In FORM, the most probable point (MPP) i.e. $\boldsymbol{U}^*$ search is the major issue to approximate P_f. Generally, the main effort of FORM is to approximate the reliability index (i.e. β) using MPP as $\beta = \boldsymbol{U}^*$. Monte Carlo simulation (MCS) generally attains accurate solutions for engineering problems if a sufficient number of simulations are used. However, high computational effort is often required for complex engineering problems due to the expensive computation of limit state function based on the MCS. Commonly, the steepest descent is applied as the first iterative formula for MPP search in structural

S.-P. Zhu and B. Keshtegar, *Structural Reliability Analysis*, Structural Integrity 31,
https://doi.org/10.1007/978-3-031-75197-4_3

reliability analysis. The FORM-based steepest descent method is typically used for the good balance between accuracy and efficiency for the reliability assessment by MPP (i.e. $\boldsymbol{U}^*$) search on the failure surface. In FORM, the P_f is approximated by reliability index i.e. $\beta = \boldsymbol{U}^*$.

Hasofer and Lind [6] used an iterative method to determine β. Subsequently, Rackwitz and Fiessler [7] enhanced this method by giving information about the distribution of non-normal variables, which is commonly referred to as the HL-RF method. Therefore, the common steepest descent search direction (SDSD) is the HL-RF method. The improved versions of HL-RF including improved HL-RF (iHL-RF) [8], stability transformation method (STM) using chaos control (CC) [2, 3], relaxed HL-RF (RHL-RF) [9, 10] have been formulated by steepest descent search direction for enriching the convergence performances of FORM. Liu and Kiureghian [11] investigated iterative analytical optimization schemes including steepest slope descent (gradient projection method), sequential quadratic programming and augmented Lagrangian method. Unlike the analytical optimization methods, the HL-RF formulation is most usually applied for solving the structural reliability problems. The numerical instabilities of HL-RF is controlled by merit function in iHL-RF method [8] and a dynamical step size given from 0 to 1 adjusted by Armijo rule, second-order finesses [10] or sufficient descent criterion [9]. The discrete dynamic system using the chaos theory with the stability transformation method is applied in STM [3], chaos control (CC) [2], and self-adaptive CC [12]. The step sizes (chaos control factors) of these modified versions for FORM using STM formulations are considered smaller than 1. Consequently, the iHL-RF, STM, CC, and RHL-RF may produce stable results compared to HL-RF but these methods are more computationally inefficient to obtain the stabilization for both linear and nonlinear limit state function (LSF).

The HL-RF method was enhanced using intervening variables and considering adaptive two-point approximation [13]. Santosh et al. [14] enhanced HL-RF by using an inexact step size adjusted by Armijo rules. Yang [2] analyzed the FORM algorithm with the help of a nonlinear transformation with the CC method. The CC (or STM) is more robust than HL-RF method [3]. Santos et al. [15] formulated another version of iHL-RF method by generating a step size using a differentiable merit function given from Wolfe conditions. The improved versions of FORM-based steepest descent search direction methods including STM (or CC) [3], RHL-RF [10] and non-negative constraint method (NNCM) [16] have been applied to enhance the performances of HL-RF formulation. These improved versions of FORM-based SDSD are enhanced by a step size less than one. In RHL-RF [10] and NNCM [15], the step size is adjusted using second-order curve but in STM [3], the step size is selected as a constant small positive value as step size $\ll 1$.

In this chapter, several FORM-based methods formulated by the steepest descent search direction vector such as HL-RF, improved HL-RF, CC, RHL-RF and NNCM are presented for evaluating MPP. Several examples are collected for showing the abilities of these methods.

3.2 HL-RF Method

Generally, the HL-RF method is commonly utilized in FORM by considering its simplicity and efficiency. The FORM-based HL-RF is a more efficient method for convex or moderate-nonlinear problems [17], because it is computed the MPP with step sizes equal to 1. The iterative formulation of HL-RF scheme is given by the following relation for searching the MPP [16].

$$\boldsymbol{U}_{k+1}^{HL} = \frac{\nabla^T g(\boldsymbol{U}_k)\boldsymbol{U}_k - g(\boldsymbol{U}_k)}{\nabla^T g(\boldsymbol{U}_k)\nabla g(\boldsymbol{U}_k)}\nabla g(\boldsymbol{U}_k) \tag{3.1}$$

where $\boldsymbol{U}$ is the standard normal variable, which is determined as follows [3],

$$\boldsymbol{U} = \frac{\boldsymbol{X} - \boldsymbol{\mu}_X^e}{\boldsymbol{\sigma}_X^e} \tag{3.2}$$

where $\boldsymbol{\mu}_X^e$ and $\boldsymbol{\sigma}_X^e$ are respectively the equivalent means and standard deviations (STD) of random variables $\boldsymbol{X}$. The equivalent STD and means for non-normal variables are approximated by Rosenblatt transformation i.e. $\boldsymbol{U} = \Phi^{-1}[F_X(\boldsymbol{X})]$ as follows [18],

$$\boldsymbol{\sigma}_X^e = \frac{1}{f_X(\boldsymbol{X})}\varphi\left[\Phi^{-1}\{F_X(\boldsymbol{X})\}\right] \tag{3.3}$$

$$\boldsymbol{\mu}_X^e = \boldsymbol{X} - \sigma_X^e \Phi^{-1}[F_X(\boldsymbol{X})] \tag{3.4}$$

where $F_X(\cdot)$ and $\Phi^{-1}(\cdot)$ are respectively the cumulative distribution function (CDF) and inverse standard normal CDF. $f_X(x)$ is the probability density function (PDF) of variable x and φ is normal standard PDF.$\nabla g(\boldsymbol{U}) = [\partial g/\partial U_1, \partial g/\partial U_2, \ldots, \partial g/\partial U_n]^T$ denotes gradient vector of LSF in standard normal space. The HL-RF-based SDSD in Eq. 3.1 shows that $\nabla g(\boldsymbol{U})$ is the main active factor in searching MPP. Actually, the iterative HL-RF in Eq. 3.1 is reformulated by the following relation,

$$\boldsymbol{U}_{k+1}^{HL} = \frac{\nabla^T g(\boldsymbol{U}_k)\boldsymbol{U}_k - g(\boldsymbol{U}_k)}{\nabla^T g(\boldsymbol{U}_k)\boldsymbol{\alpha}_{k+1}}\boldsymbol{\alpha}_{k+1} \tag{3.5}$$

where α_{k+1} denotes the normalized search direction (NSD) vector, which is determined as below,

$$\boldsymbol{\alpha}_{k+1} = -\frac{\nabla g(\boldsymbol{U}_k)}{\nabla g(\boldsymbol{U}_k)} \tag{3.6}$$

Generally, Eq. 3.5 of FORM is applied to determine the MPP (i.e. $\boldsymbol{U}^*$) as presented schematically in the MPP. As presented in Fig. 3.1, the MPP located on LSF has the

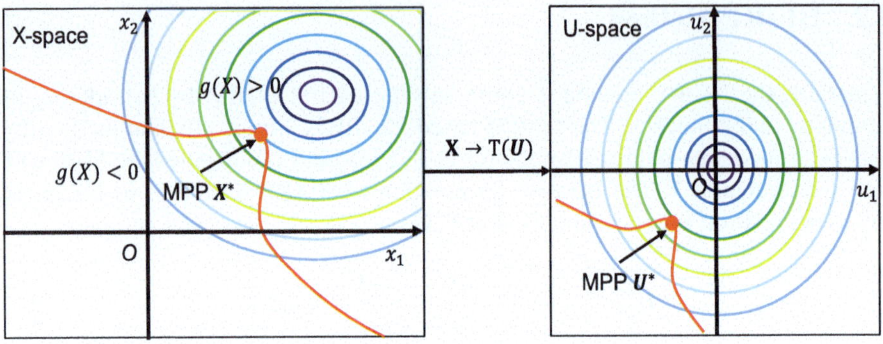

Fig. 3.1 View for transferring variables in normal standard space in the FORM method

shortest distance from its origin in standard normal space. The main iterative process of FORM is to find the MPP ($\boldsymbol{U}^*$) and then approximate P_f by β as follows,

$$P_f = \Phi(-\beta) = \Phi\left(-\boldsymbol{U}^*\right) \tag{3.7}$$

where the reliability index is determined as below,

$$\beta_{k+1} = \frac{\nabla^T g(\boldsymbol{U}_k)\boldsymbol{U}_k - g(\boldsymbol{U}_k)}{\nabla^T g(\boldsymbol{U}_k)\boldsymbol{\alpha}_{k+1}} \tag{3.8}$$

The iterative process of FORM-based HL-RF for computing the reliability index is summarized as below steps,

Step 1: Define LSF $g(\boldsymbol{X})$, give stopping criterion (ε), statistical properties of basic random variables ($\boldsymbol{\mu}$ and $\boldsymbol{\sigma}$), set $k = 1$ and $\boldsymbol{X}_1 = \boldsymbol{\mu}$;
Step 2: Normalize the random variables by transferring from $\boldsymbol{X}$-space to $\boldsymbol{U}$-space based on Eqs. 3.3 and 3.4;
Step 3: Iterate the HL-RF for finding MPP as below framework;

(i) Compute $g(\boldsymbol{U}_k)$, $\nabla g(\boldsymbol{U}_k)$;
(ii) Determine $\boldsymbol{\alpha}_{k+1} = -\frac{\nabla g(\boldsymbol{U}_k)}{\|\nabla g(\boldsymbol{U}_k)\|}$;
(iii) Compute $\boldsymbol{U}_{k+1}^{HL} = \frac{\nabla^T g(\boldsymbol{U}_k)\boldsymbol{U}_k - g(\boldsymbol{U}_k)}{\nabla^T g(\boldsymbol{U}_k)\boldsymbol{\alpha}_{k+1}}\boldsymbol{\alpha}_{k+1}$;
(iv) Determine $\beta_{k+1} = \frac{\nabla^T g(\boldsymbol{U}_k)\boldsymbol{U}_k - g(\boldsymbol{U}_k)}{\nabla^T g(\boldsymbol{U}_k)\boldsymbol{\alpha}_{k+1}}$.

Step 4: If $|\beta_{k+1} - \beta_k| < \varepsilon$ then stop iterative process and print $\boldsymbol{X}^* = \boldsymbol{X}_{k+1}$, $\boldsymbol{U}^* = \boldsymbol{U}_{k+1}$, $\beta = \beta_{k+1}$;
Else: $k = k + 1$ and $\boldsymbol{U}_k = \boldsymbol{U}_{k+1}^{HL}$. Compute $\boldsymbol{X}_{k+1} = T^{-1}(\boldsymbol{U}_{k+1})$ go to Step 2.

The robustness of HL-RF problems is investigated by a mathematical LSF in the first stage, and then, the albites of HL-RF method are compared for several engineering problems in the second stage. The same stopping criterion as $|\beta_{k+1} - \beta_k| <$

10^{-6} is selected for all examples. The iterations as evaluating the $\nabla g(\boldsymbol{U})$ and converged β are used to demonstrate the abilities of these FORM-based SDSD.

Example 3.1 Nonlinear limit state function

LSF in Eq. 3.9 is frequently considered to test the abilities of reliability algorithms.

$$g = x_1^3 + x_1^2 x_2 + x_2^3 - 18 \tag{3.9}$$

where, normal variables x_1 and x_2 have $\mu_1 = 10$, $\sigma_1 = 5$ and $\mu_2 = 9.9$, $\sigma_2 = 5$.

The periodic-2 solutions obtained by HL-RF algorithm as [1.04958, 1.15364] for reliability index. According to the nonlinear discrete map presented in Eqs. 3.2–3.4 or transforming random variables in standard normal space, the MPP is computed by an iterative formulation that it is depended on properties of random variables as,

$$\boldsymbol{U} = f\left(\boldsymbol{X}, \boldsymbol{\mu}_X^e, \boldsymbol{\sigma}_X^e\right) \tag{3.10}$$

Based on Eq. 3.10, the means and standard deviations (STD) of basic random variables are affected on the convergence properties of HL-RF. Furthermore, various mean and STD of variables x are as: μ and σ are discussed on the convergence iterative FORM-based HL-RF as presented results in Fig. 3.2. The results of Fig. 3.2 illustrated that the HL-RF is an unstable iterative method with instability in the reliability index such as the periodic and chaotic solutions corresponding to various μ_1 and σ_1.

The points at previous new iterations and reliability index are schematically plotted for HL-RF approach in Fig. 3.3. As seen, the normalized vector $\boldsymbol{\alpha}_{k+2}$ is parallel to $\boldsymbol{\alpha}_k$. Consequently, $\boldsymbol{\alpha}_{k+2}$ and $\boldsymbol{\alpha}_k$ are the same thus we have $\boldsymbol{U}_{k+2} = \boldsymbol{U}_k$. This means that it is obtained as a periodic solution. As seen from the presented results for Example 3.1, the HL-RF produced this explained issue as a 2-periodic solution. Therefore, HL-RF maybe an unstable formulation when the LSF has high curvature at MPP thus, the nonlinearity degree of LSF has strongly affected the stabilization of HL-RF formula.

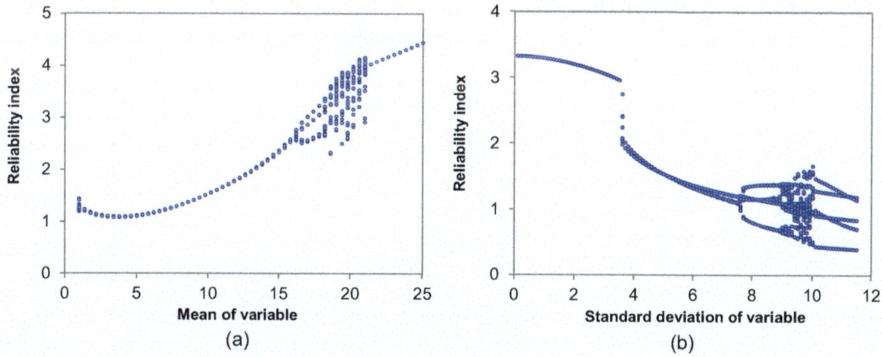

Fig. 3.2 Unstable results of HL-RF corresponding to **a** various means and **b** STD of x_1 for Example 3.1. Reprinted from [19], with permission from Elsevier

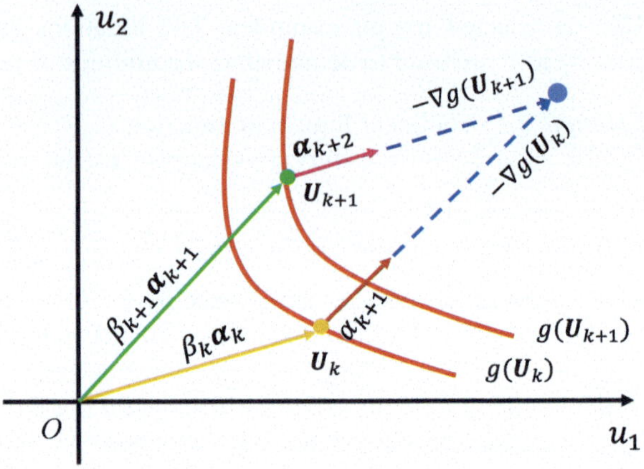

Fig. 3.3 Instability with 2-periodicperiodic solution of HL-RF for a nonlinear problem

Example 3.2 A 10-bar truss.

A 10-bar truss structure with three cross-sectional areas of A_1 applied for vertical elements, A_2 applied for horizontal elements and A_3 applied for diagonal elements as plotted in Fig. 3.4, is given by the following LSF,

$$g = d_0 - B\Delta(P, A_1, A_2, A_3, L, E) \tag{3.11}$$

As shown in Fig. 3.4, the truss structure is subjected to a load P at point 3 and point 1, which is vertically displaced at point 3 to determine its performance. This performance is calculated from the model uncertainty variable B, Young's module E and A_1, A_2, A_3, P which are presented in Table 3.1.

L (Length of vertical and horizontal span) and limited displacement (d_0) are deterministic variables in this example defined as $L = 9$and$d_0 = 0.1$ meters, respectively. The safety index is 4.4937 related failure probability of 3.5×10^{-6} using MCS

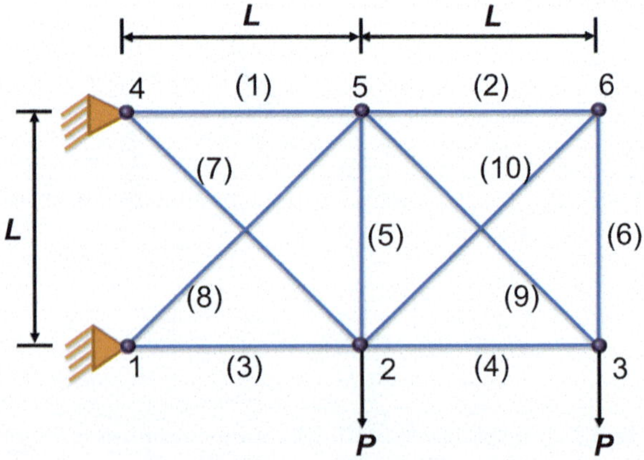

Fig. 3.4 10-bar truss structure for Example 3.2. Reprinted from [20], with permission from Elsevier

Table 3.1 Random variables for 10-bar truss example

PDF	Coefficient of variation	Mean	Variable
Gumbel	0.10	2.5×10^5	$P(N)$
Normal	0.05	10^{-2}	$A_1(\mathrm{m}^2)$
Normal	0.05	1.5×10^{-3}	$A_2(\mathrm{m}^2)$
Normal	0.05	6×10^{-3}	$A_3(\mathrm{m}^2)$
Normal	0.10	1	B
Lognormal	0.05	6.9×10^4	$E(\mathrm{Mpa})$

with 1.7×10^9 data points. The HL-RF algorithm converges to a stable result as $\beta = 4.30089$ after 25 iterations. As seen, the HL-RF is strongly more efficient than the MCS for this example by acceptable results for the real structure.

Example 3.3 Roof truss.

As shown in Fig. 3.5, A roof structure under uniform load (q) with the length of l is given as below LSF,

$$g = 0.03 - \left(\frac{ql^2}{2}\right)\left(\frac{3.81}{AcEc} + \frac{1.13}{AsEs}\right) \tag{3.12}$$

As seen, compression bars and tension bars are respectively made by reinforced concrete and steel where A_s and A_c are respectively the steel and concrete cross-section with corresponding modulus of elasticity E_s and E_c. These normal variables are presented in Table 3.2.

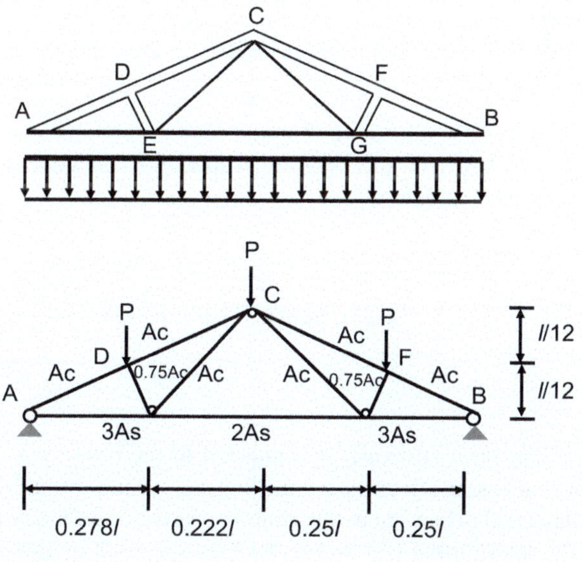

Fig. 3.5 Schematic view of roof truss example. Reprinted from [21], with permission from Elsevier

Table 3.2 Random variables for roof truss presented in Example 3.3

Random variable	q (N/m)	l (m)	Ac (m^2)	As (m^2)	Ec (Pa)	Es (Pa)
Mean	20,000	12	0.04	9.82×10^{-4}	2×10^{10}	1×10^{11}
STD	1400	0.12	0.0048	5.9852×10^{-5}	1.2×10^{9}	6×10^{9}

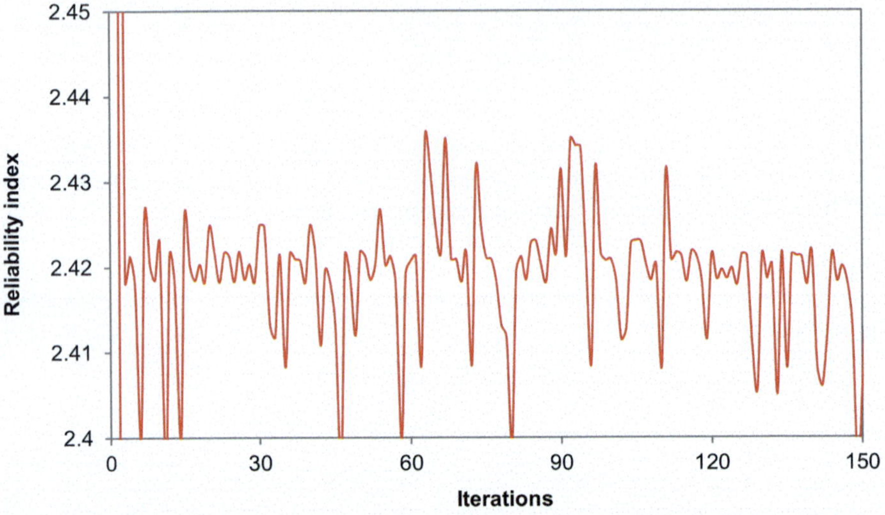

Fig. 3.6 Reliability index iteration using HL-RF for roof truss example

The results obtained by MCS with 1×10^6 samples, the reliability index and failure probability are respectively extracted as $\beta = 2.34697$ and $P_f = 9.395 \times 10^{-3}$. The results of reliability index with respect to the iterations for HL-RF is presented in Fig. 3.6 for the roof truss example. As seen, the HL-RF provides instable solutions, chaotically.

Example 3.4 Slender column.

The LSF for the buckling capacity of slender column plotted in Fig. 3.7 is computed as below,

$$
\begin{gathered}
g = 40 - \delta \\
\delta = t[(\tan(\psi) - \psi) + \frac{b^2\left(1+\sqrt{1-4\left(HEI\tan^2(\psi)\right)/b^2}\right)^2}{4HEI\tan(\psi)} \\
\psi = L\sqrt{\frac{P}{EI}}, t = \frac{H}{EI\left(\frac{P}{EI}\right)^{1.5}}
\end{gathered}
\tag{3.13}
$$

The strut structure is connected to the base by a rotary spring, which provides a non-linear effect by controlling the torque through the parameter b. The strut is designed to be used in the same way as the base. Six normal variables that are applied for determining LSF in Eq. 3.13 are specified in Table 3.3.

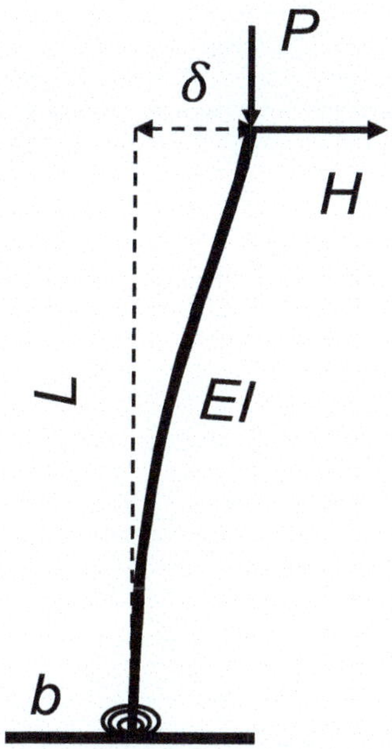

Fig. 3.7 Slender column view in Example 3.4. Reprinted from [17], with permission from Springer Nature

Table 3.3 Variables for slender column applied in Example 3.4

Description	*COV*	Mean	Variable
Horizontal load (N)	0.2	10,000	$H(x_2)$
Column of length (mm)	0.05	10,000	$L(x_4)$
Young's modulus (MPa)	0.1	20,000	$E(x_3)$
Spring parameter (N mm)	0.1	0.4167	$B(x_6)$
Moment of inertia (mm^4)	0.05	1.0417×10^{10}	$I(x_5)$
Axial load (N)	0.2	1,000,000	$P(x_1)$

As extracted results from Ref. [17], $\beta = 2.3508131$ and $\boldsymbol{U}^* = [0.5608, 1.2274, -1.2315, 1.3696, -0.5594, 0.00021]$. The HL-RF method is not converged for this example and it has a chaotic solution that its iteration of reliability index is presented in Fig. 3.8.

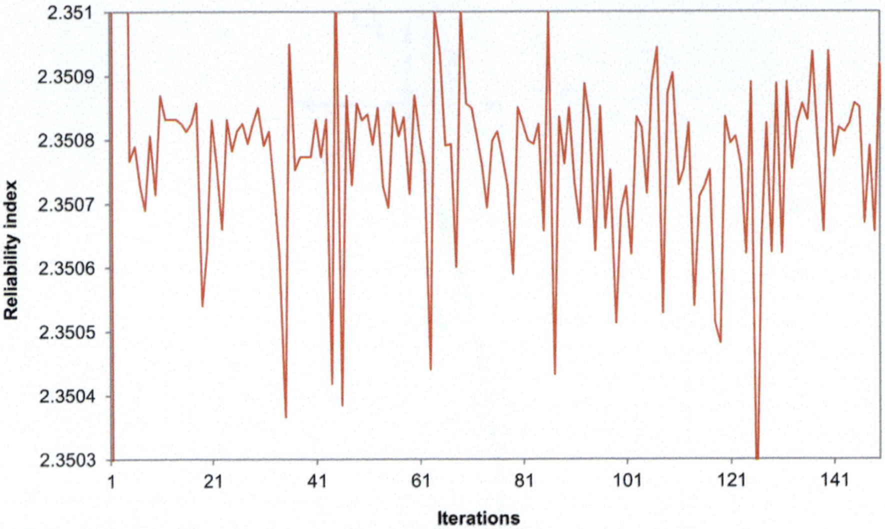

Fig. 3.8 Reliability index iteration using HL-RF for slender column example

3.3 Improve HL-RF Method

Liu and Kiureghian [8] used a merit function $m(\boldsymbol{U})$ to improve the abilities of the HL-RF method. They have shown that this procedure leads to convergence during successive iterations according to applying a step size less than 1 for HL-RF by established below,

$$\begin{gathered} g(\boldsymbol{U}^*) = 0, \boldsymbol{U}^* + \lambda^* \nabla g(\boldsymbol{U}^*) = 0 \\ \lambda^* = \nabla \frac{\nabla^T g(\boldsymbol{U}^*)\boldsymbol{U}^*}{|\nabla g(\boldsymbol{U}^*)|^2} \end{gathered} \tag{3.14}$$

in which the merit function is defined as follows,

$$m(\boldsymbol{U}) = \frac{1}{2}\left| \boldsymbol{U} - \nabla \frac{\nabla g(\boldsymbol{U})\boldsymbol{U}}{|\nabla g(\boldsymbol{U})^2|^2} \nabla g(\boldsymbol{U}) \right|^2 + \frac{1}{2} C g(\boldsymbol{U})^2 \tag{3.15}$$

where C is a positive constant value. In the improved HL-RF method, the new point is computed by step size which satisfies the merit function as follows,

$$\boldsymbol{d}_k = \frac{1}{|\nabla g(\boldsymbol{U}_k)|^2} [\nabla^T g(\boldsymbol{U}_k)\boldsymbol{U}_k \nabla g(\boldsymbol{U}_k)] \nabla g(\boldsymbol{U}_k) - \boldsymbol{U}_k \tag{3.16}$$

$$\boldsymbol{U}_{k+1} = \boldsymbol{U}_k + \lambda_k \boldsymbol{d}_k \tag{3.17}$$

The above formulation with an appropriate step size λ_k can provide a suitable reduction for merit function $m(\boldsymbol{U})$. For an acceptable iteration, it should be given as $m(\boldsymbol{U}_{k+1}) < m(\boldsymbol{U}_k)$. Therefore, the merit function in Eq. 3.15 is a scheme for selecting a step size α_k less than 1 for HL-RF iterative formulation. If $\boldsymbol{\alpha}_k < 1$ then we can be given a condition as $m(\boldsymbol{U}_{k+1}) < m(\boldsymbol{U}_k)$. If the line search direction $\boldsymbol{d}_k$ is not computed as an acceptable normalized sensitivity vector, then it is not guaranteed for giving stable results by the iterative formulation in the improved HL-RF [8].

As mentioned, the step size has affective rules on the convergence performances of the reliability methods. In the improved HL-RF approach, the convergence performances are better than the HL-RF method. They applied the Armijo line search rule for selecting the step size at every iteration of FORM formula. They assumed that for step size (1) the step size should not be selected too large for controlling the instabilities of HL-RF method and (2) The step length should not be chosen so small as to increase the risk of premature convergence. With certain conditions, the proper choice of step size should be guaranteed for giving global convergence in iHL-RF. The iterative steps for iHL-RF are presented as below,

Step 1: Define LSF $g(\boldsymbol{X})$, Give stopping criterion (ε), statistical properties of basic random variables (μ and σ), set $k = 1, \boldsymbol{X}_1 = \boldsymbol{\mu}.\boldsymbol{\alpha}_0 = 1$;
Step 2: Normalize the random variables by transferring from $\boldsymbol{X}$-space to $\boldsymbol{U}$-space based on Eqs. 3.3 and 3.4;
Step 3: Iterate the iHL-RF for finding MPP as below framework;
Compute $g(\boldsymbol{U}_k), \nabla g(\boldsymbol{U}_k)$;
Determine negative unit normal vector as $\boldsymbol{\alpha}_{k+1} = -\frac{\nabla g(\boldsymbol{U}_k)}{\|\nabla g(\boldsymbol{U}_k)\|}$;
Compute new point using HL-RF with $\boldsymbol{U}_{k+1}^{HL} = \frac{\nabla^T g(\boldsymbol{U}_k)\boldsymbol{U}_k - g(\boldsymbol{U}_k)}{\nabla^T g(\boldsymbol{U}_k)\boldsymbol{\alpha}_{k+1}}\boldsymbol{\alpha}_{k+1}$;
Determine the new merit function $m(\boldsymbol{U}) = \frac{1}{2}\left|\boldsymbol{U} - \nabla\frac{\nabla g(\boldsymbol{U})\boldsymbol{U}}{|\nabla g(\boldsymbol{U})^2|^2}\nabla g(\boldsymbol{U})\right|^2 + \frac{1}{2}Cg(\boldsymbol{U})^2$;
Check the condition of $m(\boldsymbol{U}_{k+1}) < m(\boldsymbol{U}_k)$, if satisfy then new point is HL-RF and go to step vi else, choose a small step size and update new point as $\boldsymbol{U}_{k+1} = \boldsymbol{U}_k + \lambda_k \boldsymbol{d}_k$;
Determine reliability index as $\beta_{k+1} = ||\boldsymbol{U}_{k+1}||$.
Step 4: If $|\beta_{k+1} - \beta_k| < \varepsilon$ then stop iterative process and print $\boldsymbol{X}^* = \boldsymbol{X}_{k+1}, \boldsymbol{U}^* = \boldsymbol{U}_{k+1}, \beta = \beta_{k+1}$;
Else: $k = k + 1$ and $\boldsymbol{U}_{k+1} = \boldsymbol{U}_{k+1}^{iHL}$. Compute $\boldsymbol{X}_{k+1} = T^{-1}(\boldsymbol{U}_{K+1})$ go to Step 2.

Example 3.5 Nonlinear LSF in Eq. 3.4.

The ability of the above algorithm is discussed for reliability analysis by presented Example 3.4 which are applied in HL-RF method. The nonlinear problem presented in Eq. 3.9 is given for first problem of evaluating the iHL-RF while the reliability index iterative history is presented in Fig. 3.9. As seen, the iHL-RF produced the stable results of $\beta = 2.2983$ after 102 iterations. This method solved the drawbacks of HL-RF and it can be introduced as improved version of FORM for this example.

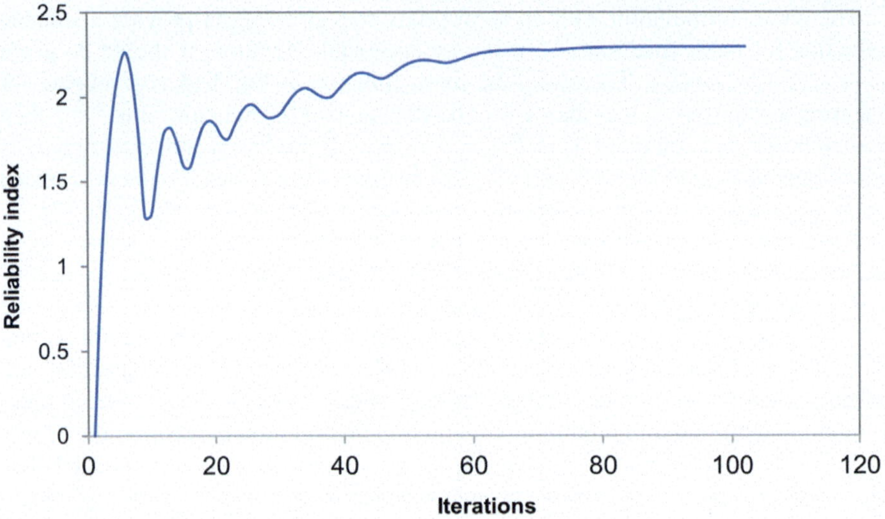

Fig. 3.9 Reliability index iteration using iHL-RF for nonlinear example

Example 3.6 A 10-bar truss.

As LSF in Eq. 3.11, this method is accurately converged with a reliability index of 4.3006 after 39 iterations. As well as HL-RF method, iHL-RF method provides stable results but it is more inefficient than HL-RF method for this truss problem by selecting the step size less than 1 for some iterations.

Example 3.7 Roof truss.

The reliability result for this problem is presented in Fig. 3.10 where the reliability index converged to 2.4138 after 136 iterations. By giving a scheme as a merit function for computing the step size can be controlled the instability of FORM-based HL-RF method that is strongly shown in this example.

Example 3.8 Slender column.

In this example the reliability index given by the iHL-RF method is presented. As seen as well as other examples, unlike HL-RF, iHL-RF is converged after 73 iterations to the safety index of 2.3507. As seen from the results presented in Fig. 3.11, the iHl-RF method is more inefficient than HL-RF but HL-RF formulation may provide unstable results for nonlinear problems. The iHL-RF uses a step size of less than 1 and a merit function as a guide for selecting the step size, which greatly improves the robustness of the HL-RF method. The efficiency of the iHL-RF method is one of the disadvantages of this method for computing reliability index and also this method for reliability problems with several local minima may converge to local optimum points that it may be not an acceptance solution for FORM.

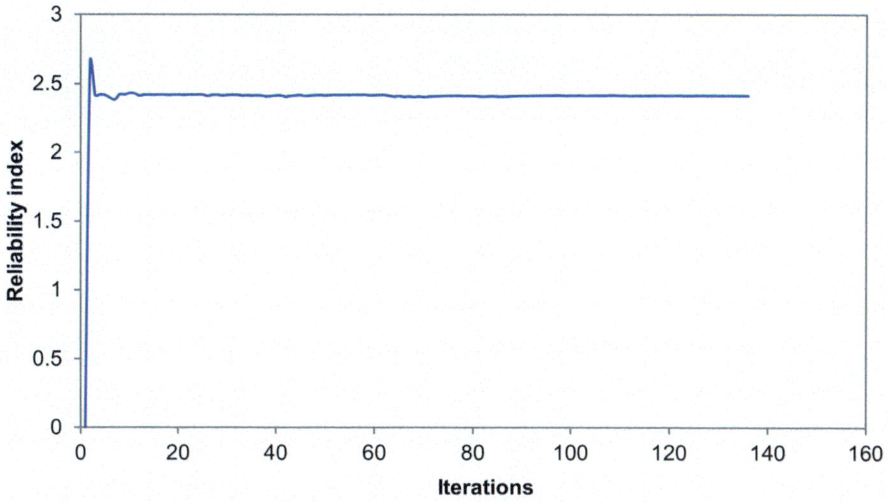

Fig. 3.10 Reliability index iteration using iHL-RF for Roof truss example

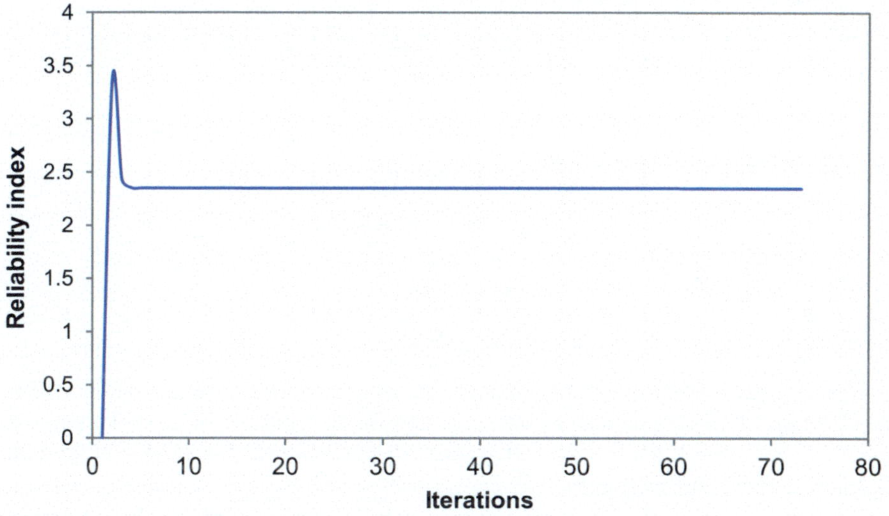

Fig. 3.11 Reliability index iteration using iHL-RF for Slender column example

3.4 STM (CC) Method

The STM-based CC can be formulated using a chaotic attractor given from a nonlinear dynamical map which is a scheme for approximating fixed points for MPP. Temporarily for highly nonlinear problems, the STM can also provide a stabilization for iterative FORM. By using STM, oscillation, chaos or bifurcation of a dynamical iterative process can be controlled [3]. Generally, for n-dimensional basic random variables applied in a dynamical system, a discrete dynamical map can be defined as below,

$$\boldsymbol{U}_{k+1} = f(\boldsymbol{U}_k, \boldsymbol{p}) \quad \boldsymbol{U} \in R^n,\ \boldsymbol{p} \in R^m, \quad k = 1, 2, 3, \ldots \tag{3.18}$$

where, $f(\boldsymbol{U}_k, \boldsymbol{p})$ is a differentiable function given by the iterative method, $\boldsymbol{U} \in R^n$ represents n-variable vector and $\boldsymbol{p}$ denotes the control vector applied in the discrete dynamical system. It is too hard to determine $\boldsymbol{p}$ for a nonlinear system because it is difficult to compute the new point $\boldsymbol{U}_{k+1}$. An appropriate discrete dynamical system with a nonlinear transferring map for giving the stabilization which is defined by STM for FORM-based formulation as below,

$$\boldsymbol{U}_{k+1} = \boldsymbol{U}_k + \lambda \boldsymbol{C}\left[f(\boldsymbol{U}_k) - \boldsymbol{U}_k\right] \tag{3.19}$$

in which, $0 < \lambda < 1$, $\boldsymbol{C}$ named involutory matrix with $n \times n$-element where 1 or -1 is applied in one element in each row and column and then others are given 0. In STM, $\boldsymbol{C}$ is usually selected unit matrix as $\boldsymbol{C} = \boldsymbol{I}$. The STM in FORM-based on Eq. 3.19 is expressed as follows,

$$\boldsymbol{U}_{k+1}^{STM} = \boldsymbol{U}_k^{STM} + \lambda\left[\boldsymbol{U}_{k+1}^{HL} - \boldsymbol{U}_k^{STM}\right] \tag{3.20}$$

where, $f(\boldsymbol{U}_k) = \boldsymbol{U}_{k+1}^{HL}$, thus, discrete dynamical map i.e. $f(\boldsymbol{U}_k) = f\left(\boldsymbol{X}_k, \boldsymbol{\mu}_X^e, \boldsymbol{\sigma}_X^e\right)$ is defined by the following relations,

$$\boldsymbol{U}_{k+1} = \boldsymbol{U}_k + \lambda \boldsymbol{d}_k \tag{3.21}$$

$$\boldsymbol{d}_k = \frac{\nabla^T g(\boldsymbol{U}_k)\boldsymbol{U}_k - g(\boldsymbol{U}_k)}{\nabla^T g(\boldsymbol{U}_k)\nabla g(\boldsymbol{U}_k)} \nabla g(\boldsymbol{U}_k) - \boldsymbol{U}_k \tag{3.22}$$

If $\lambda = 1$, then the dynamical system has not any control for stabilization. In $\lambda = 1$, the STM and HL-RF schemes as the same approach. Yang [3] suggested a smaller value for the control factor as $\lambda = 0.1$, thus, STM has a low-convergence rate with high efforts for computations.

Recently, the performance measure approach applied in reliability-based design optimization has been modified by CC formulation [2]. The modified CC methods have been extended using a control factor with less than one. By the above relation the steps for computing the MPP-based CC algorithm are presented as below:

Step 1: Define LSF $g(\boldsymbol{X})$, Give stopping criterion (ε), statistical properties of basic random variable (μ and σ), set $k = 1$, $\boldsymbol{X}_1 = \boldsymbol{\mu}$, $\lambda = 0.1$;
Step 2: Normalize the random variables by transferring from $\boldsymbol{X}$-space to $\boldsymbol{U}$-space based on Eqs. 3.3 and 3.4;
Step 3: Iterate the CC for finding MPP as below framework;
Compute $g(\boldsymbol{U}_k)$, $\nabla g(\boldsymbol{U}_k)$;

Determine negative unit normal vector as $\boldsymbol{\alpha}_{k+1} = -\frac{\nabla g(\boldsymbol{U}_k)}{\|\nabla g(\boldsymbol{U}_k)\|}$;

Compute new point using HL-RF with $\boldsymbol{U}_{k+1}^{HL} = \frac{\nabla^T g(\boldsymbol{U}_k)\boldsymbol{U}_k - g(\boldsymbol{U}_k)}{\nabla^T g(\boldsymbol{U}_k)\boldsymbol{\alpha}_{k+1}}\boldsymbol{\alpha}_{k+1}$;

Determine the new point by $\boldsymbol{U}_{k+1} = \boldsymbol{U}_k + \lambda \boldsymbol{d}_k$, where, $\boldsymbol{d}_k = \frac{\nabla^T g(\boldsymbol{U}_k)\boldsymbol{U}_k - g(\boldsymbol{U}_k)}{\nabla^T g(\boldsymbol{U}_k)\nabla g(\boldsymbol{U}_k)}\nabla g(\boldsymbol{U}_k) - \boldsymbol{U}_k$;

Determine reliability index as $\beta_{k+1} = ||\boldsymbol{U}_{k+1}||$.

Step 4: If $|\beta_{k+1} - \beta_k| < \varepsilon$ then stop the iterative STM and print $\boldsymbol{X}^* = \boldsymbol{X}_{k+1}$, $\boldsymbol{U}^* = \boldsymbol{U}_{k+1}$, $\beta = \beta_{k+1}$;

Else: $k = k + 1$ and $\boldsymbol{U}_{k+1} = \boldsymbol{U}_{k+1}^{STM}$. Compute $\boldsymbol{X}_{k+1} = T^{-1}(\boldsymbol{U}_{k+1})$ go to Step 2.

It can be seen that the method is as simple as the HL-RF method and differs from iHL-RF in that its application to reliability problems is based on simple relationships. The main difference between this method and the HL-RF method is the use of a constant chaos control factor. The chaos control factor is similar to weight, and considering the previous results, we give about 90% of the iterative relations, as shown in Fig. 3.12. As can be seen from the figure, the HL-RF method provides unstable bi-periodic results, whereas the CC method provides stable results by giving a scalar factor that participates in the previous point, and the new search direction is not parallel to the previous search direction or the normalized sensitivity vector.

The capability of this method is evaluated by several problems which are presented for evaluation of performance convergences as robustness and efficiency. We compare the number of iterations and the convergence results for evaluating the efficiency and robustness, respectively.

Example 3.9 High-nonlinear problem.

LSF is measured as [18],

$$g(\boldsymbol{X}) = x_1^3 + x_1^2 x_2 + x_2^3 - 18 \tag{3.23}$$

where, normal variables x_1 and x_2 have $\mu_1 = 10$, $\mu_2 = 9.9$ and $\sigma_1 = \sigma_2 = 5$. Extracted results. Using Wolfe conditions, results from Ref. [18] showed $\beta =$

Fig. 3.12 Schematic view for an iteration of the CC method for a nonlinear problem

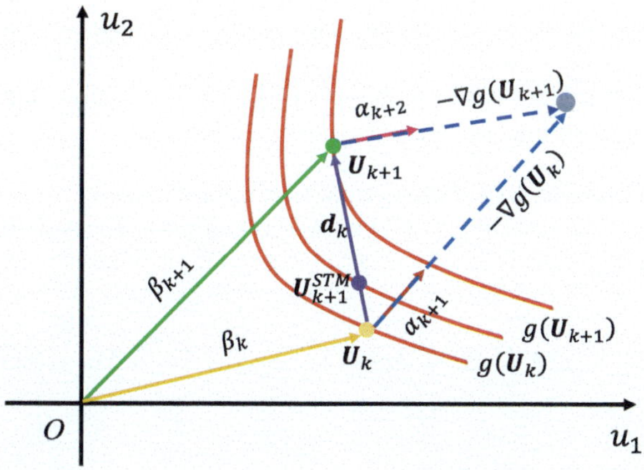

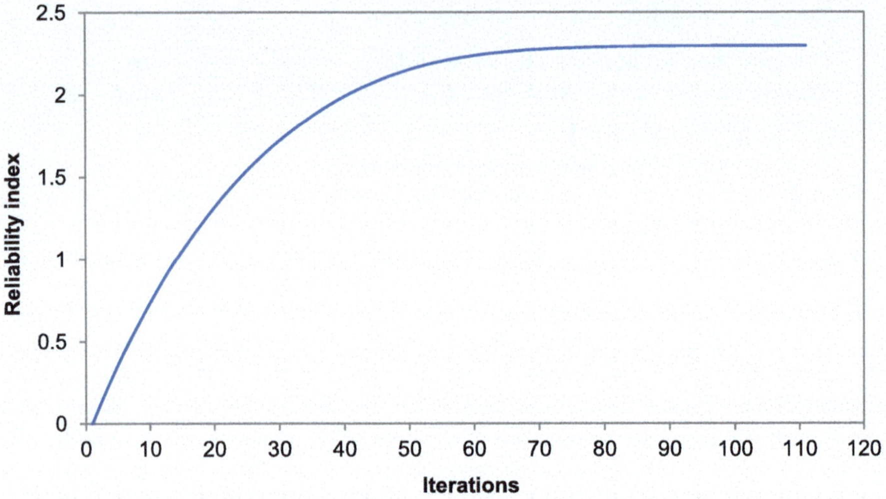

Fig. 3.13 Reliability index iteration using CC for nonlinear example

2.298252 which was obtained by modified HL-RF with a conjugate descent approach. With 10^6 samples applied in MCS, the reliability index is $\beta = 2.5139$ with related $P_f = 0.007975$.

The STM-based CC is converged for this example with MPP and reliability index of $\boldsymbol{X}^* = [1.6864, 1.9690]$ ($\boldsymbol{X}^* = [-1.6627, -1.5862]$) and $\beta = 2.2979$ which are captured after 111 iterations as presented results in Fig. 3.13. The results are closely in agreement with extracted results from literature and it is shown that the CC is a robust formulation of FORM but it is more efficient.

Figure 3.14 illustrates the iteration-based MPP search for the HL-RF and CC methods in the nonlinear example. As can be seen from the figure, the CC method does not show any chaotic results but converges to the MPP based on a line determined by the search direction, which has a scalar factor of 0.1. This clearly shows that the results of the CC method are stable if a small value of the chaos control factor is chosen, and therefore the stability of the iterations depends heavily on the chaos control factor, which is almost given a very small value. Therefore, it is an inefficient method for practical reliability problems.

Example 3.10 Non-normal nonlinear problem.

The LSF of Eq. 3.24 with artificial noise terms defined for this example as below [15],

$$g(\boldsymbol{X}) = x_1 + 2x_2 + 2x_3 + x_4 - 5x_5 - 5x_6 + 0.001\sum_{i=1}^{6}\sin(100x_i) \tag{3.24}$$

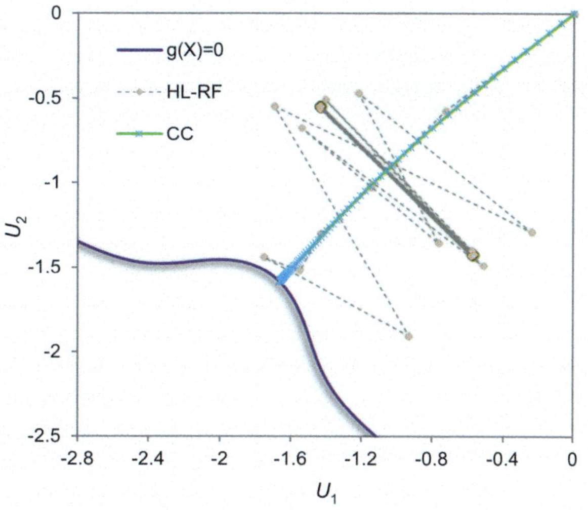

Fig. 3.14 Iteration for MPP search in normal standard space using CC and HL-RF methods for nonlinear example

The Lognormal variables $x_1 - x_6$ defined as $x_1 - x_4$ with the mean of 120 and STD of 12;x_5 with the mean of 50 and STD of 15; and x_6 with the mean of 40 and STD of 12.

Extracted results from Ref. [18] showed $\beta = 2.348257$ and $\boldsymbol{X}^* = [117.35, 115.28, 115.28, 117.35, 83.52, 55.64]$. Using 10^6 samples applied in MCS, it is obtained that $\beta = 2.2576$.

This problem has moderate nonlinearity the extracted results from Ref. [18] showed that the HL-RF method provided an unstable solution but the CC method is accurately converged with the results agreement with literature as the MPP of $\boldsymbol{U}^* = [-0.17933, -0.35565, -0.35565, -0.17933, 1.89933, 1.26054]$, $\boldsymbol{X}^* = [117.2875, 115.2426, 115.2426, 117.2875, 83.6383, 55.4696]$ and reliability index of $\beta = 2.3403$ after 91 iterations. The iterative reliability index of this problem is presented in Fig. 3.15. It is demonstrated from Fig. 3.15 that CC algorithm converged to stable MPP. It can be concluded that the CC method has improved the instability of FORM-based HL-RF formulation. However, it provides inefficient results. In addition, in the CC method, a constant chaos control factor is given, so in applicable engineering problems, some problems may arise, such as unstable iteration of the FORM formula given by CC.

Example 3.11 A Conical Structure.

A conical structure with axial load P and bending moment M is given for given by the following relation performance function [19],

$$g = 1 - \frac{\sqrt{3(1 - v^2)}}{\pi\, E\, t^2 \cos^2\alpha} \times \left(\frac{P}{0.66} + \frac{M}{0.41\, r_1}\right) \tag{3.25}$$

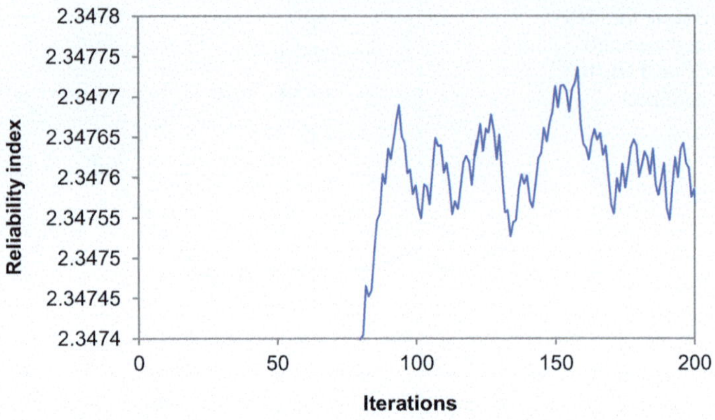

Fig. 3.15 Reliability index iteration using CC for non-normal nonlinear example

The LSF is computed by buckling failure mode for this conical structure buckling failure mode is computed by Poisson's ratio i.e. $\nu = 0.3$ and six normal and non-normal variables as presented in Table 3.4.

By MCS with 10^7 samples, the reliability index is captured as 4.4163. Figure 3.16 illustrates the iterative histories of the reliability index for the STM-based CC with formulated by a dynamical discrete map. The results of FORM algorithms illustrated that STM is converged to unstable solutions as chaos. The nonlinear dynamical map by CC formulation with chaos control factor of 0.1 cannot guarantee the stability of the FORM-based MPP search by STM formulation for this example.

Example 3.12 A steel frame structure.

A frame plotted in Fig. 3.17 by the following LSF as lateral displacement is given as,

$$g = 0.05 - \Delta \tag{3.26}$$

Table 3.4 Random variables information for conical structure of Example 3.3

Description	PDF	COV	Mean	Variables
Young's modulus	Lognormal	0.05	70,000	E (MPa)
Slop angle	Normal	0.02	0.524	α(rad)
Thickness	Normal	0.05	0.0025	t (m)
Internal radius	Normal	0.025	0.9	r_1(m)
Axial load	Gumbel	0.08	70,000	P (N)
Bending moment	Gumbel	0.08	80,000	M (N · m)

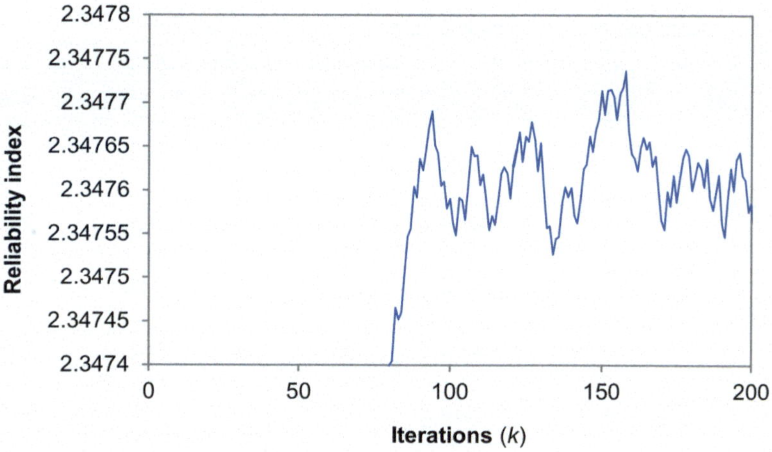

Fig. 3.16 Iterations of reliability index using CC for conical structure example

Fig. 3.17 The steel frame structure for Example 3.12. Reprinted from [19], with permission from Elsevier

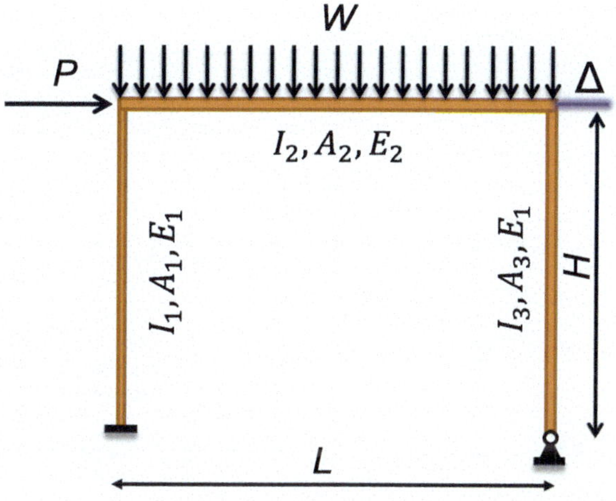

This frame has height H and span L with two supports as clamped (at left column) and simply (at right column) boundaries that it is subjected to distributed load W and lateral load P. The statistical details of random variables are arranged in Table 3.5.

The failure probability (related safety index) is obtained by MCS using 1.2×10^6 samples as $P_f = 1.08 \times 10^{-3}$ ($\beta = 3.0673$). As seen from the iterative histories for the reliability index using STM presented in Fig. 3.18, the STM is not a robust method as chaotic results for STM based on this applicable engineering problem. Therefore, the STM with a smaller chaos control factor may give a stable result for highly nonlinear LSFs but for some application problems in the reliability analysis, it should be adjusted for providing a stable convergence in the structural reliability analysis.

Table 3.5 Random variables information for steel frame of Example 3.12

Variable	Mean	STD	PDF	Variable	Mean	STD	PDF
A_1 (m^2)	91×10^{-4}	9.1×10^{-4}	Lognormal	E_1 (kg/m^2)	2.38×10^{10}	1.9×10^9	Normal
A_2 (m^2)	53.8×10^{-4}	5.38×10^{-4}	Lognormal	E_2 (kg/m^2)	2.17×10^{10}	1.74×10^9	Normal
A_3 (m^2)	106×10^{-4}	10.6×10^{-4}	Lognormal	H (m)	4.5	0.36	Normal
I_1 (m^4)	80.9×10^{-6}	8.09×10^{-6}	Lognormal	L (m)	6	0.48	Normal
I_2 (m^4)	83.6×10^{-6}	8.36×10^{-6}	Lognormal	P (kg)	2000	350	Gumbel
I_3 (m^4)	112.6×10^{-6}	11.26×10^{-6}	Lognormal	W (kg/m)	4500	630	Gumbel

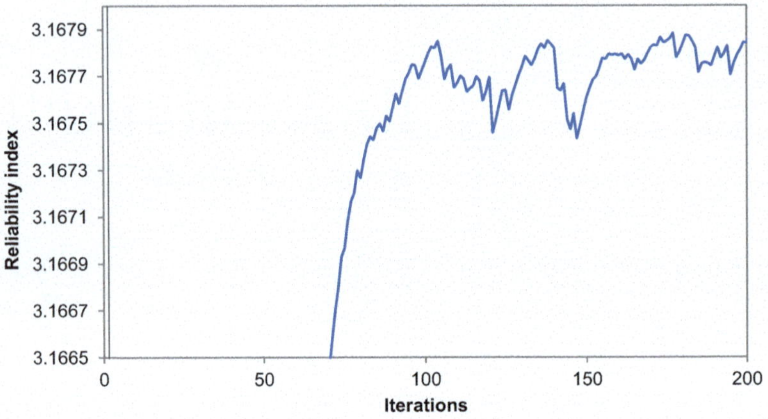

Fig. 3.18 Iterations of reliability index using CC for steel frame example

3.5 RHL-RF Method

The results from the above section illustrated that HL-RF yielded to unstable results as chaos and periodic. The STM with parameters of $\xi = 0.1$ exhibited an unstable chaotic solution but it is converged for the highly nonlinear performance function. However, the STM is an inefficient method in highly convex reliability problems and it also provided unstable chaotic solutions for some applicable engineering problems. The iHL-RF where the step size is computed based on a merit function provided stable results for applicable engineering problems but, it is an inefficient formulation. The dynamic adjusting the step size as well as the chaos control factor is an essential challenge in the steepest descent search direction. The relaxed formulation of the FORM is similar to the iHL-RF method it can be given from the relaxed iterative approach. In this section, a relaxed dynamical approach of the FORM is presented and its formulation is extended by the steepest descent search direction.

The convergence instability of iterative FORM algorithms can be controlled by a relaxed approach as well as STM, commonly [3]. As a linear algebraic equation by relation $\boldsymbol{AX} = \boldsymbol{b}$, the formulation is applied with the below format,

$$X_i^{k+1} = \frac{b_i - \sum_{j=1}^{i-1} a_{ij} X_j^{k+1} - \sum_{j=i+1}^{n} a_{ij} X_j^k}{a_{ii}} \quad (i = 1, 2, 3, \ldots, n) \tag{3.27}$$

The relaxed form of the above relation form is rewritten by the following relation to control the convergence of Eq. 3.27.

$$X_i^{k+1} = (1-\alpha) X_i^k + \alpha \left(\frac{b_i - \sum_{j=1}^{i-1} a_{ij} X_j^{k+1} - \sum_{j=i+1}^{n} a_{ij} X_j^k}{a_{ii}} \right) \quad (i = 1, 2, 3, \ldots, n) \tag{3.28}$$

where, a_{ij} denotes the element of matrix A, α is the relaxed factor applied for controlling instability of the Gauss–Seidel method. According to the above relations, the robustness of HL-RF iterative algorithm is improved for solving the nonlinear limit state function as below,

$$\boldsymbol{U}_{k+1} = (1 - \alpha_k)\, \boldsymbol{U}_k + \alpha_k\, \boldsymbol{U}_{k+1}^{HL} \tag{3.29}$$

where $\boldsymbol{U}_{k+1}^{HL}$ is the point obtained by HL-RF formula see Sect. 3.2. α_k denotes the relaxed factor as a positive less than one i.e. $0 < \alpha_k \leq 1$. It is considered that $\boldsymbol{d}_k$ should be satisfied conditions $\nabla^T g(\boldsymbol{U}_k)\ \boldsymbol{d}_k = 0$ for robust solutions this we have $\boldsymbol{U}_k^T \boldsymbol{d}_k < 0$ [8]. Therefore, vector $\boldsymbol{d}_k$ is determined as,

$$\boldsymbol{d}_k = \boldsymbol{U}_{k+1}^{HL} - \boldsymbol{U}_k \tag{3.30}$$

The relaxed factor should be selected as an acceptable value not too large (FORM may provide unstable solutions) nor too small (FORM may slowly converge). Therefore, it should be applied line search rules including the Armijo, Minimization, Goldstein, Limited minimization, Wolfe and Strong Wolfe rules [19] to control the instability of an iterative FORM formula. However, all inexact line search methods as presented above require high computational costs or they may provide a non-convergence issue. Meanwhile, relaxed factors without inexact rules can be computed for SRA with efficient results.

The second-order polynomial function by one extremum can be applied based on the interpolating technique for approximation of the relaxed factor. The optimum value of a second-order polynomial function is relaxed to relaxed factor that is the extremum of the polynomial function. Formerly, a novel approach for approximating the relaxed factor is presented in this section. It is given a relaxed factor as below relation,

$$h(\alpha) = A + B\alpha + C\alpha^2 \tag{3.31}$$

where h is the unconstrained reliability function interpolated by second-order polynomial function. On the other hand, the unconstrained reliability function is given based on FORM formulation as below relation [18],

$$h_1(\boldsymbol{U}) = \frac{\boldsymbol{U}^T\boldsymbol{U}}{2} + \lambda g(\boldsymbol{U}) \tag{3.32}$$

where λ is the penalty factor which is the optimum condition of $\boldsymbol{U} + \lambda \nabla g(\boldsymbol{U}) = 0$. By using Eq. 3.32, the penalty factor is computed as,

$$\lambda = -\frac{\nabla^T g(\boldsymbol{U})\boldsymbol{U}}{\nabla^T g(\boldsymbol{U})\nabla g(\boldsymbol{U})} \tag{3.33}$$

Thus, we have,

$$h_1(\boldsymbol{U}) = \frac{\boldsymbol{U}^T\boldsymbol{U}}{2} - \frac{\nabla^T g(\boldsymbol{U})\boldsymbol{U}}{\nabla^T g(\boldsymbol{U})\nabla g(\boldsymbol{U})} g(\boldsymbol{U}) \tag{3.34}$$

By the Taylor's series, the reliability function in Eq. 3.34 at point ($\boldsymbol{U}_{k+1}$) is approximated as,

$$h_1(\boldsymbol{U}_{k+1}) = h_1(\boldsymbol{U}_k) + \alpha_k \nabla^T h_1(\boldsymbol{U}_k)\boldsymbol{d}_k \tag{3.35}$$

By Eqs. 3.34 and 3.35, The coefficients A, B and C are given by,

$$\alpha = 0 \Rightarrow h(\alpha) = h_1(\boldsymbol{U}_k) \Rightarrow A = h_1(\boldsymbol{U}_k) \tag{3.36}$$

It is considered that the relaxed factor is the optimum of the previous iteration ($\alpha = 0$); by differentiating Eq. 3.36 corresponding with α, and then it is related to the differential of Eq. 3.33, we have,

$$\left.\begin{array}{l} \frac{dh(\boldsymbol{U}_{k+1})}{d\alpha} = \nabla^T h_1(\boldsymbol{U}_k)\boldsymbol{d}_k = -\boldsymbol{d}_{k-1}^T\boldsymbol{d}_k \\ \left.\frac{dh(\alpha)}{d\alpha}\right|_{\alpha=0} = B + 2C\alpha|_{\alpha=0} = B \end{array}\right\} \Rightarrow B = -\boldsymbol{d}_{k-1}^T\boldsymbol{d}_k \tag{3.37}$$

For $\alpha = 1$, then $h(\alpha) = h_1(\boldsymbol{U}_{k+1})$. Thus, we will have,

$$A + B + C = h_1(\boldsymbol{U}_{k+1}) \tag{3.38}$$

Substitute Eqs. 3.36 and 3.37 into Eq. 3.38, C is determined as follows,

$$C = h_1(\boldsymbol{U}_{k+1}) - h_1(\boldsymbol{U}_k) + \boldsymbol{d}_{k-1}^T\boldsymbol{d}_k \tag{3.39}$$

Put A, B and C in Eq. 3.31, we will have,

$$h(\alpha) = h_1(\boldsymbol{U}_k) - \alpha\, \boldsymbol{d}_{k-1}^T \boldsymbol{d}_k + \alpha^2\left[h_1(\boldsymbol{U}_{k+1}) - h_1(\boldsymbol{U}_k) + \boldsymbol{d}_{k-1}^T \boldsymbol{d}_k\right] \tag{3.40}$$

The extremum of Eq. 3.40 is related to the relaxed factor, thus,

$$\begin{aligned} \frac{dh(\alpha)}{d\alpha} &= B + 2\alpha_k^{opt} C = 0 \\ \alpha_k^{opt} &= -\frac{B}{2C} = \frac{\boldsymbol{d}_{k-1}^T \boldsymbol{d}_k}{2\left[h_1(\boldsymbol{U}_{k+1}) - h_1(\boldsymbol{U}_k) + \boldsymbol{d}_{k-1}^T \boldsymbol{d}_k\right]} \end{aligned} \tag{3.41}$$

$h_1(\boldsymbol{U}_k)$ and $h_1(\boldsymbol{U}_{k+1})$ denote the unconstrained reliability problems at previous and new points, which can be estimated as below relations,

$$h_1(\boldsymbol{U}_k) = \frac{\boldsymbol{U}_k^T \boldsymbol{U}_k}{2} - \frac{\nabla^T g(\boldsymbol{U}_k)\boldsymbol{U}_k}{\nabla^T g(\boldsymbol{U}_k)\nabla g(\boldsymbol{U}_k)} g(\boldsymbol{U}_k) \tag{3.42}$$

$$h_1(\boldsymbol{U}_{k+1}) = \frac{\boldsymbol{U}_{k+1}^{T\,HL} \boldsymbol{U}_{k+1}^{HL}}{2} - \frac{\nabla^T g(\boldsymbol{U}_k)\boldsymbol{U}_{k+1}^{HL}}{\nabla^T g(\boldsymbol{U}_k).\nabla g(\boldsymbol{U}_k)} g\left(\boldsymbol{U}_{k+1}^{HL}\right) \tag{3.43}$$

The relaxed factor is selected as $0 < \alpha \le 1$ and it should be omitted the negative values. If for $\alpha = 1$, the results are followed to HL-RF method. The second-order interpolation provides a relax between 0 and 1 with a dynamic process. By RHL-RF method, it is computed a larger relaxed factor than CC. Consequently, RHL-RF may be more efficient than the CC algorithms. The iterative procedure of this method is presented as below steps:

Step 1: Define LSF $g(\boldsymbol{X})$, give stopping criterion (ε), statistical properties of basic random variables (μ and σ), set $k = 1$, $\boldsymbol{X}_1 = \boldsymbol{\mu}$,$\boldsymbol{\alpha}_1 = 1$;
Step 2: Normalize the random variables by transferring from $\boldsymbol{X}$-space to $\boldsymbol{U}$-space based on Eqs. 3.3 and 3.4;
Step 3: Iterate the RHL-RF for finding MPP as below framework;

(i) Compute $g(\boldsymbol{U}_k)$, $\nabla g(\boldsymbol{U}_k)$;
(ii) Determine negative unit normal vector as $\boldsymbol{\alpha}_{k+1} = -\frac{\nabla g(\boldsymbol{U}_k)}{\nabla g(\boldsymbol{U}_k)}$;
(iii) Compute new point using HL-RF with $\boldsymbol{U}_{k+1}^{HL} = \frac{\nabla^T g(\boldsymbol{U}_k)\boldsymbol{U}_k - g(\boldsymbol{U}_k)}{\nabla^T g(\boldsymbol{U}_k)\boldsymbol{\alpha}_{k+1}} \boldsymbol{\alpha}_{k+1}$;.
(iv) Determine search direction vector $\boldsymbol{d}_k = \boldsymbol{U}_{k+1}^{HL} - \boldsymbol{U}_k$;
(v) For $k > 2$, determine the value of $h_1(\boldsymbol{U}_k)$ and $h_1(\boldsymbol{U}_{k+1})$ and then, relaxed factor $\alpha_k^{opt} = -\frac{B}{2C} = \frac{\boldsymbol{d}_{k-1}^T \boldsymbol{d}_k}{2\left[h_1(\boldsymbol{U}_{k+1}) - h_1(\boldsymbol{U}_k) + \boldsymbol{d}_{k-1}^T \boldsymbol{d}_k\right]}$;
(vi) Compute the new point $\boldsymbol{U}_{k+1}^{RHL} = (1 - \alpha_k)\boldsymbol{U}_k + \alpha_k\, \boldsymbol{U}_{k+1}^{HL}$.

Step 4: If $|\beta_{k+1} - \beta_k| < \varepsilon$ then stop RHL-RF method and print $\boldsymbol{X}^* = \boldsymbol{X}_{k+1}$, $\boldsymbol{U}^* = \boldsymbol{U}_{k+1}$. $\beta = \beta_{k+1}$;

Else: $k = k + 1$ and $\boldsymbol{U}_{k+1} = \boldsymbol{U}_{k+1}^{RHL}$. Compute $\boldsymbol{X}_{k+1} = T^{-1}(\boldsymbol{U}_{k+1})$ go to Step 2.

The results of the previous iteration method are directly applied for approximating the relaxed factor. The formulation of this method is as simple as HL-RF but it is formulated based on an exact line search, unlike iHL-RF method. Based on computational efforts for approximate the optimum relaxed factor, the RHL-RF method has a complicated formulation compared to HL-RF and STM. Abilities for efficiency and robustness are examined using several examples as presented in below.

Example 3.13 Nonlinear limit state function.

The limit state function in Eq. 3.9 is given as the first example for evaluating the RHL-RF method that its LSF is as $g = x_1^3 + x_1^2 x_2 + x_2^3 - 18$, where the normal variables of x_1 and x_2 have $\mu_1 = 10$, σ_1=5 and $\mu_2 = 9.9$, $\sigma_2 = 5$.

The iterations for the reliability index are presented for nonlinear examples using RHL-RF method. As seen in Fig. 3.19, stable convergence is obtained for this problem with a reliability index of 2.2982 after 39 iterations. The stability of the nonlinear LSF is satisfied based on the RHL-RF while it is strongly more efficient than the STM method. The dynamical procedure for computing the step size accurately computes the search direction vector for this example.

Iteration for searching the MPP in the normal standard space is plotted for RHL-RF method in Fig. 3.20. It can be given that the search direction is computed on a straight line after three iterations and at the final iteration, the length of the search direction vector is computed as a small value. However, the HL-RF has been successfully searched for 1–3 first iterations, but for high nonlinearity of the limit state function it lost its stability for iterations more than 4.

Example 3.14 Two-degree of dynamic system.

A two-degree dynamic system as presented in Fig. 3.21 is defined below LSF [19],

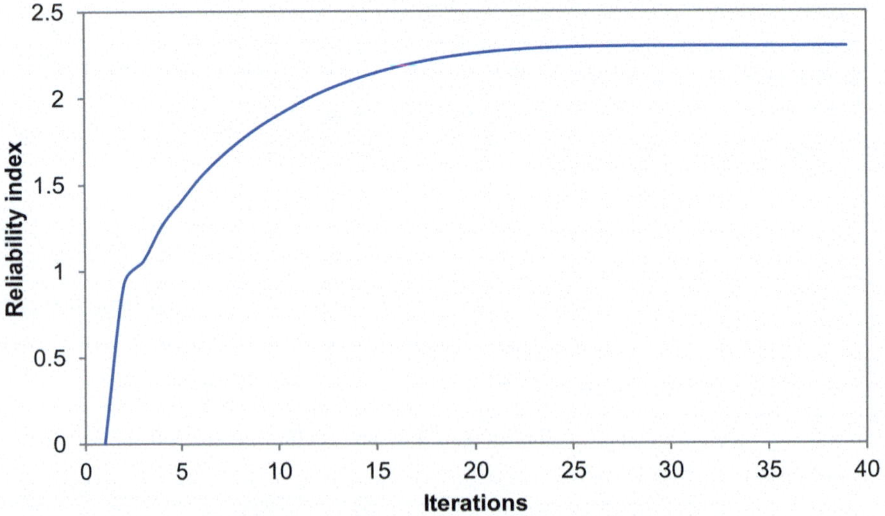

Fig. 3.19 Reliability index iteration using RHL-RF for nonlinear example

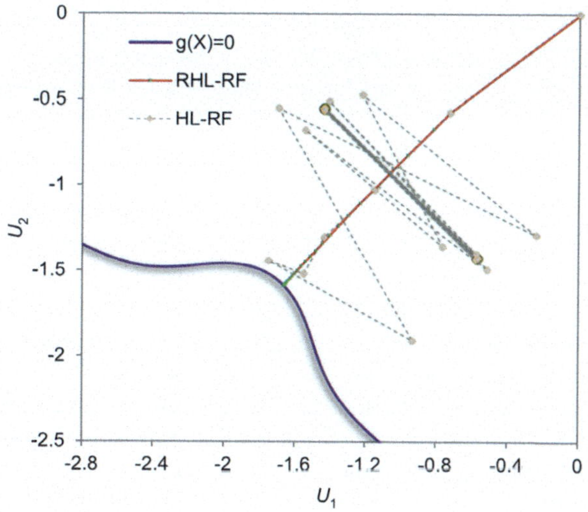

Fig. 3.20 Iteration for MPP search in normal standard space using RHL-RF and HL-RF methods for nonlinear example

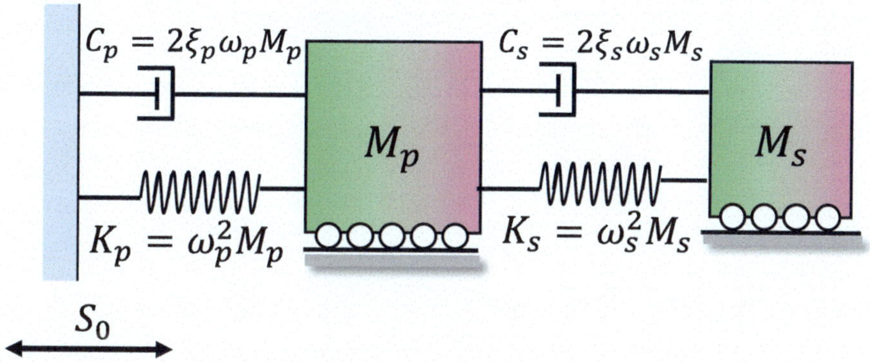

Fig. 3.21 Two-degree dynamic system for Example 3.14. Reprinted from [22], with permission from Elsevier

$$g = F_s - K_s \times P(E[x_s^2])^{1/2} \tag{3.44}$$

where Pis3, F_s is the force capacity and mean-square displacement computed as below,

$$E[x_s^2] = \frac{\pi S_0}{4\xi_s\omega_s}\left[\frac{\xi_a\xi_s}{\xi_p\xi_s(4\xi_a^2 + \theta^2) + \gamma\xi_a^2} \times \frac{\left(\xi_p\omega_p^3 + \xi_s\omega_s^3\right)\omega_p}{4\xi_a\omega_a^4}\right] \tag{3.45}$$

where mass ratio i.e.γ is $\frac{M_s}{M_p}$; average damping ratio i.e.ξ_a is $\frac{\xi_p+\xi_s}{2}$; average frequency i.e.ω_a is $\frac{\omega_p+\omega_s}{2}$; tuning parameter i.e.$\theta$ is $\frac{\omega_p-\omega_s}{\omega_a}$; S_0 is the white noise.

Eight lognormal variables are tabulated in Table 3.6 for this example.

The identical results from Ref. [19] showed that reliability index of 2.00496. As given results from Fig. 3.22, the periodic-2 reliability index as $\beta = [4.21701, 4.98074]$ is captured by HL-RF method. However, RHL-RF provides stable results as reliability index of 2.0149 which is obtained after 85 iterations. This problem has an acceptable result with the results extracted from the literature. As seen from the results presented in Fig. 3.22, the RHL-RF provides chaotic solutions in the iteration procedure but at the final iteration, this formulation gives its stability for iterations. By second-order fitting of the LSF-based step size in the RHL-RF method, the search direction vector may be computed inaccurately direction in MPP search processes that it is shown in the results of Fig. 3.22 for iterations $4 < k < 39$, while the stability of the RHL-RF can be given based on the dynamical step size at final iteration. It provided stable results but the efficiency of the RHL-RF method is dependent on the search direction vector which is directly applied to computing the step size.

Table 3.6 Random variables information for dynamic system in Example 3.14

variable	M_s	M_p	K_s	K_p	ξ_s	ξ_p	F_s	S_0
Mean	0.01	1	0.01	1	0.02	0.05	15	100
STD	0.001	0.1	0.001	0.2	0.01	0.02	1.5	10

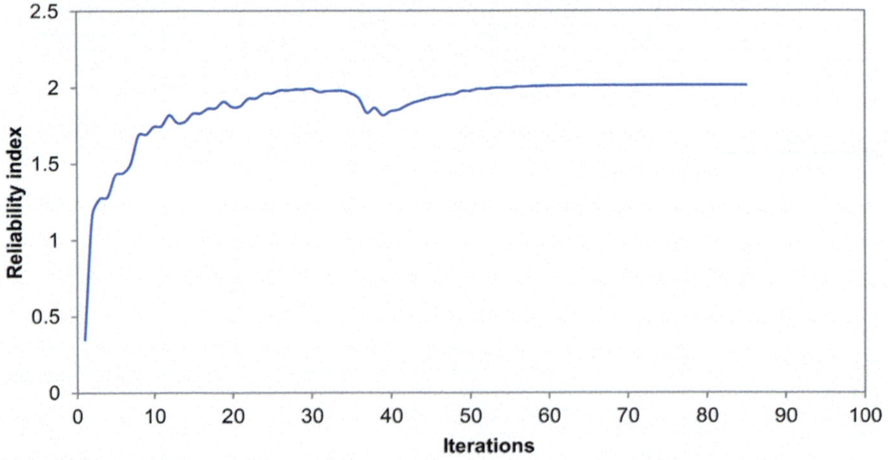

Fig. 3.22 Reliability index iteration using RHL-RF for dynamic system Example 3.14

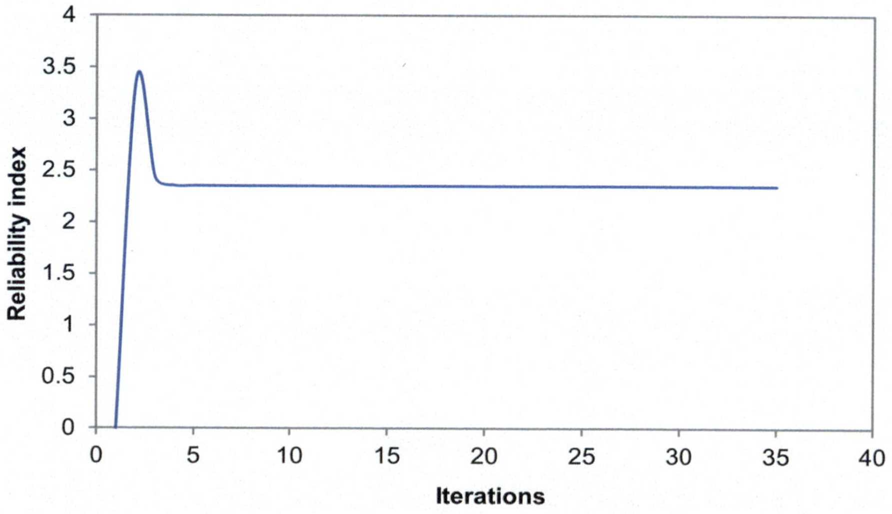

Fig. 3.23 Reliability index iteration using RHL-RF for Slender Column

Example 3.15 Slender Column.

A slender column in buckling failure mode applied in Example 3.15, Sect. 3.2 is applied to illustrate the abilities of FORM-based RHL-RF formulation. This problem involved six independent random variables. The reliability index is given based on the RHL-RF method as $\beta = 2.3507$ after 35 iterations and the convergence for the reliability index is presented in Fig. 3.23. The stability convergence for this problem is given from the results presented in Fig. 3.23 and the smooth results for the reliability index with a high-convergence rate us obtained using RHL-RF method. This approach can provide stable results for some engineering cases with moderate LSF, but its ability is strongly dependent on the search direction vector computed based on the HL-RF formulation. Thus, the step size is used for controlling the instability of the iterative formulation.

Example 3.16 Roof truss.

The roof truss problem given in Sect. 3.2 with LSF in Eq. 3.12 is selected for this example. As seen, the HL-RF method is not converged in this problem but the ability of RHL-RF method should be examined in this structure, and the convergence of the reliability index for this example is plotted in Fig. 3.24. The result of Fig. 3.24 showed that the RHL-RF method is converged for this problem (i.e. $\beta = 2.4173$ after 136 iterations), however, its stability is omitted several times in the search process of MPP. The chaotic results are obtained for some cases of iterations but the new and previous normalized selectivity vector is not located on the same direction. The RHL-RF method can improve the stability of MPP search for problems with the periodic solution obtained by HL-RF method, but for some applicable engineering problems with chaotic results which are given by HL-RF method the RHL-RF may be

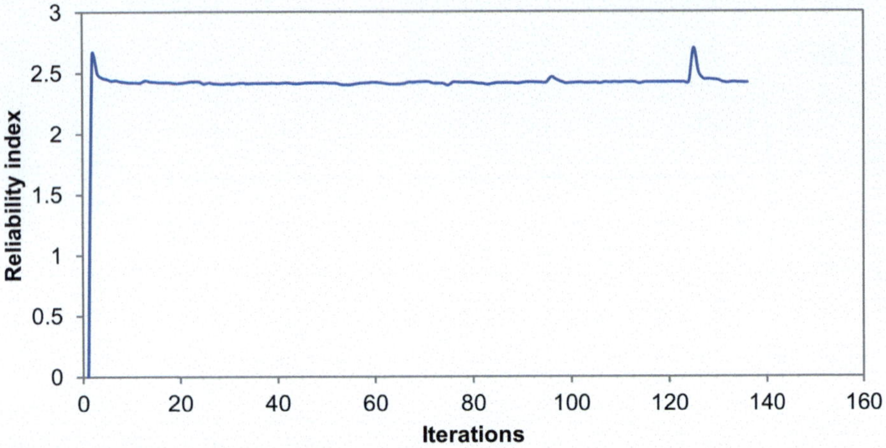

Fig. 3.24 Reliability index iteration using RHL-RF for Roof truss

confused for computing appropriate search direction. Therefore, it may require more computational burden for MPP search and give stable results by using the dynamical step size.

3.6 Non-negative Constraint Method

An iterative approach-based FORM formulation is provided by computing a search direction vector and dynamical step sizes in Non-negative constraint method. In this approach, the LSF is transformed into a real and a non-negative function that its values at all positions in the original space give the real-positive value. The change in constraint is the key point in the non-negative constraint (NNC) method [16]. In all FORM iterative formulations, the non-negative constraint as LSF is given as real and positive values. By NNC technique, the challenge of approximating an appropriate penalty factor at every iteration can be solved. The step size is computed self-adaptively that is a simple contribution of iHL-RF method. The formulation of this method is described in this section. By Considering the unconstrained function for searching the MPP, we have,

$$h(\boldsymbol{U}) = \frac{\boldsymbol{U}^T\boldsymbol{U}}{2} + \lambda g(\boldsymbol{U}) \tag{3.46}$$

This is a constrained problem of the FORM which is given based on the minimum distance of a point on limit state function thus the minimum distance ($\frac{\boldsymbol{U}^T\boldsymbol{U}}{2}$) is defined as objective and the LSF ($g(\boldsymbol{U})$) is as subjective of the probabilistic optimization model as below relation,

$$\begin{cases} \min \frac{U^T U}{2} \\ subjecttog(U) = 0 \end{cases} \quad (3.47)$$

where λ > 0 denotes the penalty factor. Generally, computing the step size and adjustment of λ are the main challenges in the FORM-based SDSD. As seen from the above sections, HL-RF is an unstable formulation for high-nonlinear problems while the improved versions solved the robustness of the FORM compared to HL-RF method, but they are inefficient methods. Based on Eq. 3.24, the large values of λ (corresponding to penalty factor) can stop the process may be not achieved. In the NNC method, the LSF i.e. g is given as a real and positive value thus it can be given a minimum of $g(U)$ at MPP to follow the zero. Therefore, Eq. 3.47 is refined as below in the NNC method,

$$\begin{cases} \min \frac{U^T U}{2} \\ subjecttog^2(U) = 0 \end{cases} \quad (3.48)$$

And we have,

$$h(U) = \frac{U^T U}{2} + \lambda g^2(U) \quad (3.49)$$

It can be concluded that $g = 0$ at MPP thus $g^2 = 0$ at MPP. Consequently, by using the non-negative h(**U**) in Eq. 3.47 and obtaining an appropriate positive $\lambda = 10^6$, the minimization of constraint function for MPP search is guaranteed at MPP. In this method, the iterative formulation is given as below relation which is extracted from the unconstraint optimization model of Eq. 3.47.

$$U_{k+1} = U_k + \alpha_k d_k \quad (3.50)$$

where d_k denotes the search direction vector computed by negative gradient vector as presented in Eq. 3.49 as follows,

$$d_k = -[U_k + 2\lambda g(U_k)\nabla g(U_k)] \quad (3.51)$$

The step size α_k is given from Taylor expansion of the function g^2 as below relations,

$$g^2(U) = g^2(U_k) + \nabla^T g^2(U_k)(U_{k+1} - U_k) \approx 0 \quad (3.52)$$

Based on Eq. 3.48, it can be conducted that $U_{k+1} - U_k = \alpha_k d_k$ thus by replacing Eq. 3.49 by Eq. 3.50. and rearranging Eq. 3.50, we give the below relation,

$$g^2(U_k) + 2\nabla^T g(U_k) g(U_k) \alpha_k d_k \approx 0 \quad (3.53)$$

Thus, the step size is computed as below relation,

$$\alpha_k = -\frac{g(\boldsymbol{U}_k)}{2\nabla^T g(\boldsymbol{U}_k)\boldsymbol{d}_k} \tag{3.54}$$

For this scheme, the iterative formulated is given on the above relations thus it can provide an iterative FORM-based NNC as below steps,

Step 1: Define LSF $g(\boldsymbol{X})$, give stopping criterion (ε), statistical properties of basic random variables (μ and σ), set $k = 1$ and $\boldsymbol{X}_1 = \boldsymbol{\mu}$, $\lambda = 10^6$ and $\alpha_1 = 1$;
Step 2: Normalize the random variables by transferring from $\boldsymbol{X}$-space to $\boldsymbol{U}$-space based on Eqs. 3.3 and 3.4;
Step 3: Iterate the NNC for finding MPP as below framework;

(i) Compute $g(\boldsymbol{U}_k)$, $\nabla g(\boldsymbol{U}_k)$;
(ii) Determine as $\boldsymbol{d}_k = -[\boldsymbol{U}_k + 2\lambda g(\boldsymbol{U}_k)\nabla g(\boldsymbol{U}_k)]$;
(iii) Determine the step size as $\alpha_k = -g(\boldsymbol{U}_k)/\left(2\nabla^T g(\boldsymbol{U}_k)\boldsymbol{d}_k\right)$;
(iv) Compute the new point as $\boldsymbol{U}_{k+1} = \boldsymbol{U}_k + \alpha_k \boldsymbol{d}_k$.

Step 4: If $|\beta_{k+1} - \beta_k| < \varepsilon$ then stop iterative formulation of NNC and print $\boldsymbol{X}^* = \boldsymbol{X}_{k+1}$,$\boldsymbol{U}^* = \boldsymbol{U}_{k+1}$, $\beta = \beta_{k+1}$;

Else: $k = k + 1$ and Compute $\boldsymbol{X}_{k+1} = T^{-1}(\boldsymbol{U}_{k+1})$ go to Step 2.
As seen from this framework, the NNC method is a simple approach compared to iHL-RF and RHL-RF, but its formulation is directly computed based on a dynamical step size, unlike HL-RF and CC methods. The NNC method may be applied for engineering reliability problems with simple relation, but it should be improved for updating step size in future research works. The NCC method is compared with four reliability problems that are given as below,

Example 3.17 Nonlinear mathematical performance function.

$$g = x_1^3 + x_1^2 x_2 + x_2^3 - 18 \tag{3.55}$$

where, normal x_1 and x_2 variables have $\mu_1 = 10$, $\mu_2 = 9.9$ and $\sigma_1 = \sigma_2 = 5$. It converged to the reliability index of 2.3014 after 27 iterations as presented results for iterations of the reliability index in Fig. 3.25. The stability for convergence is presented in Fig. 3.25 for this example, which obtaina stable result with smooth iterations. Therefore, the NNC method accurately computes the search direction with an appropriate step size.

The iterations for MPP search in normal standard space are presented in Fig. 3.26 for NNC method. Unlike the HL-RF method, the NNC iterative formulation has a good ability to searching the MPP. The MPP is obtained based on a smooth line with linear forms as seen in Fig. 3.26 with red color. This method can provide the stabilization of the MPP with an appropriate step size.

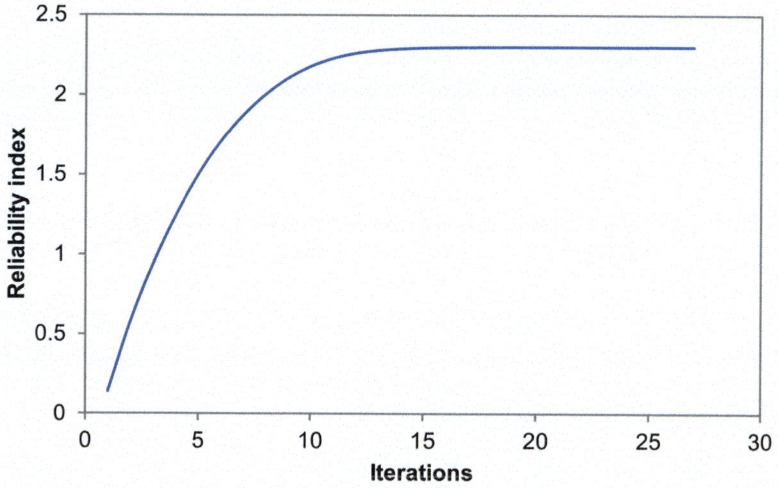

Fig. 3.25 Reliability index iteration using NNC for nonlinear problem

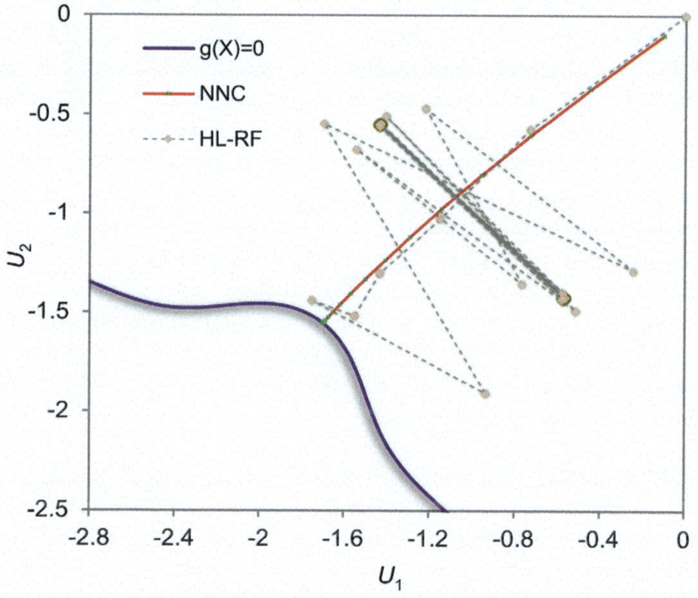

Fig. 3.26 Iteration for MPP search in normal standard space using NNC and HL-RF methods for nonlinear example

Example 3.18 A nonlinear non-normal problem.

A nonlinear second-order LSF, which is taken from a pipeline performance generated by the response-surface function as below [8],

$$\begin{aligned} g = 1.1 - 0.00115x_1x_2 + 0.00157x_2^2 \\ +0.00117x_1^2 + 0.0135x_2x_3 - 0.0705x_2 - 0.00534x_1 \\ -0.0149x_1x_3 - 0.0611x_2x_4 + 0.0717x_1x_4 - 0.226x_3 \\ +0.0333x_3^2 - 0.558x_3x_4 + 0.998x_4 - 1.339x_4^2 \end{aligned} \tag{3.56}$$

where, x_1-x_4 are random variables simulated as x_1 with type-II largest distribution (mean = 10and STD = 5), x_2 and x_3 with normally distribution (means = 25and0.8, STD = 5and0.2, respectively); and x_4 with lognormal distribution (mean = 0.0625and STD = 0.0625).

By extracted results from Ref. [8], the reliability index is $\beta = 1.36$ and MPP is $\boldsymbol{X}^* = [15.09, 25.027, 0.8653, 0.03582]$. Using the RHL-RF given from Ref. [19], $\beta = 1.33053$ and $\boldsymbol{X}^* = [14.905, 25.067, 0.85956, 0.046061]$. The NNC method is converged to the reliability index of 1.4704 and the MPP of $\boldsymbol{U}^* = [1.2915, -0.0371, 0.1622, 0.683]$ and $\boldsymbol{X}^* = [14.8721, 24.8147, 0.8324, 0.0781]$ after 21 iterations. The converged results based on the NNC are followed by the results of Refs. [8, 19]. However, it is some differences with the MPP and the reliability index is captured based on FORM-based NNC formulation and the results of Ref. [8]. This can be suspended as a non-negative LSF in the NNC formulae that compute the search direction and step size. This approach converges to the MPP local minimum of the problem, ignoring the global nature of it being the minimum of the reliability index in the NNC formulation of the problem.

The reliability index given from the NNC method with respect to the iterations is plotted in Fig. 3.27 for nonlinear no-normal problems. It is obvious that the NNC method provides stable results with smooth iteration, more efficiently. However, the convergent performances of this method should be examined for convex and high-dimensional non-normal problems.

Example 3.19 A vehicle disc brake system.

A brake disc with friction materials and back plates is shown in Fig. 3.28, the RSM is applied for approximation of the failure domain that is defined as follows,

$$\begin{aligned} g = 0.046287 + 0.20458\mu - 0.059821p - 0.00036549\,h_1 - 0.010037h_2 \\ +0.013836h_3 + 0.24308\mu p - 0.0037884\mu h_1 + 0.0023358\mu h_2 \\ -0.016981\mu h_3 + 0.029287ph_1 - 0.015872ph_2 - 0.0028333ph_3 \\ +0.0007175h_1h_2 - 0.00046158h_1h_3 - 0.00036486h_2h_3 - 0.39076\mu^2 \\ -0.015986p^2 - 0.0011936h_1^2 + 0.000269h_2^2 + 0.00062638h_3^2 - 0.01 \end{aligned} \tag{3.57}$$

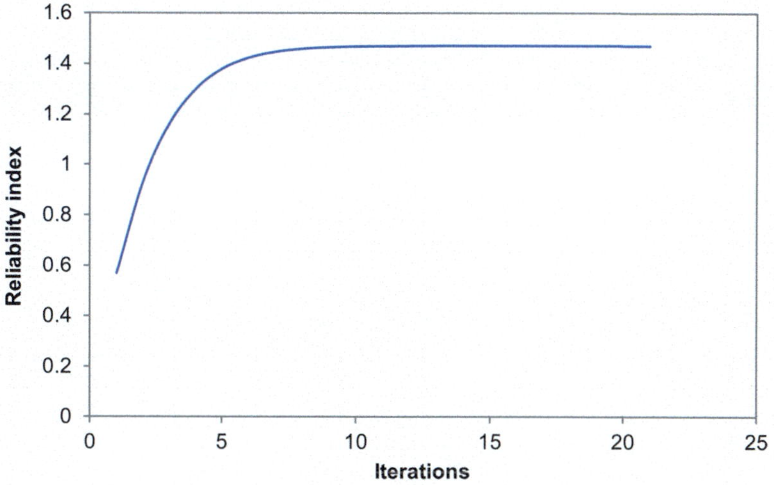

Fig. 3.27 Reliability index iteration using NNC for nonlinear non-normal problem

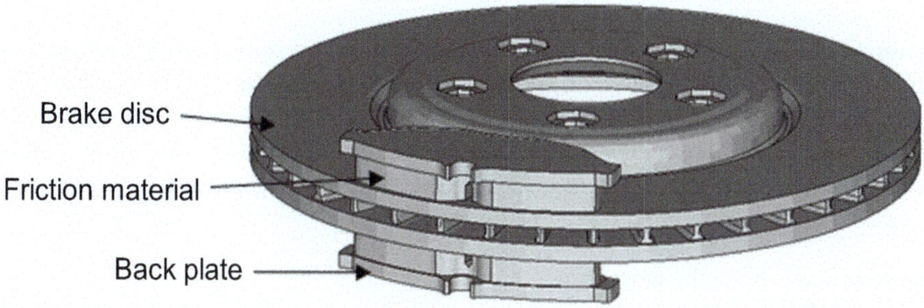

Fig. 3.28 Disc brake for Example 3.19. Reprinted from [23], with permission from Elsevier

where p is the brake pressure, μ is the friction coefficient, h_1 is the thickness of friction material, h_2 is the disc thickness and h_3 is the back plate thickness. The random variables as recorded in Table 3.7 for Example 3.19.

The results given by MCS with 1.2×10^6 samples are obtained as $\beta = 2.5849$ and the related failure probability of 0.0049. By using the iterative method of NNC, the reliability index and failure probability are computed as 2.2416 and 0.0125 after

Table 3.7 Information of random variables for disc brake applied in Example 3.19

Variable	μ	h_1 (mm)	h_2 (mm)	h_3 (mm)	p (MPa)
Mean	0.4	15	20	10	0.5
CoV	0.35	0.1	0.1	0.1	0.25
PDF	Normal	Gumbel	Gumbel	Gumbel	Lognormal

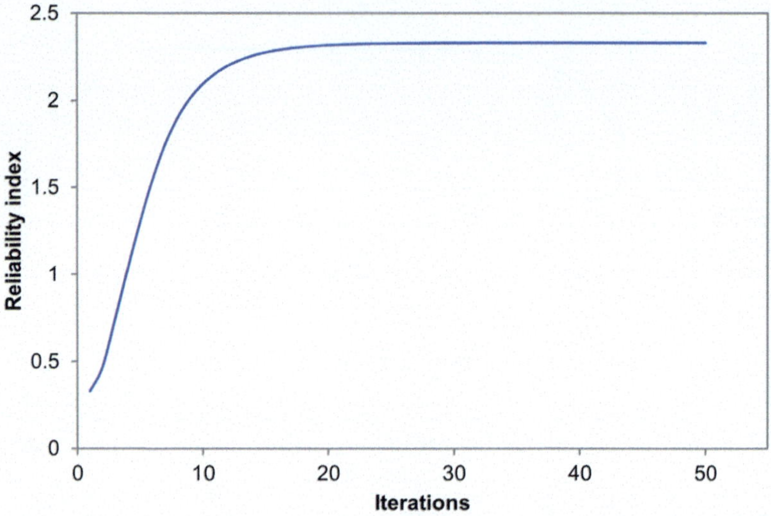

Fig. 3.29 Reliability index iteration using NNC for vehicle disc brake problem

51 iterations. The NCC method-based FORM formulation converges, more robustly. The efficiency of this method is moderate for this example by 51 iterations for searching MPP as shown in Fig. 3.29 for reliability index iterations. By linear approximating the performance function at MPP in the FORM formula, the error of this problem showed that it has a high-concave LSF for this problem which is regressed by second-order polynomial response surface method.

Example 3.20 A steel frame.

The steel frame problem used as Example 3.20 in Sect. 3.3 is given for the final problem that is used by the LSF $g = 0.05 - \Delta$ for showing the abilities of NNC iterative formula. In this example, lateral displacement is determined using the finite element model by random variables presented in Table 3.5.

The result of NNC-based FORM formulation is presented for the reliability index in Fig. 3.30 for the steel frame example. It should be pointed out that the NCC is a robust iterative procedure which the NNC method guarantees stable results for FORM-based SDSD but the accuracy and efficiency of this method are missing for several problems as well as the steel frame example. The converged reliability index using NNC method is 3.2268 which is obtained after 38 iterations that showed that the NNC has more efficiency for convergence of this problem. This is a kind of convex example that this method is shown best capability for reliability analysis of convex reliability problems but the convergence performances are strongly dependent on the nonlinearity degree of the LSF at MPP.

As well as other studied problems for evaluating the ability of NNC method, the iterative procedure of the NCC is without any numerical instabilities such as periodic or chaotic solutions. The penalty coefficient in this method is one of the key

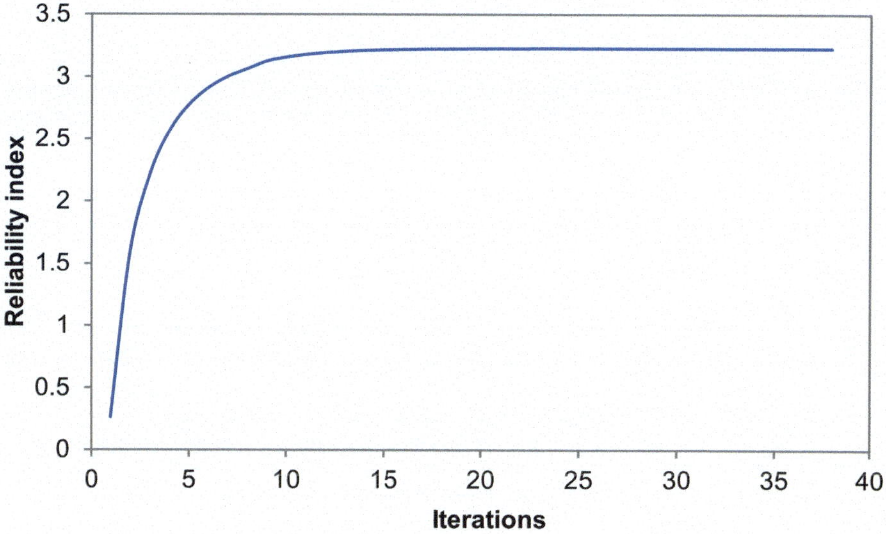

Fig. 3.30 Reliability index iteration using NNC for steel frame problem

factors for giving stable results. If it is selected a too small value, inefficient results may be obtained using this method but for a too large penalty factor λ, it may obtain inaccurate line search and the stability of this method may be omitted. For this reason, the reliability index of the steel frame problems is investigated for various penalty factors that captured results are cited in Fig. 3.31. As seen from the results of Fig. 3.31, the premature results are obtained based on the smaller penalty factor. On the other hand, this method is not valid for the penalty factor of $100 < \lambda < 600$. Therefore, the capability of this method depended on acceptable selection of the penalty function which is different for every problem. That it can be given an accurate result based on the penalty factor of more than 10^5 for this problem.

3.7 Discussion

In order to better go about understanding the capabilities of a method, including globality, robustness and efficiency, it should be illustrated based on several examples of numerical methods. These capabilities are defined as follows,

(1) Globality: related to the global results of the FORM-based formulation that it is related to the minimum reliability index when the LSF is related to zero. FORM is a probabilistic optimization model that it should provide a minimum distance form origin to limit state surface in normal standard space. Therefore, its globality is important for convergence properties. An iterative formulation

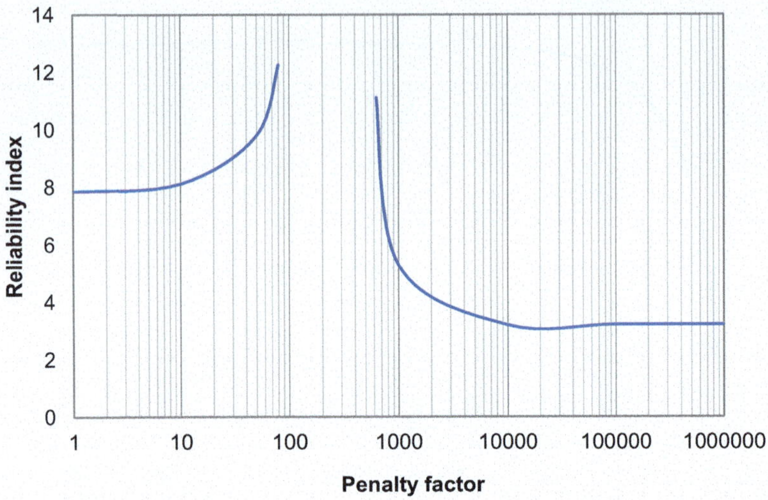

Fig. 3.31 Reliability index corresponding to various penalty factors in NNC method for steel frame problem

provides a stable result but it may be not global optimum condition. There may be converged, prematurely or locally;

(2) Robustness: give a stable solution using an iterative formulation which is applied for MPP search. The robust results with stable solutions are one of the main convergent properties in the analytical reliability analysis method. Today by developing the computer service and extending their abilities for doing computational methods with high-process, it is important that a method has a robust formulation for giving the global MPP;

(3) Efficiency: the efforts of a FORM algorithm for finding the MPP are related to the number of iterations or a number of call functions for determining the MPP. In all algorithms, we used the numerical differential approach for computing the gradient vector by the finite difference method. Consequently, the times and efforts should be cited for computing the iterative formulation of FORM so that it can be related to their iterations.

The above abilities are discussed in this part based on three examples that these examples are solved by HL-RF, iHL-RF STM, RHL-RF and NNC. The parameters of STM are given as a chaos control factor of $\xi = 0.1$ and involutory matrix of $\boldsymbol{C} = \boldsymbol{I}$ in the iHL-RF method, the parameter C is computed as below relations,

$$C > \frac{\beta}{|\nabla g(\boldsymbol{U})|} \tag{3.58}$$

Thus, we have,

$$C = 1.5\frac{\beta}{|\nabla g(U)|} + 3 \tag{3.59}$$

Equation 3.57 is dynamically applied for improved HL-RF method for giving adjusting guidance for selecting the step size as below relation,

$$\alpha_k = \begin{cases} \alpha_k m(U_k) < m(U_{k-1}) \\ 0.95\alpha_k m(U_k) \geq m(U_{k-1}) \end{cases} \tag{3.60}$$

The step size method is dynamically computed for RHL-RF as well as the NNC formulation while the penalty factor for NCC method is given 10^4 for all problems. The stopping criterion is given as $|\beta_{k+1} - \beta_{k+1}| < 10^{-6}$ for all problems.

Example 3.21 A vehicle side-impact.

The LSF for this problem is given from the response surface method by the following second-order polynomial function [12],

$$\begin{aligned} g = {} & 0.489x_3x_7 + 0.843x_5x_6 - 0.0432x_9x_{10} \\ & +0.0556x_9x_{11} + 0.000786x_{11}^2 - 0.75 \end{aligned} \tag{3.61}$$

where $i = 1 \sim 7, x_i \sim N(1, 0.005), i = 8 \sim 9, x_i \sim N(0.3, 0.006), x_{10} \sim N(0, 10.0)$.

In this example, the HL-RF method provides unstable solutions, i.e., 4-step reliability indices, while the other steepest descent methods, which are improved versions of HL-RF, converge robustly. As presented results in Fig. 3.32, the reliability index is obtained using the STM ($\beta = 3.4975$), iHL-RF ($\beta = 3.4975$), RHL-RF ($\beta = 3.4975$) and NCC ($\beta = 3.5317$) methods after 124, 79, 28 and 23 iterations, respectively. As seen, the NNC method provides a different reliability index from other methods while it is more efficient than others. The RHL-RF provides the global with robust results, more efficiently. The STM and iHL-RF methods are inefficient approaches among others. Consequently, the search direction in the steepest descent-based FORM formula is a vital challenge for developing a method.

Example 3.22 Two-degree dynamic system.

A two-degree dynamic system is given by the LSF of $g = Fs - Ks \times P(E[x_s^2])^{1/2}$ [19] whose random variables are presented in Table 3.6. The all methods presented in this chapter are applied to solve this example and the results of the reliability index is plotted in Fig. 3.33.

As seen from Fig. 3.33, the HL-RF method provided periodic-2 as $\beta = [4.21701, 4.98074]$ solutions. But, the other methods provided stable results as,

(i) Using iHL-RF method; $\beta = 2.0164$ captured after 97 iterations with 9808 call functions;
(ii) using RHL-RF method; $\beta = 2.0163$ captured after 69 iterations with 6995 call functions;

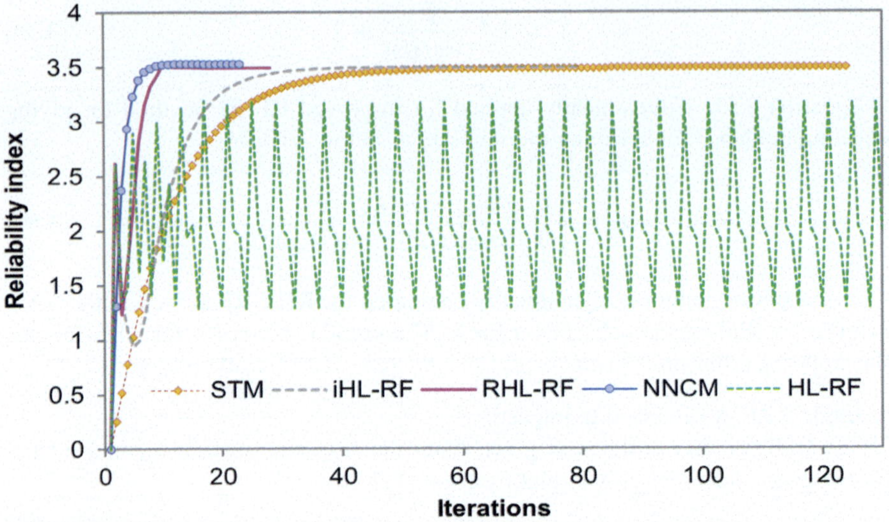

Fig. 3.32 The convergence for reliability index of Example 3.21 using steepest descent search direction methods

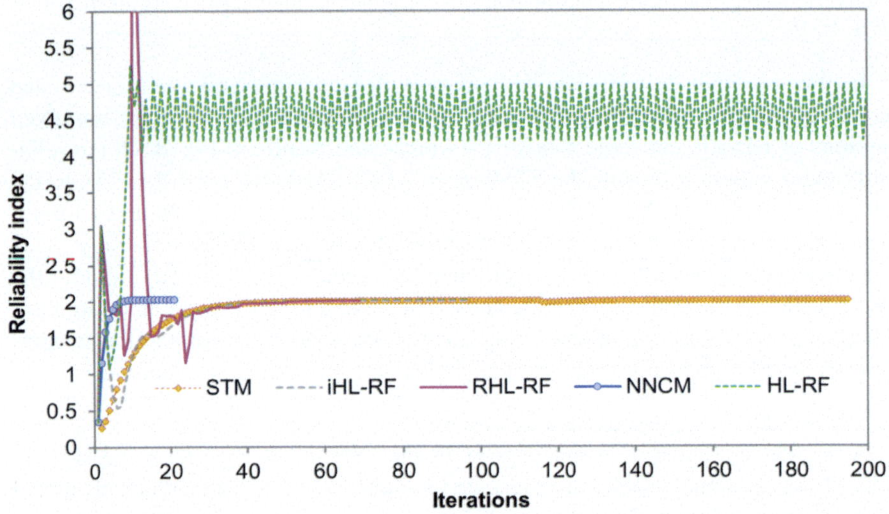

Fig. 3.33 The convergence for reliability index of Example 3.22 using steepest descent search direction methods

(iii) using STM method; $\beta = 2.0164$ captured after 195 iterations with 19,905 call functions;

(iv) using NNC method; $\beta = 2.0319$ which is captured after 21 iterations and 2408 call functions.

The NNC method is robust and quite efficient. Relaxed HL-RF is more efficient than iHL-RF and STM. The STM has the last efficiency among other methods. The nonlinearity degree of the LSF is neglected in the STM while it is strongly affected in the formulation of the NCC method. The step size of RHL-RF is computed with an inaccurate line search at the first iteration but it is funded the right way at the final iterations.

Example 3.23 A steel frame.

This example is also compared with the results of several methods in this chapter the LSF of the steel frame is given as $g = 0.05 - \Delta$. The studied methods formulated by the steepest descent approach are used for the reliability analysis of this problem and their reliability iterative process is plotted in Fig. 3.34. It can be concluded that the STM and HL-RF provided unstable results as chaotic solutions while the other methods provided stable results. The stability of the NNC is provided based on no-negative LSF while its globality is an issue for this iterative FORM formulation. The reliability index using NNC method is strongly obtained with a fast convergence rate but it is not a global result. It can be concluded that NNC method is fast but its results may be provided, prematurely. The slow convergence rate is obtained for both iLH-RF and the RHL-RF method thus it cannot be selected as an acceptable method with efficient formulation, global convergence and robust results but for applicable engineering problems the NCC is a fast scheme with simple formulation. The iHL-RF and RHL-RF reliability methods require guidance on calculating the step size. The STM is very simple but it is converged very slowly. Therefore, the STM is not suggested as an applicable FORM method for engineering problems due to low convergence rates and unstable results. The HL-RF may provide a fast convergence rate but it is not a robust method.

3.8 Summary

This chapter presents the FORM-based steepest descent search direction. The iterative formulation of FORM is established based on the steepest descent sensitivity vectors that almost all analytical methods of FORM are enhanced by a step size less than 1. In this chapter, all collected the FORM formulas are developed as an improved version of the HL-RF method named as the steepest descent method. The main improvement of these approaches is similar to the relaxed approach that is applied to improving the robustness of FORM-based HL-RF. The HL-RF method is the first method that we try to show its abilities and convergence feedback including chaotic and periodic results. By using HL-RF method, we try to collect several approaches of FORM named (i) improve HL-RF, (ii) chaos control method, (iii) relaxed HL-RF, (iv) Non-negative constraint method that they are formulated by a refined HL-RF formulation for improving gradient line search direction based on a step size.

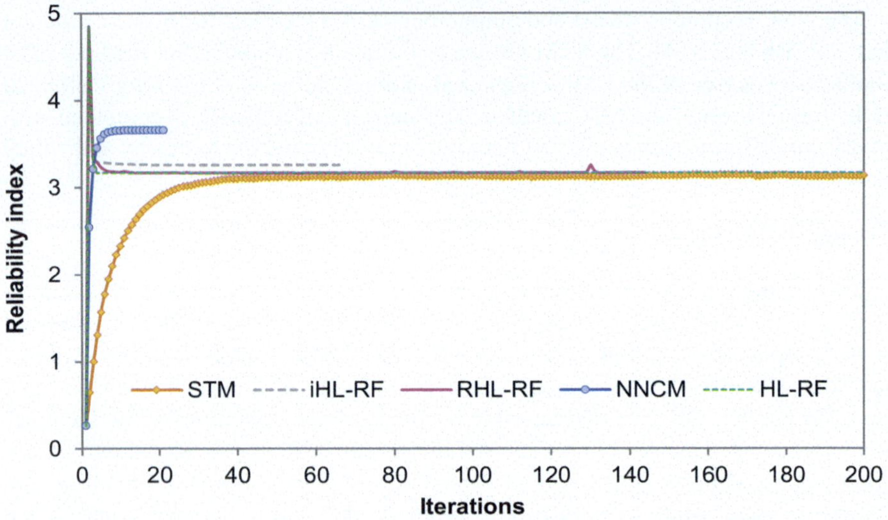

Fig. 3.34 The convergence for reliability index of Example 3.23 using steepest descent search direction methods

References

1. O. Ditlevsen, H.O. Madsen, *Structural Reliability Methods* (Wiley, New York, 1996), p.178
2. Z. Meng, G. Li, B.P. Wang, P. Hao, A hybrid chaos control approach of the performance measure functions for reliability-based design optimization. Comput. Struct. **146**, 32–43 (2015)
3. D. Yang, Chaos control for numerical instability of first order reliability method. Commun. Nonlinear Sci. Numer. Simul. **15**(10), 3131–3141 (2010)
4. I. Lee, K. Choi, L. Du, D. Gorsich, Inverse analysis method using MPP-based dimension reduction for reliability-based design optimization of nonlinear and multi-dimensional systems. Comput. Methods Appl. Mech. Eng. **198**(1), 14–27 (2008)
5. A. Der Kiureghian, H.Z. Lin, S.J. Hwang, Second-order reliability approximations. J. Eng. Mech. **113**(8), 1208–1225 (1987)
6. J.J. Kim, Realizing the creep and damage effect on masonry panel design based on reliability analysis. Materials **17**(11), 2643 (2024)
7. C. Hou, Q. Xu, Y. Li, Z. Sun, Reliability analysis of geosynthetic-reinforced slopes under rainfall infiltration. Geotext. Geomembr. **52**(1), 156–165 (2024)
8. P.L. Liu, A. Der Kiureghian, Optimization algorithms for structural reliability. Struct. Saf. **9**(3), 161–177 (1991)
9. B. Keshtegar, I. Lee, Relaxed performance measure approach for reliability-based design optimization. Struct. Multidiscip. Optim. **54**, 1439–1454 (2016)
10. B. Keshtegar, Z. Meng, A hybrid relaxed first-order reliability method for efficient structural reliability analysis. Struct. Saf. **66**, 84–93 (2017)
11. P.L. Liu, A.D. Kiureghian, Finite element reliability of geometrically nonlinear uncertain structures. J. Eng. Mech. **117**(8), 1806–1825 (1991)
12. B. Keshtegar, P. Hao, Z. Meng, A self-adaptive modified chaos control method for reliability-based design optimization. Struct. Multidiscip. Optim. **55**, 63–75 (2017)
13. W. Liping, R. Grandhi, Efficient safety index calculation for structural reliability analysis. Comput. Struct. **52**(1), 103–111 (1994)
14. T. Santosh, R. Saraf, A. Ghosh, H. Kushwaha, Optimum step length selection rule in modified HL-RF method for structural reliability. Int. J. Press. Vessel. Pip. **83**(10), 742–748 (2006)

15. S.R.d. Santos, L.C. Matioli, A.T. Beck, New optimization algorithms for structural reliability analysis. Comput. Model. Eng. Sci. **83**(1), 23–55 (2012)
16. M.A. Roudak, M. Karamloo, Establishment of non-negative constraint method as a robust and efficient first-order reliability method. Appl. Math. Model. **68**, 281–305 (2019)
17. B. Keshtegar, Stability iterative method for structural reliability analysis using a chaotic conjugate map. Nonlinear Dyn. **84**, 2161–2174 (2016)
18. B. Keshtegar, M. Miri, Introducing conjugate gradient optimization for modified HL-RF method. Eng. Comput. **31**(4), 775–790 (2014)
19. B. Keshtegar, Chaotic conjugate stability transformation method for structural reliability analysis. Comput. Methods Appl. Mech. Eng. **310**, 866–885 (2016)
20. B. Keshtegar, A hybrid conjugate finite-step length method for robust and efficient reliability analysis. Appl. Math. Model. **45**, 226–237 (2017)
21. C. Luo, S.P. Zhu, B. Keshtegar, W. Macek, R. Branco, D. Meng, Active Kriging-based conjugate first-order reliability method for highly efficient structural reliability analysis using resample strategy. Comput. Methods Appl. Mech. Eng. **423**, 116863 (2024)
22. C. Luo, S.P. Zhu, B. Keshtegar, X. Niu, O. Taylan, An enhanced uniform simulation approach coupled with SVR for efficient structural reliability analysis. Reliab. Eng. Syst. Saf. **237**, 109377 (2023)
23. B. Xia, H. Lü, D. Yu, C. Jiang, Reliability-based design optimization of structural systems under hybrid probabilistic and interval model. Comput. Struct. **160**, 126–134 (2015)

Chapter 4
Convergence Control for First Order Reliability Methods

Abstract In this chapter, the convergence control for achieving the stability of first order reliability method (FORM)-based steepest formulation is provided by improved HL-RF iterative formulation. Three conditions as (i) merit condition, (ii) angle condition and (iii) sufficient descent condition are presented for controlling instability of the FORM-based improved HL-RF and relaxed FORM approach. These conditions are applied for a suitable framework for monitoring the stability of FORM-based steepest descent method which are dynamically applied for adjusting step size. The step size using an accelerated inexact inner loop is computed for achieving the stable results of iterative FORM formulated by steepest descent methods.

Keywords Merit condition · Angle condition · Sufficient descent condition · Improved HL-RF

4.1 Introduction

The global convergences as robust iterations and efficiency are two major performances for First Order Reliability Method (FORM)-based iterative methods developed by steepest descent search direction. These convergent abilities are dissuaded in Chap. 3. However, a relaxed factor, step size, or chaos control factor are the major differences in the improved version of the HL-RF (steepest descent method without control). The guidance for computing the appropriate search direction is the main challenge in this chapter. By computing an accurate line search direction as well as the steepest decent vector and the step size, the robust convergence of the steepest descent methods of the FORM formula is given based on the efficient computational burden. The HL-RF which is one of the oldest steepest descent methods may be converged to unstable results for some cases of structural reliability analysis or LSFs with high-nonlinearity functions at Most Probable Point (MPP). However, the improved version of the HL-RF is robustly converged but they are inefficient approaches. By an improved version of the FORM formulation, the iterative formulation based on the steepest descent method is given as follows [1],

S.-P. Zhu and B. Keshtegar, *Structural Reliability Analysis*, Structural Integrity 31,
https://doi.org/10.1007/978-3-031-75197-4_4

$$\boldsymbol{U}_{k+1} = \boldsymbol{U}_k + \alpha_k \boldsymbol{d}_k \tag{4.1}$$

where, $\boldsymbol{d}_k = \boldsymbol{U}_{k+1}^{HL} - \boldsymbol{U}_k$ relates to the search direction vector, α_k is the step size which is given $\alpha_k > 0$. The steepest descent direction is generally the line search, since the opposite of the gradient direction of Limit State Function (LSF) or the sensitivity vector direction in the FORM-based HL-sRF is located on the steepest descent direction, thus the MPP search is efficiently computed. On the other hand, computing the accurate step size can be given by the following optimization model [2],

$$\alpha_{k,opt} = min\big[f(\boldsymbol{U}_k + \alpha_k \boldsymbol{d}_k)\big] \tag{4.2}$$

By decreasing $\boldsymbol{d}_k$, we give a suitable line search for FORM iterative optimization model. The MPP search by using a stable convergence depends on the value of step size. The line search method as Wolfe conditions is extended in the mathematical optimization problems as follows,

$$\begin{aligned} & f(\boldsymbol{U}_k + \alpha_k \boldsymbol{d}_k) \le f(\boldsymbol{U}_k) + \sigma_0 \alpha_k \nabla^T f(\boldsymbol{U}_k) \boldsymbol{d}_k \nabla^T f(\boldsymbol{U}_k \boldsymbol{U}_k + \alpha_k \boldsymbol{d}_k) \\ & \boldsymbol{d}_k \ge \sigma_1 \nabla^T f(\boldsymbol{U}_k) \boldsymbol{d}_k \end{aligned} \tag{4.3}$$

The strong Wolfe conditions are given for fast conversance, more preferably. In this rule, the below conditions should be satisfied,

$$\begin{aligned} & f(\boldsymbol{U}_k + \alpha_k \boldsymbol{d}_k) \le f(\boldsymbol{U}_k) + \sigma_0 \alpha_k \nabla^T f(\boldsymbol{U}_k) \boldsymbol{d}_k \\ & \left|\nabla^T f(\boldsymbol{U}_k \boldsymbol{U}_k + \alpha_k \boldsymbol{d}_k) \boldsymbol{d}_k\right| \le \sigma_1 \left|\nabla^T f(\boldsymbol{U}_k) \boldsymbol{d}_k\right| \end{aligned} \tag{4.4}$$

where, σ_0 and σ_1 are two positive constants as $0 < \sigma_0 < 0.5$ and, $\sigma_0 < \sigma_1 < 0.5$ the FORM-based iHL-RF algorithms which are satisfy the Wolfe rule conditions are more computational inefficient. On the other hand, the mathematical search direction conditions such as Wolfe and strong Wolfe conditions presented in Eqs. 4.3 and 4.4 are applied for unconstrained optimization problems. However, the optimization model of the FORM-based MPP search is a constrained probabilistic problem as shown in the below model [3],

$$\begin{cases} \min \dfrac{\boldsymbol{U}^T \boldsymbol{U}}{2} \\ subject\ to\ g(\boldsymbol{U}) = 0 \end{cases} \tag{4.5}$$

The constraint of the optimization model is LSF, thus, the application of some methods of line search direction method is more computational efforts for FORM method. Another type of inexact lines search direction is rule of Armijo conditions which are applied for determining the step size as below,

$$\alpha_k = \max\big[S_k, \gamma S_k, \gamma^2 S_k, \ldots, \gamma^n S_k\big], n = 1, 2, \ldots \tag{4.6}$$

The step size given form the above Ss should be satisfied the below condition,

$$f(\boldsymbol{U}_k + \alpha_k \boldsymbol{d}_k) - f(\boldsymbol{U}_k) \le \sigma \alpha_k \nabla^T f(\boldsymbol{U}_k)\boldsymbol{d}_k \tag{4.7}$$

where, $S_k = -\frac{\nabla^T f(\boldsymbol{U}_k)\boldsymbol{d}_k}{L|\boldsymbol{d}_k|^2}$ and $0 < \gamma < 1, 0 < \sigma \langle 0.5, L \rangle 0$.

The Armijo-based line search is a variant of Goldstein's rule that it is one of the most efficient step length selection rules. In structural reliability analysis applied for computing the step size in iHL-RF, Santosh et al. [4] used the Armijo for determining the step size. Zhang and Der Kinreghian [3] improved the HL-RF method by applying the Armijo rule to enhance the merit function instability. Keshtegar and Chakraborty [5] improved the finite step size method of FORM formulation based on the Armijo rule using a dynamical explicit step size. This condition is applied for computing the self-adaptive step length of the directional stability transformation method-based FORM formulation by Bin et al. [6]. This line search may improve the conveyance rate of the iHL-RF method presented by Liu and Der Kinreghian [1]. However, it is an inefficient approach for formulation of the steepest descent condition of FORM applied in [4]. On the other hand, the two-point approximation-based step size as the optimum value of dynamical step size is developed by Keshetagar and Miri [7]. However, the iHL-RF presented in Ref. [7] has some feedback as slow convergence for convex problems and inefficient convergence as well as the iHL-RF by giving the results presented in Chap. 3. The initial step size and adjusting coefficient in the finite step length method are given based on Armijo line search rules by Huang et al. [8], which are optimized to enhance the applicability of the algorithm emphatically for highly nonlinear functions. An improved HL-RF-based FORM iterative formulation has been presented based on the modified Armijo line search rule which has been applied for interpolation of step size. The Armijo line search guarantees the global convergence ability for FORM-based steepest descent condition but with a more complicated formulation for computing the optimum step size. The determining process of step size is of vital importance in the structural reliability analysis. The appropriate step size provides a suitable line search for MPP as close as possible to obtain an optimal point for computing the reliability index. Nevertheless, the optimum step size is a main effort in the FORM-based steepest descent search direction that it should satisfy two below conditions,

(1) The step size should not be selected a too large value, because it may give an unstable result as chaotic or periodic solutions.
(2) The step size should not be selected a too small value, because it may increase the risk of premature convergence or the iteration of the algorithm may be very slow with inefficient results.

The performances of the steepest descent method with regard to efficiency and robustness strongly depend on the selection of step size. The Armijo rule can used for determining the step size in the iteration procedure but it is time-consuming to select the step size. Thus, it should be changed the search direction vector or step size for giving an efficient reliable result which is the main challenge for this investigation. In this chapter, three applicable line search methods are discussed named as (1) the merit function proposed by Liu and Der Kinreghian [1], (2) the sufficient descent condition, and (3) the angle condition, the merit function in this current work has some modification for improving the ability of FORM formulation. Finally, these three conditions are applied for the relaxed formulation of the FORM-based steepest descent search direction. Finally, these line search conditions are compared together through several examples.

4.2 Merit Function Condition

Liu and Kiureghian [1] introduced a merit function ($m(\boldsymbol{U})$) for a guidance of selecting step size and it applied in the improved HL-RF method. The step size is given less than 1 by satisfying the condition as $m(\boldsymbol{U}_{k+1}) < m(\boldsymbol{U}_k)$, in which the merit function is defined as below:

$$m(\boldsymbol{U}_k) = \frac{1}{2}\left|\boldsymbol{U}_k - \frac{\nabla^T g(\boldsymbol{U}_k)\boldsymbol{U}_k}{|\nabla g(\boldsymbol{U}_k)|^2}\nabla g(\boldsymbol{U}_k)\right|^2 + \frac{1}{2}Cg(\boldsymbol{U}_k)^2 \tag{4.8}$$

where C is a positive constant value i.e. $C > 0$. The parameter C is computed as below relations,

$$C > \frac{|\boldsymbol{U}_k|}{|\nabla g(\boldsymbol{U}_k)|} \tag{4.9}$$

Thus, we have

$$C = 1.5\frac{|\boldsymbol{U}_k|}{|\nabla g(\boldsymbol{U}_k)|} + 3 \tag{4.10}$$

Equation 4.8 is dynamically applied for improved HL-RF method for giving an adjusting. For an acceptable iteration, it should be given as $m(\boldsymbol{U}_{k+1}) < m(\boldsymbol{U}_k)$. Therefore, the merit function is a scheme for selecting a step size α_k less than 1. For $k \to \infty$ then it should be given $\lim\limits_{k\to\infty} m(\boldsymbol{U}_k) = 0$. The most decreasing of the merit function is an acceptable line search to MPP search since it is a non-negative value that its optimum is 0. This is the main effort of applying the merit function in this chapter. Note that the optimum results are given based on the merit function through the search MPP on the LSF by satisfying the$g = 0$. Therefore, the new point is a fixed point by $\boldsymbol{U}_{k+1} \approx \boldsymbol{U}_k$. It can be conducted from the merit function presented in Eq. 4.8 that for $k \to \infty$, we have $m(\boldsymbol{U}_k) = 0$ thus $g(\boldsymbol{U}_k)^2 = 0$ and then $g = 0$. By this condition, it can be shown that,

$$\left|\boldsymbol{U}_k - \frac{\nabla^T g(\boldsymbol{U}_k)\boldsymbol{U}_k}{|\nabla g(\boldsymbol{U}_k)|^2}\nabla g(\boldsymbol{U}_k)\right|^2 = 0 \Rightarrow \boldsymbol{U}_k = \frac{\nabla^T g(\boldsymbol{U}_k)\boldsymbol{U}_k}{|\nabla g(\boldsymbol{U}_k)|^2}\nabla g(\boldsymbol{U}_k) \tag{4.11}$$

And $U_{k+1} \approx U_k$ at $g(U_k) = 0$. Thus the U_{k+1} is a fixed point. Thus, a mean effort is defined based on the merit function as the acceptable step is given based on merit function when the lowest decorating value for $m(U_{k+1})$ is captured.

Example 4.1 For showing the instability results by using the performances of the merit function, it is used a highly non-normal performance function as below [9],

$$g(X) = x_1^4 + x_2^2 - 50 \tag{4.12}$$

where x_1 and x_2 are basic random variables. The x_1 is Lognormal variable with the mean value of 5 and a standard deviation of 1, and x_2 is a Gumbel random variable with a mean of 10 and a standard deviation of 10. Based on FORM and Monte Carlo Simulation (MCS) with 1.2×10^6 samples, the reliability indexes extracted from Ref. [9] are $\beta_{FORM} = 3.259315$ and $\beta_{MCS} = 3.5009$, respectively. This example is solved based on the HL-RF formulation to illustrate the merit function abilities. The merit function for iteration results of the HL-RF method is plotted in Fig. 4.1. It can be concluded that the strong reduction of the merit function is obtained at iterations 1–4 but the stability of the FORM formulation is loosed for $k > 5$.

By appalling the iHL-RF (improved HL-RF) method, this problem is solved and the results of iHL-RF are presented in Fig. 4.2 for merit function. It can be seen that the merit function followed to zero at final iterations but for the initial iteration which needs control on the instability, after several iterations, it is given a suitable line search using the merit function to determine the stability.

For improved the efficiency of the iHl-RF, it provides stable results as reliability index of 3.2593 with iterations of 147. It is an inefficient method thus, it supposed that if $m(U_{k+1}) < m(U_k)$ then we have the acceptable search direction. Consequently, the step size can be increased when it computed $g(U_{k+1})^2 < g(U_k)^2$. Thus, this condition in the reliability analysis is improved as below steps.

First, it establishs a framework to find step size for determining the new point as $U_{k+1} = U_k + \alpha_k d_k$ where, $d_k = \frac{\nabla^T g(U_k)U_k - g(U_k)}{\|\nabla g(U_k)\|^2} \nabla g(U_k) - U_k$ for iteration k. By changing the step size α_k, the new point is changed. By this assumption, we have below steps for adjusting the step size using merit function.

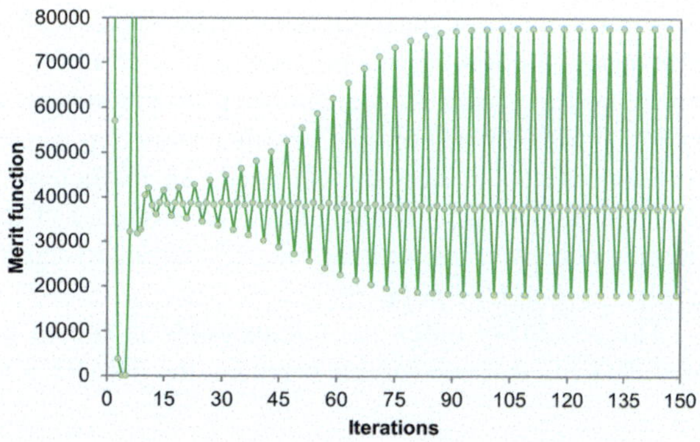

Fig. 4.1 Merit function with respect to iteration for $g(X) = x_1^4 + x_2^2 - 50$ when $\alpha_k = 1$

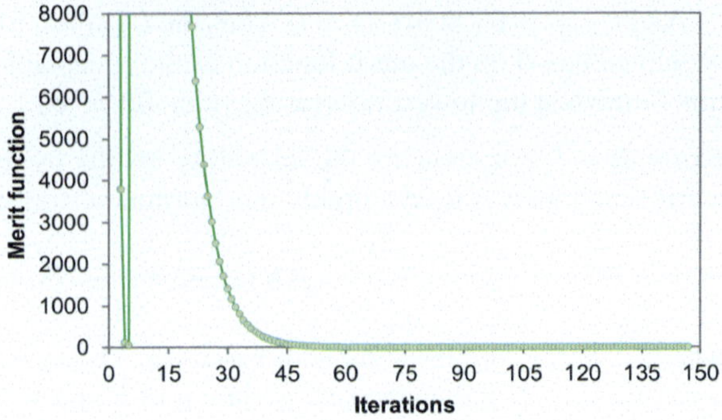

Fig. 4.2 Merit function with respect to iteration of iHL-RF for $g(\boldsymbol{X}) = x_1^4 + x_2^2 - 50$ by using $\alpha_k = \alpha$

Step 1: Give $\boldsymbol{d}_k$ and $\boldsymbol{U}_k$ as fixed vectors and set $m = 1$;
Step 2: Try to satisfy the merit condition as If $m(\boldsymbol{U}_k) < m(\boldsymbol{U}_{k-1})$, then do

(i) $\alpha_m = 2\gamma^m \sigma$ (4.13)

(ii) where, $0.9 < \gamma < 1$, for applicable selection $\gamma = 0.95$ and $0.5 \leq \sigma \leq 1$. As the HL-RF method, it is related to 0.5, but in this investigation, it gives $\sigma = 0.75$;
(iii) Determine the new point as $\boldsymbol{U}_{k+1} = \boldsymbol{U}_k + \alpha_m \boldsymbol{d}_k$;
(iv) Compute the LSF at new points as $g(\boldsymbol{U}_{k+1})$;
(v) If $g(\boldsymbol{U}_{k+1})^2 < g(\boldsymbol{U}_k)^2$ then $\alpha_{k+1} = \alpha_m$ and go to Step 3; else $m = m + 1$ and go to step 1;

Step 3: Determine new point as $\boldsymbol{U}_{k+1} = \boldsymbol{U}_k + \alpha_m \boldsymbol{d}_k$.

This iterative simple framework for determining step size using the merit function is presented as the flowchart in Fig. 4.3. It can be seen, that the step size is computed based on an iterative process when the merit condition is obtained based on the previous iterations, thus, it is efforted to find a step size more than 1 until $g(\boldsymbol{U}_{k+1})^2 < g(\boldsymbol{U}_k)^2$. This inexact step size is reduced by a nonlinear form as $\alpha_m = \gamma^{m+k}\alpha_k$ when it is a nonlinear LSF and the merit condition is not satisfied.

Example 4.1 in Eq. (4.12) is solved based on the presented inexact step size that is determined based on the merit function by an inexact line search presented in Fig. 4.2. The merit function for this problem is plotted in Fig. 4.3 which is given by inexact line search approach. As seen, the merit function is not any instabilities and $m(\boldsymbol{U}_k) < m(\boldsymbol{U}_{k-1})$ for almost iterations. The merit function is generally an acceptable condition for computing the step size. It needs an inefficient computational evaluation for determining the LSF when $m(\boldsymbol{U}_k) < m(\boldsymbol{U}_{k-1})$. This method is computationally inefficient comparing with the HL-RF method to determine the LSF at new points in $g(\boldsymbol{U}_{k+1})^2 < g(\boldsymbol{U}_k)^2$ loop.

The reliability index for this example is shown in Fig. 4.4 for $g(\boldsymbol{X}) = x_1^4 + x_2^2 - 50$ LSF. As seen for this problem, the iHL-RF formulation is provided a robust

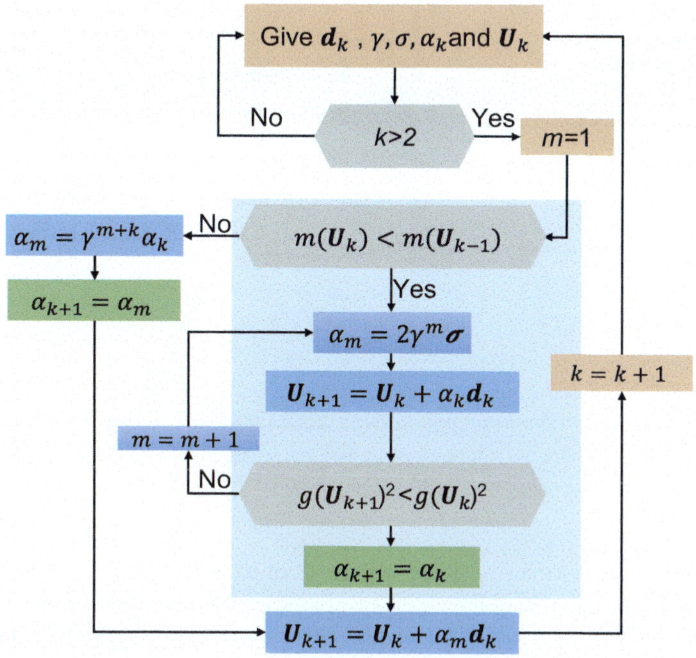

Fig. 4.3 Inexact step size selection using merit function procedure

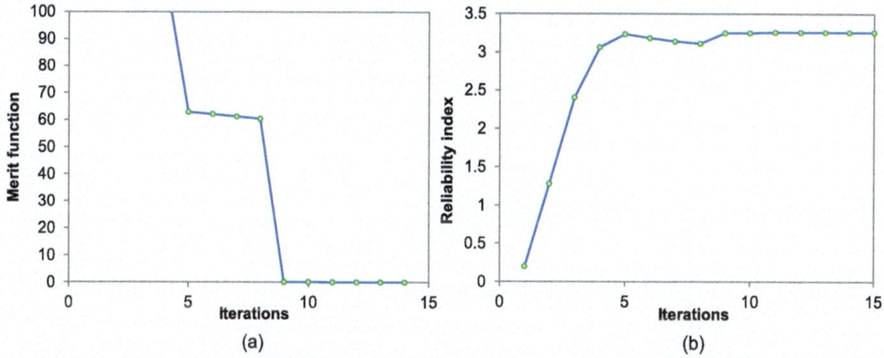

Fig. 4.4 Iteration of iHL-RF **a** Merit function;**b** reliability index for $g(X) = x_1^4 + x_2^2 - 50$ with inexact line search

formulation by the inexact line search after 15 iterations with a converged reliability index of 3.2593 which is computed based on the 933 call function.

Example 4.2 Normal nonlinear function

LSF is given by the following performance function,

$$g(X) = x_1^3 + x_1^2 x_2 + x_2^3 - 18 \tag{4.14}$$

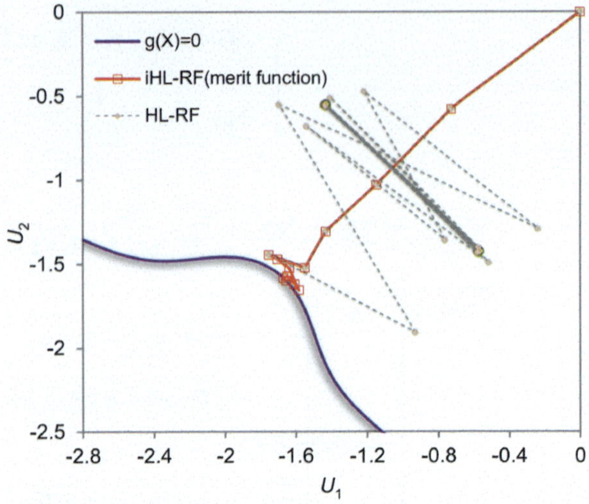

Fig. 4.5 Iteration for MPP search in normal standard space using iHL-RF given by inexact step size and HL-RF methods for Example 4.2

where x_1 and x_2 are normal variables with random parameters $\mu_1 = 10, \sigma_1 = 5$ and $\mu_2 = 9.9, \sigma_2 = 5$. The iHL-RF with inexact lines search computed by merit function is applied to analyze this problem. The iterative of the iHL-RF method is presented in Fig. 4.4 which is plotted in normal standard space for this problem. As seen, the MPP has computed accurately with a related reliability index of 2.2973 obtained after 13 iterations and 308 call functions that are closely in agreement with the results of the Stability Transformation Method (STM) and Relaxed HL-RF (RHL-RF) methods presented in Chap. 2. It can be concluded that the efficiency of iHL-RF is strongly improved while stable results are obtained (Fig. 4.5).

4.3 Sufficient Descent Condition

Al-Baali [10] proposed the sufficient descent condition for the first time for the convergence property of the Fletcher-Reeves conjugate gradient method [11] and then Gilbert and Nocedal applied this condition for obtaining global convergence of conjugate gradient methods. For a real and positive constant of $c > 0$, the sufficient decent condition is defined as follows,

$$\boldsymbol{d}_k \nabla^T f(\boldsymbol{U}_k) \leq -c\nabla^T f(\boldsymbol{U}_k)\nabla f(\boldsymbol{U}_k) \tag{4.15}$$

In the conjugate gradient-based descent method, the positive parameter c is suggested as $c = 7/8$ by Hager and Zhang [12]. The sufficient decent condition is an inexact line search that the globality for iterative process of an optimization method is provided when there is satisfied $\boldsymbol{d}_k \nabla^T f(\boldsymbol{U}_k) \leq -c\nabla^T f(\boldsymbol{U}_k)\nabla f(\boldsymbol{U}_k)$ for all iterations. Generally, in exact line search, we supposed that $\boldsymbol{d}_k \nabla^T f(\boldsymbol{U}_k) = 0$ for finding the step size or search direction vector but in inexact line search, a domain of the step size can be computed that they can provide an accurate line search for FORM-iterative formulation. As we know the reliability model is an unconcentrated optimization relation, thus application of the inexact condition for searching the step size is very difficult. For applying the sufficient condition in the FORM formulation which is computed the MPP by an iterative process, it is determined the $\boldsymbol{d}_k$ and $\nabla f(\boldsymbol{U}_k)$ as below relation,

$$\boldsymbol{d}_k = \boldsymbol{U}_{k+1}^{HL} - \boldsymbol{U}_k \tag{4.16}$$

$$\nabla f(\boldsymbol{U}_k) = \boldsymbol{U}_{k-1} - \boldsymbol{U}_k \tag{4.17}$$

where

$$\boldsymbol{U}_{k+1}^{HL} = \frac{\nabla^T g(\boldsymbol{U}_k) U_k - g(\boldsymbol{U}_k)}{\nabla^T g(\boldsymbol{U}_k) \nabla g(\boldsymbol{U}_k)} \nabla g(\boldsymbol{U}_k) \tag{4.18}$$

The $\boldsymbol{U}_{k-1}$ and $\boldsymbol{U}_k$ are the previous converged points obtained based on an iterative formulation. Thus, we have

$$[\boldsymbol{U}_{k+1}^{HL} - \boldsymbol{U}_k]^T [\boldsymbol{U}_{k-1} - \boldsymbol{U}_k] \langle -c |\boldsymbol{U}_{k-1} - \boldsymbol{U}_k|^2 \tag{4.19}$$

Based on the triangular condition, we can conduct that

$$[\boldsymbol{U}_{k+1}^{HL} - \boldsymbol{U}_k]^T [\boldsymbol{U}_{k-1} - \boldsymbol{U}_k] < |\boldsymbol{U}_{k+1}^{HL} - \boldsymbol{U}_k| |\boldsymbol{U}_{k-1} - \boldsymbol{U}_k| \tag{4.20}$$

By replacing Eq. 4.20 in Eq. 4.19, we can obtain the sufficient descent condition based on the below simple formulation,

$$|\boldsymbol{U}_{k+1}^{HL} - \boldsymbol{U}_k| \langle c |\boldsymbol{U}_k - \boldsymbol{U}_{k-1}| \tag{4.21}$$

The above relation is almost applied for iterative formulation based on the HL-RF without control. However, the HL-RF method is an unstable iterative formulation as results presented in Chap.3. Therefore, an appropriate lined search by given a suitable step size α_k can be satisfied the sufficient descent condition as presented in Eq. 4.21. Thus, it changes as below,

$$|\boldsymbol{U}_{k+1} - \boldsymbol{U}_k| \langle c |\boldsymbol{U}_k - \boldsymbol{U}_{k-1}| \tag{4.22}$$

And

$$\boldsymbol{U}_{k+1} = \boldsymbol{U}_k + \alpha_k [\boldsymbol{U}_{k+1}^{HL} - \boldsymbol{U}_k] \tag{4.23}$$

Equation 4.22 is applied to determine the step size. Unlike the merit function condition, in sufficient descent condition, the previous results of the iterative formulation are applied for computing the step size while in the merit function, we need to compute the LSF in the inner loop as an inexact line search. This method is shown by the below steps for computing the step size.

Step 1: Set $c = 7/8$ and $m = 0$. The results of HL-RF method $\boldsymbol{U}_{k+1}^{HL}$ and previous points $\boldsymbol{U}_k$, $\boldsymbol{U}_{k-1}$ in the iteration process;
Step 2: Compute $|\boldsymbol{U}_k - \boldsymbol{U}_{k-1}|$;
Step 3: Check the sufficient descent condition as below,

(i) Determine step size as $\alpha_m = 2c^m\sigma$;
where, $0.5 \leq \sigma \leq 1$ that based on the HL-RF method, it is related to 0.5;
(ii) Compute the new point based on step size as $\boldsymbol{U}_{k+1} = \boldsymbol{U}_k + \alpha_m[\boldsymbol{U}_{k+1}^{HL} - \boldsymbol{U}_k]$;
(iii) Determine $|\boldsymbol{U}_{k+1} - \boldsymbol{U}_k|$;
(iv) Check conditions as $|\boldsymbol{U}_{k+1} - \boldsymbol{U}_k| < c|\boldsymbol{U}_k - \boldsymbol{U}_{k-1}|$, if satisfied then $\alpha_k = \alpha_m$ and the sufficient decent satisfies, got to step 4; else $m = m + 1$ and go to step 3.

Step 4: Compute the new point as $\boldsymbol{U}_{k+1} = \boldsymbol{U}_k + \alpha_k[\boldsymbol{U}_{k+1}^{HL} - \boldsymbol{U}_k]$.

By using the above steps, the procedure for computing the step size using sufficient decent condition is shown in Fig. 4.4. As seen, this method is simpler than the FORM with merit function. The inexact line search-based steep descent search direction can be used for computing a largest step size. The globality of the FORM-based sufficient descent conation for k-iteration is simply conducted based on the Eq. 4.22. It can be given a parameter t as $0 < t < c < 1$, thus we have for k-th iteration as,

$$|\boldsymbol{U}_{k+1} - \boldsymbol{U}_k| = t|\boldsymbol{U}_k - \boldsymbol{U}_{k-1}| \leq t \cdot t|\boldsymbol{U}_{k-1} - \boldsymbol{U}_{k-2}| = \ldots \leq t^k|\boldsymbol{U}_1 - \boldsymbol{U}_0| \quad (4.24)$$

Thus, we have

$$|\boldsymbol{U}_{k+1} - \boldsymbol{U}_k| \leq t^k|\boldsymbol{U}_1 - \boldsymbol{U}_0| \quad (4.25)$$

It can be concluded that $\lim_{k\to\infty} t^k \approx 0$ because $0 < t < 1$. Thus, it can be given for k iteration of FORM which conducted by the sufficient descent condition as $|\boldsymbol{U}_{k+1} - \boldsymbol{U}_k| \approx 0$. This means that the $\boldsymbol{U}_{k+1}$ is a fixed point and the results of FORM is given or the iterative scheme is converged, slowly. Thus, it provides stable results for FORM-based iteration but the globality of this method should be discussed (Fig. 4.6).

Example 4.1 that is applied in Sect. 4.2 by the following LSF $g(\boldsymbol{X}) = x_1^4 + x_2^2 - 50$ is given for first discussion of the FORM-based iHL-RF method enhanced by the sufficient decent condition that it is named as iHL-RF (Sufficient Descent Condition (SDC)) method. The converged reliability index of 3.2593 is given after 25 iterations with 316 call functions. As seen, this method is accurately converged with the same

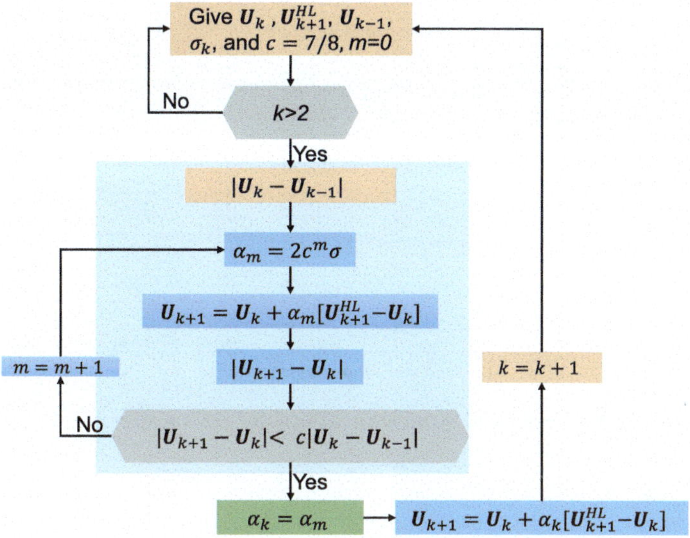

Fig. 4.6 Inexact step size selection using sufficient decent condition

results given from the literature. However, the efficiency of the iHL-RF is strongly enhanced while it provided robust results for MPP as $\boldsymbol{X}^* = [2.6475, 0.9351]$. This inexact search condition provides a fixed point at final iteration based on the computation of sufficient descent condition applied in this current investigation.

The reliability index with respect to iterations obtained by iHL-RF (SDC) is presented in Fig. 4.4 for this example. As seen for the results of this problem, the smooth curve for reliability index is obtained. This is means, the iHL-RF (SDC) provides the stable MPP and accurate line search is obtained. On the other hand, a simple inexact step size is defined that it is more efficient than the improved versions of steepest descent method such as SDC and RHL-RF (Fig. 4.7).

The step size obtained based on the inexact line search of sufficient decent condition is plotted in Fig. 4.8 for Example 4.1 as a nonlinear non-normal problem. It can be seen that, the same step size is not obtained for every iteration and they change between 1 and 0.15 thus, this provides a dynamical procedure for computing the step size.

Example 4.2 applied in Sect. 4.2 is also selected for the next problem which is used for evaluating performances of the iHL-RF (SDC). The reliability index as shown in Fig. 4.9 is obtained as 2.2983 after 18 iterations and 234 call functions. As seen from the results of Fig. 4.9, the straight line for the reliability index is given at the final iterations. The globality of the FORM iterative formulation is conducted based on the sufficient descent condition while this result is accurate.

The convergence process for the MPP search 9 of Example 4.2 in normal standard space is plotted in Fig. 4.10 for HL-RF and the iHL-RF method given by the inexact line search using sufficient descent condition. The iHL-RF (SDC) is iterated as same

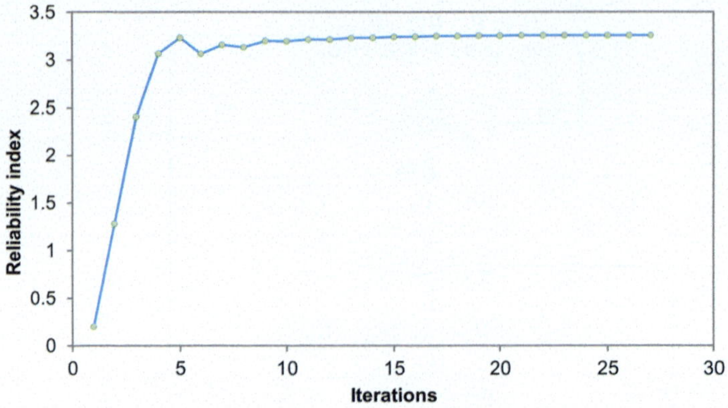

Fig. 4.7 Reliability index computed by sufficient descent for $g(X) = x_1^4 + x_2^2 - 50$

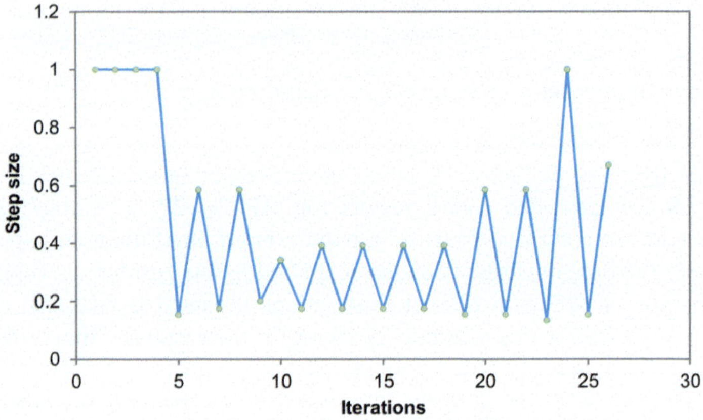

Fig. 4.8 Inexact step size obtained by sufficient descent condition for $g(X) = x_1^4 + x_2^2 - 50$

as HL-RF for iterations less than 5. However, the stability of the HL-RF is loosed for $k > 5$ but the iHL-RF (SDC) is controlled instability of FORM formulation for this example.

4.4 Angle Condition

A metaheuristic condition is proposed in this condition for controlling the instability of the FORM formulation. This criterion named as angle condition. The angle format is also used for the inverse reliability method for Reliability Based Design Optimization (RBDO) using the performance measure approach. The hybrid mean

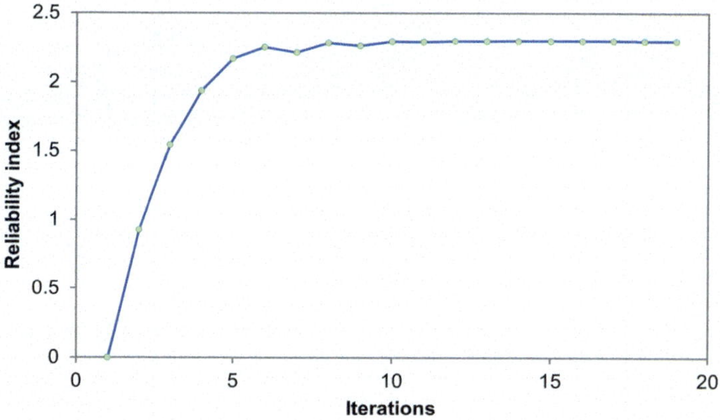

Fig. 4.9 Reliability index computed by sufficient descent with inexact line search for $g(\boldsymbol{X}) = x_1^3 + x_1^2 x_2 + x_2^3 - 18$

Fig. 4.10 Iteration for MPP search in normal standard space using iHL-RF (SDC) given by inexact step size and HL-RF methods for $g(\boldsymbol{X}) = x_1^3 + x_1^2 x_2 + x_2^3 - 18$

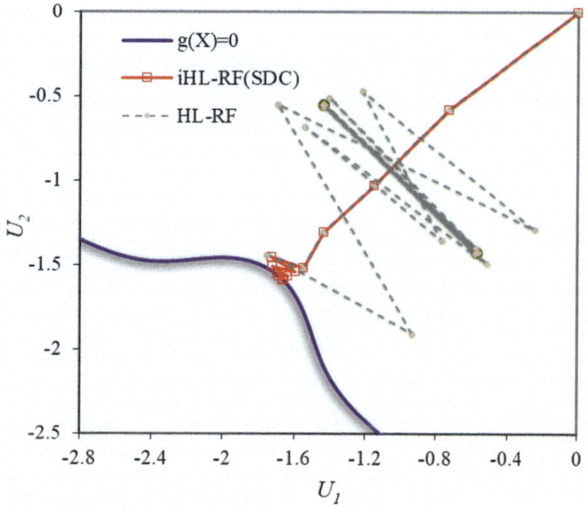

value method has been combined as the conjugate mean value and the advanced mean value by using an angle condition which is computed as below,

$$\varsigma_{k+1} = (\boldsymbol{n}_{k+1} - \boldsymbol{n}_k) \cdot (\boldsymbol{n}_k - \boldsymbol{n}_{k-1}) \tag{4.26}$$

where, $\boldsymbol{n} = \frac{U}{|U|}$ (normalized vector) and ς is a criterion for function type of the LSF at point $\boldsymbol{U}_{k+1}$. It can be given based on three consecutive iterations of the mean value method the ς as below:

$sign(\varsigma_{k+1}) > 0$, then the LSF is a convex performance function
$sign(\varsigma_{k+1}) \leq 0$, then the LSF is a concave performance function

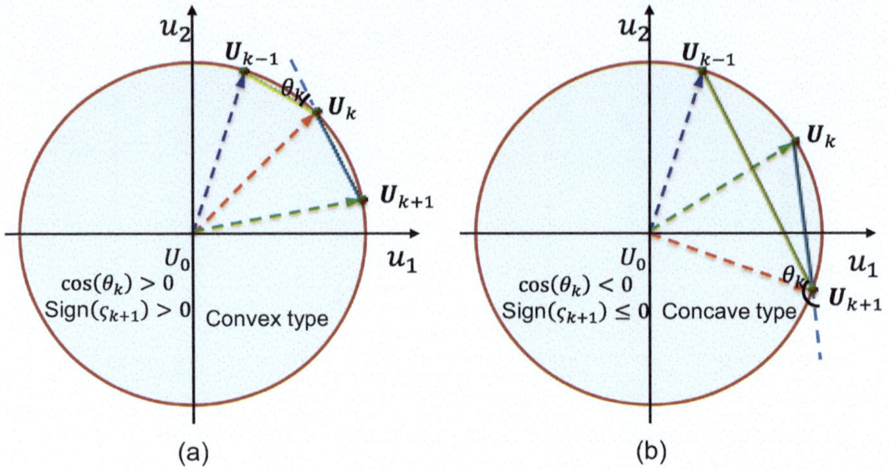

Fig. 4.11 Convexity criterion for performance functions **a** convex type; **b** concave type

The criterion ς shows the performance function type and $\boldsymbol{n}_k$ is the normalized steepest descent direction for the performance function at point $\boldsymbol{U}_k$. This condition is plotted in Fig. 4.11 for advanced mean value method as inverse reliability iterative formulation.

The convexity criterion is a way to provide the hybrid reliability method that can be used in performance measure approach applied in the reliability-based design optimization as several methods including; hybrid mean value (the Advanced Mean Value (AMV) combined with the conjugate mean value), hybrid modified mean value [13] where AMV is combined with chaos control method and adaptive chaos control method. It can be seen for Fig. 4.11, ς_{k+1} is cosine of the angle θ_k between $\boldsymbol{n}_{k+1} - \boldsymbol{n}_k$ and $\boldsymbol{n}_k - \boldsymbol{n}_{k-1}$. If the $Sign(\varsigma_{k+1})$ as shown in Fig. 4.11a is larger than zero (convex problem), the iterative point of the AMV method is applied in the hybrid methods while, the iterative AMV formula may be oscillated for concave performance measure function as shown Fig. 4.11b for $Sign(\varsigma_{k+1}) < 0$. Then, the hybrid formula of MPP search using the convexity criterion is applied. The angle type criterion as a positive or negative factor which is computed based on $Sign(\varsigma_{k+1})$ is applied for the hybrid formulation of inverse FORM iterative scheme in the RBDO problems. However, the angle condition is a novel inexact line search in structural reliability analysis that can be computed based on the result of three points as below relation,

$$\theta_{k+1} < c\theta_k \tag{4.27}$$

where, c is a positive value less than 1, as $0.9 \leq c < 1$. θ_{k+1} angle between the new point $\boldsymbol{U}_{k+1}$ and point $\boldsymbol{U}_k$. θ_{k+1} is determined as below,

$$\theta_{k+1} = cos^{-1}\left(\frac{\boldsymbol{U}_{k+1}}{|\boldsymbol{U}_{k+1}|}.\frac{\boldsymbol{U}_k}{|\boldsymbol{U}_k|}\right) \tag{4.28}$$

As well as above, θ_k as angle between points $\boldsymbol{U}_{k-1}$ and $\boldsymbol{U}_k$, is computed by the following relation,

$$\theta_k = cos^{-1}\left(\frac{\boldsymbol{U}_{k-1}}{|\boldsymbol{U}_{k-1}|}\cdot\frac{\boldsymbol{U}_k}{|\boldsymbol{U}_k|}\right) \tag{4.29}$$

The schematic view of this condition is presented in Fig. 4.12. This condition is similar to the sufficient descent condition but the instability is controlled by angles between sequence points. Consequently, the direction of line search is considered in this condition, unlike the sufficient descent condition in Eq. 4.22. This criterion provides convexity results for all iterations of the performance function based on a suitable direction.

By the angle condition is presented in Eq. 4.27, it can be conducted that the stable results are given when $\theta_{k+1} \approx 0$. This means that we have a fixed point and the new point is located on previous point of $\boldsymbol{U}_k$. Therefore, we have $\frac{\boldsymbol{U}_{k+1}}{|\boldsymbol{U}_{k+1}|}\cdot\frac{\boldsymbol{U}_k}{|\boldsymbol{U}_k|} \approx 1$, thus $\boldsymbol{U}_{k+1} \approx \boldsymbol{U}_k$. By the angle condition, it is supposed that $\theta_1 < \theta_2 < \theta_3 < \ldots\theta_k < \theta_{k+1}$ thus for $k \to \infty$, $\theta_{k+1} \to 0$. This means that stability of FORM is obtained based on angle condition. The iterative procedure for angle condition for computing new point is given as below steps.

Step 1: Give, $0.9 \leq c < 1$, $0.9 < \gamma < 1$ and $m = 0$. The results of HL-RF method $\boldsymbol{U}_{k+1}^{HL}$ and previous points $\boldsymbol{U}_k, \boldsymbol{U}_{k-1}$ in the iteration process;

Step 2: Compute $\theta_k = \cos^{-1}\left(\frac{\boldsymbol{U}_{k-1}}{|\boldsymbol{U}_{k-1}|}\cdot\frac{\boldsymbol{U}_k}{|\boldsymbol{U}_k|}\right)$;

Step 3: Check the angle condition as below:

(i) Determine step size as $\alpha_m = 2\gamma^m\sigma$.
where, $0.5 \leq \sigma \leq 1$ that based on the HL-RF method, it is related to 0.5;

(ii) Compute the new point based on step size as $\boldsymbol{U}_{k+1} = \boldsymbol{U}_k + \alpha_m[\boldsymbol{U}_{k+1}^{HL} - \boldsymbol{U}_k]$;

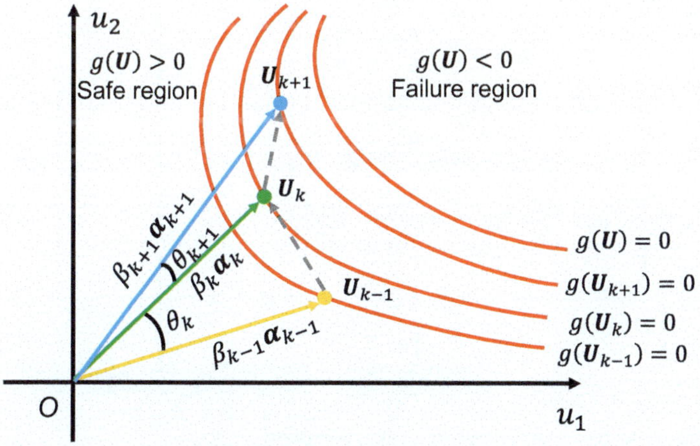

Fig. 4.12 Angle between the points for three iterations of FORM

(iii) Determine $\theta_{k+1} = \cos^{-1}\left(\frac{U_{k+1}}{|U_{k+1}|} \cdot \frac{U_k}{|U_k|}\right)$;
(iv) Check condition as $\theta_{k+1} < c\theta_k$, if satisfied then $\alpha_k = \alpha_m$ and got to Step 4; else $m = m + 1$ and go to Step 3;

Step 4: Compute the new point $U_{k+1} = U_k + \alpha_k[U_{k+1}^{HL} - U_k]$.

The framework for detraining the inexact step size using angle criterion is presented in Fig. 4.13. As seen, this method is similar to the sufficient descent condition but the search direction is considered in this current criterion.

For comparing its abilities for both efficiency and robustness of this method, the Example 4.1 i.e. $g(X) = x_1^4 + x_2^2 - 50$ and Example 4.2 i.e. $g(X) = x_1^3 + x_1^2 x_2 + x_2^3 - 18$ given from sect. 4.4–sect. 4.2 as nonlinear non-normal and highly nonlinear problems are examined. The factors γ and σ are respectively given as $\gamma = 0.98$ and $\sigma = 0.5$.

The reliability index for Example 4.1 using the angle condition is plotted in Fig. 4.14. As seen in this figure as well as iHL-RF (SDC), the smooth convergence is obtained based on this improved HL-RF method enhanced by inexact step size with angle condition named iHL-RF (Adaptive Conjugate (AC)). The reliability index is given as $\beta = 3.2593$ which is captured after 20 iterations with 245 call functions for $g(X) = x_1^4 + x_2^2 - 50$.

The inexact step size with dynamical properties and the angles between the iterative points of the non-normal nonlinear LSF are presented in Fig. 4.15a and b, respectively. It can be concluded that the angle at the final iterations is limited to zero, thus a fixed point is obtained based on the proposed condition. The step size is

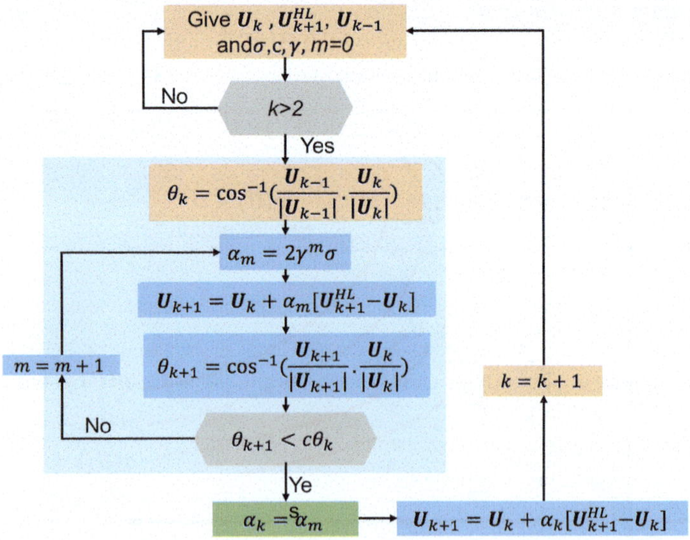

Fig. 4.13 Inexact step size selection using angle condition

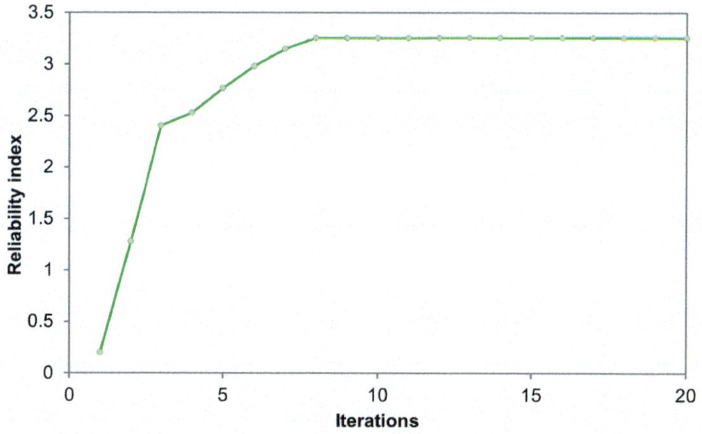

Fig. 4.14 Reliability index computed by angle condition for $g(\boldsymbol{X}) = x_1^4 + x_2^2 - 50$

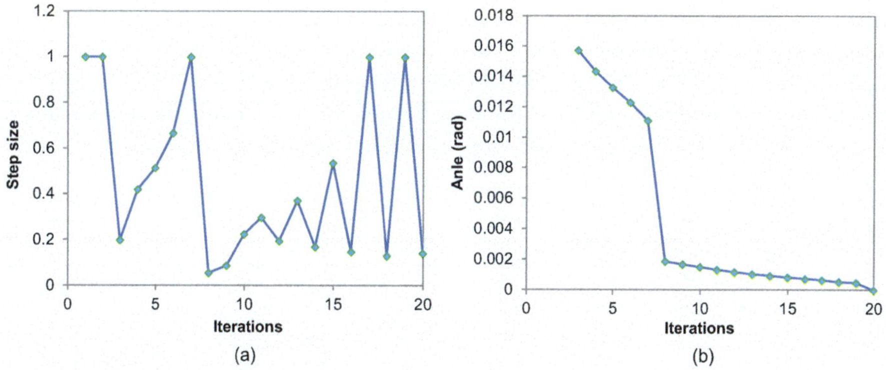

Fig. 4.15 Results of iterations for iHL-RF (AC) for $g(\boldsymbol{X}) = x_1^4 + x_2^2 - 50$ **a** inexact step size and **b** angle between sequence points

dynamically computed in intervals of 0–1 while for some iterations, a smaller value to control of the instability of FORM formulation.

The results of iHL-RF (AC) are obtained for Example 4.2 that convergence process of the reliability index is given in Fig. 4.16. As well as the sufficient descent condition, the iHL-RF by using inexact line search provides the stable results as reliability index of 2.2983 which is obtained after 19 iterations and 245 call functions. It can be conducted that the global convergence is obtained which closely agrees well with the result of the sufficient decent condition with efficient computational cost.

The iterations as scatterplots of the MPP search for this nonlinear example are plotted in Fig. 4.17. As seen from the results of this figure, the distance between the new and previous points is reduced by increasing the iteration which is controlled based on the angle of sequence points. Finally, we can have a very small step size at

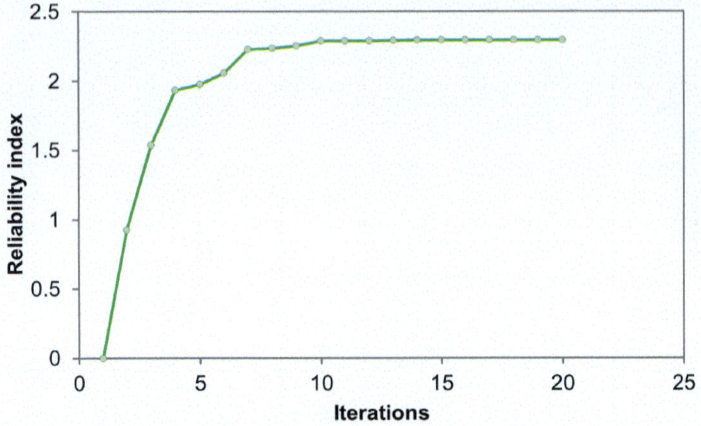

Fig. 4.16 Reliability index computed by angle condition for $g(\boldsymbol{X}) = x_1^3 + x_1^2 x_2 + x_2^3 - 18$

the final iteration as 2.83×10^{-5}. By comparing the iterations of the iHL-RF using sufficient descent condition and the angle condition in the Fig. 4.10 and Fig. 4.17, the iterations of the iHL-RF (AC) are different from the iterations of iHL-RF (SDC). The angle condition is considered the direction of the line search, more accurately. It is searched the MPP in a straight line without any significant chaos.

Fig. 4.17 Iteration for MPP search in normal standard space using iHL-RF (AC) for $g(\boldsymbol{X}) = x_1^3 + x_1^2 x_2 + x_2^3 - 18$

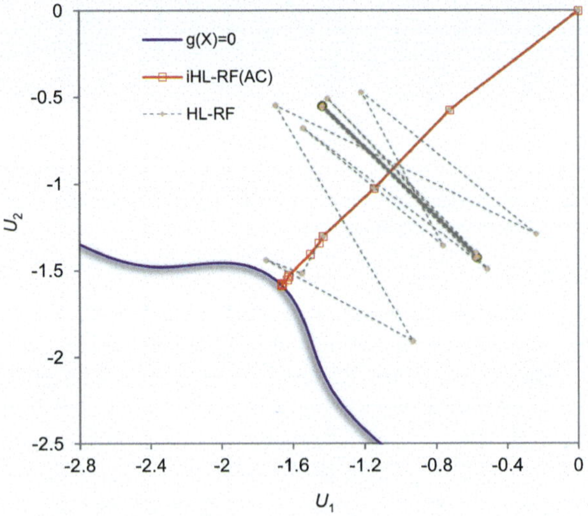

4.5 Relaxed Method-Based Angle Condition

It can be supposed that if the line search direction is accurately computed, then we can increase the step size along the search direction vector. The relaxed formulation is an acceptable strategy for increasing the step size along the suitable line search which is computed based on angle condition. The superior efficiency condition it is may be obtained based on the hybrid method for moderately nonlinear LSF. The performances of the iHL-RF may be not changed using the relaxed formulation for highly nonlinear LSFs. The relaxed formulation is presented in Chap. 2 as below,

$$\boldsymbol{U}_{k+1} = (1 - \alpha_k)\boldsymbol{U}_k + \alpha_k \boldsymbol{U}_{k+1}^{HL} \tag{4.30}$$

The step size α_k should satisfy the angle condition that it is given for the first iteration as $\alpha_1 = 2\sigma$. It can be given a step size of more than 1 for this problem when it selects $\sigma > 0.5$. On the other hand, the globality of this method is similar to HL-RF computed by the angle condition. However, when the inexact step size is greater than 1, it may increase the efficiency of the steepest descent condition for some problems as real engineering applications. The steps for MPP search and reliability index computation are presented for relaxed FORM formulation coupled by the angle condition named (RHL-RF (AC)) as below,

Step 1: Define the L $g(\boldsymbol{X})$., Give the statistical properties of random variables (μ and σ), stopping criterion (ε) and two factors for adjusting step size $0.5 < \sigma < 1$, $0.9 \le \gamma < 1$ and $0.9 < c < 0.95$. Let $k = 1$, choose $\boldsymbol{X}_1 = \boldsymbol{\mu}$;
Step 2: Normalize the random variables by transferring from $\boldsymbol{X}$-space to $\boldsymbol{U}$-space;
Step 3: Iterative process of RHL-RF (AC) as below,

(i) Compute $g(\boldsymbol{U}_k)$, $\nabla_u g(\boldsymbol{U}_k)$ and set $\alpha_0 = 2\sigma$ and $m = 0$;
(ii) Determine the new point based on HL-RF method as below;

$$\boldsymbol{U}_{k+1}^{HL} = \frac{\nabla^T g(\boldsymbol{U}_k)\boldsymbol{U}_k - g(\boldsymbol{U}_k)}{\nabla^T g(\boldsymbol{U}_k)\nabla g(\boldsymbol{U}_k)} \nabla g(\boldsymbol{U}_k) \tag{4.31}$$

(iii) Compute the new point $\boldsymbol{U}_{k+1}^{RHL} = (1 - \alpha_k)\boldsymbol{U}_k + \alpha_k \boldsymbol{U}_{k+1}^{HL}$;
(iv) Check the angle condition as below for $k \ge 3$;
(v) Give $\boldsymbol{U}_{k+1}^{HL}$ with previous points $\boldsymbol{U}_k, \boldsymbol{U}_{k-1}$;
(vi) Compute $\theta_k = cos^{-1}\left(\frac{\boldsymbol{U}_{k-1}}{|\boldsymbol{U}_{k-1}|} \cdot \frac{\boldsymbol{U}_k}{|\boldsymbol{U}_k|}\right)$ and $\theta_{k+1} = cos^{-1}\left(\frac{\boldsymbol{U}_{k+1}^{RHL}}{|\boldsymbol{U}_{k+1}^{RHL}|} \cdot \frac{\boldsymbol{U}_k}{|\boldsymbol{U}_k|}\right)$;
(vii) If $\theta_{k+1} < c\theta_k$, then $\alpha_m = 2\sigma$, else.

   (a) $m = m + 1$;
   (b) Determine step size as $\alpha_m = 2\gamma^m \sigma$;
   (c) Compute the new point as $\boldsymbol{U}_{k+1} = (1 - \alpha_m)\boldsymbol{U}_k + \alpha_m \boldsymbol{U}_{k+1}^{HL}$;
   (d) Updating angle as $\theta_{k+1} = \cos^{-1}\left(\frac{\boldsymbol{U}_{k+1}}{|\boldsymbol{U}_{k+1}|} \cdot \frac{\boldsymbol{U}_k}{|\boldsymbol{U}_k|}\right)$ and go to step vii;

(viii) Set inexact step size is $\alpha_k = \alpha_m$;

(ix) Compute new point using step size as $\boldsymbol{U}_{k+1} = (1 - \alpha_k)\boldsymbol{U}_k + \alpha_k \boldsymbol{U}_{k+1}^{HL}$;

Step 4: If $|\boldsymbol{U}_{k+1} - \boldsymbol{U}_k| < \varepsilon$ then stop and print $\boldsymbol{X}^* = \boldsymbol{X}_{k+1}, \boldsymbol{U}^* = \boldsymbol{U}_{k+1}. \beta = \beta_{k+1}$
Else $k = k + 1$ and compute $\boldsymbol{X}_{k+1} = T^{-1}(\boldsymbol{U}_{k+1})$ go to Step 2.

The flowchart of the RHL-RF (AC) is presented in Fig. 4.18. It is shown that an inner loop is applied for adjusting the relaxed factor for satisfy the angle conditions. The simple formulation is applied for determining relaxed factor and the finding MPP. The major difference between this method and angle condition is to select a step size more than 1 for the FORM formulation. In this method we have an inexact step size more than 1 that it may be controlled based on the parameter σ.

For nonlinear LSF presented in Example 4.2, several σ are given for evaluating the convergence abilities of the RHL-RF (AC) method. The reliability index and step size are used for comparing the results of the RHL-RF (AC) with different adjusting factors of step size σ. These results are given for $c = 0.9$ and $\gamma = 0.98$ that the

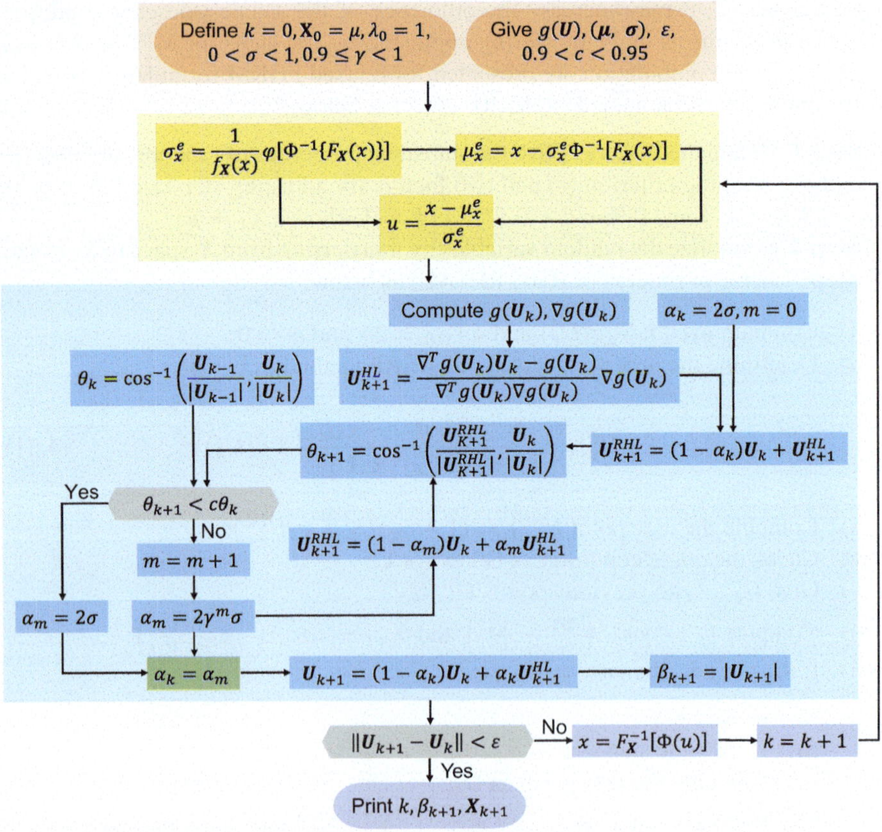

Fig. 4.18 Flowchart of the RHL-RF (AC) for reliability analysis

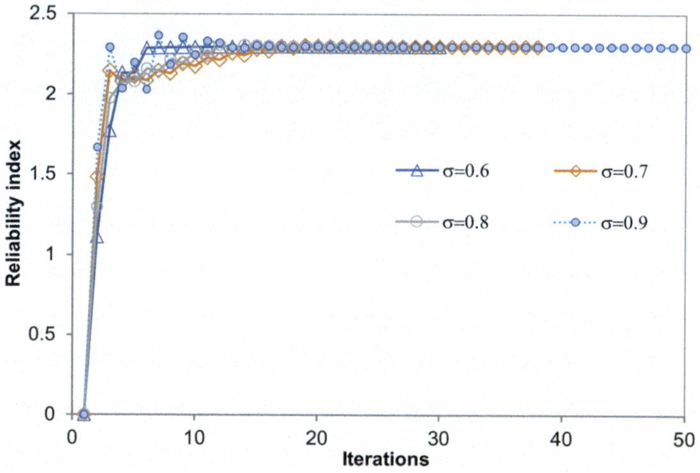

Fig. 4.19 Reliability index using RHL-RF (AC) for different values of parameter σ for $g(\boldsymbol{X}) = x_1^3 + x_1^2x_2 + x_2^3 - 18$

reliability index for different $\sigma = 0.6, 0.7, 0.8, 0.9$ are presented in Fig. 4.19 and the related step size is given for different adjusting factors applied in Eq. 4.19 for $g(\boldsymbol{X}) = x_1^3 + x_1^2x_2 + x_2^3 - 18$ example.

It can be concluded that, by increasing the adjusting factor, the iterations for convergence are also increased. The same reliability index is given based on different values of σ. The nonlinearity format of this LSF needs to provide stable results twice more than the method with angle condition for providing stability. This method provides the globality for FORM iteration formulation and it can introduce the angle condition as a novel convergence controlling framework that in this method based on the line search approach, controls the instabilities of FORM based on the step size of more than 1.

As seen from the results in Fig. 4.20, the step size varies within interval 0–2 σ with RHL-RF (AC). This angle condition is a good rule as a golden step that provides a step size based on the results of previous points. The nonlinearity of this example is high at MPP thus, the proposed method efforts are for giving the stabilization.

On the other hand, it is may create a new question about the adjusting factor σ as why it considered a value less than 0.5.

Based on this important quarry, it shows that, if the σ is given a value less than 0.5 then it may be provided a premature convergence and the RHL-RF (AC) may converge inaccurately. The results of the reliability index for σ less than 0.5 as 0.1, 0.2, and 0.4 are presented in Fig. 4.21 which is provided by the RHL-RF (AC) with parameters of $c = 0.9$ and $\gamma = 0.98$. It can be seen that the efficiency of the iHL-RF (AC) and the RHL-RF (AC) are not improved but this relaxed FORM formulation with inexact search direction provides the inefficient results for $\sigma < 0.5$. For $\sigma = 0.1$, the RHL-RF (AC) is prematurely converged. For all studied adjusting factors, we obtain the stable results without the instabilities as chaotic and periodic

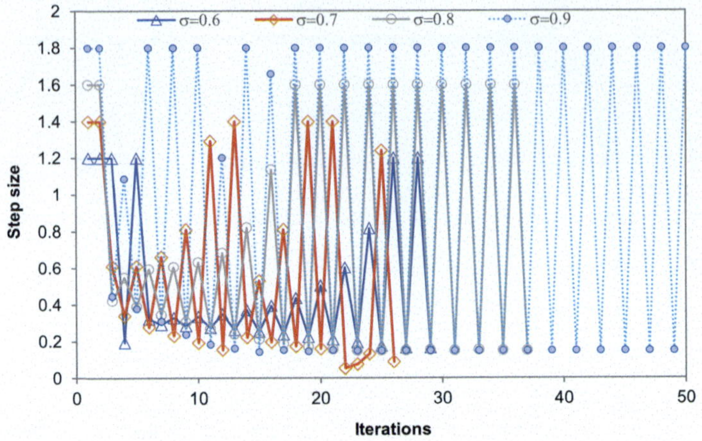

Fig. 4.20 Step size using RHL-RF (AC) for different values of parameter σ for $g(\boldsymbol{X}) = x_1^3 + x_1^2 x_2 + x_2^3 - 18$

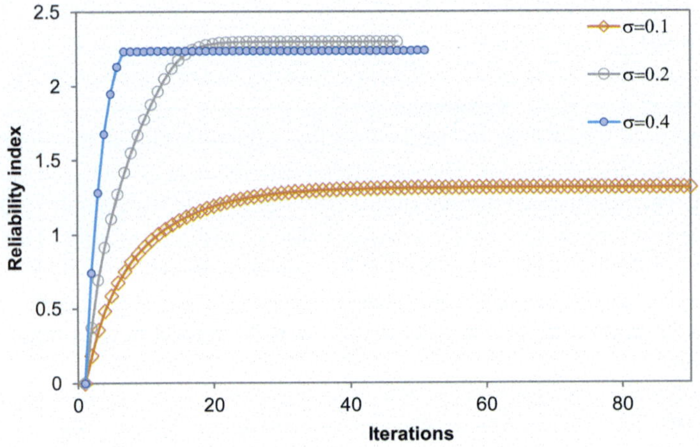

Fig. 4.21 Reliability index using RHL-RF (AC) for values of parameter σ less than 0.5 for nonlinear example

solutions but selecting the step size less than 0.1 is given the inaccurate convergence, more inefficiently. In this condition, we obtained an accurate step size as too small that it provided slowly iterations with premature convergence.

There is a suitable step size for angle condition then the global convergence with accurate MPP search is obtained while the adjusting factor can control the maximum value of the step size. The maximum step size depends on the nonlinearity of the performance function. It is suggested that the below comments for choosing the maximum step size in the RHL-RF (AC).

a. For linear LSFs, it can be given as $0.8 \leq \sigma < 1$;
b. For moderate nonlinear LSFs, it can be given as $0.7 \leq \sigma < 0.8$;
c. For nonlinear LSFs, it can be given as $0.6 \leq \sigma < 0.7$;
d. For highly nonlinear LSFs, it can be given as $0.5 \leq \sigma < 0.6$;

the above comments can be provided the accurate results with stable convergence while its efficiency is controlled.

4.6 Discussion

This chapter focuses on the inexact line search approaches that three conditions are discussed for searching the MPP. The main effort is to find a direction which is used for new point. The inexact step size is presented for obtaining the suitable line search direction of the steepest decent condition. The stability of the FORM is guaranteed based on the angle and sufficient decent condition but for giving the appropriate results as global convergency, it should be selected an appropriate step size. The step size in this chapter is computed with inexact line search that the conditions of globality should be evaluated based on three rules as merit function, sufficient descent and angle properties.

For comparing the methods presented in this chapter, we give four structural examples as (1) a vehicle disc brake system, (2) a two-degree freedom primary-secondary dynamic system, (3) a steel frame, and (4) a conical structure. The LSF of these examples are the nonlinear properties. Also in these all problems, the same stopping criterion is given as $|\beta_{k+1} - \beta_k| < 10^{-6}$ and factor $\gamma = 0.98$ but, factor c for SDC is $7/8$ and for AC is 0.9. The σ is also given as 0.5 for iHL-RF (SDC) and iHL-RF (AC) and for each example of the RHL-RF (AC), it compared two values for σ as 0.6 and 0.8. Thus, we have five algorithms as below for comparing the engineering structures.

(1) iHL-RF (Merit Condition (MC)) related to iHL-RF that its instability is controlled by the merit function;
(2) iHL-RF (SDC) related to iHL-RF method that its instability is controlled by sufficient decent conditions with parameters of $\gamma = 0.98$, $c = 7/8$;
(3) iHL-RF (AC) related to the iHL-RF method that its instability is controlled by angle conditions with parameters of $\gamma = 0.98$, $c = 0.9$;
(4) RHL-RF (AC-0.6) related to RHL-RF method that its instability is controlled by angle conditions with parameters of $\gamma = 0.98$, $c = 0.9$, $\sigma = 0.6$;
(5) RHL-RF (AC-0.8) related to the RHL-RF method that its instability is controlled by angle conditions with parameters of $\gamma = 0.98$, $c = 0.9$, $\sigma = 0.8$.

From the convergence results, the reliability index is used for checking the stability while call functions and iteration for computing the gradient vector are applied for showing the efficiency of the algorithms.

Example 4.3 A vehicle disc brake system

A brake disc as shown in Fig. 4.21, is expressed by the following LSF:

$$
\begin{aligned}
g = {} & 0.046287 + 0.20458\mu - 0.059821p - 0.00036549h_1 - 0.010037h_2 \\
& +0.013836h_3 + 0.24308\mu p - 0.0037884\mu h_1 + 0.0023358\mu h_2 \\
& -0.016981\mu h_3 + 0.029287ph_1 - 0.015872ph_2 - 0.0028333ph_3 \\
& +0.0007175h_1h_2 - 0.00046158h_1h_3 - 0.00036486h_2h_3 - 0.39076\mu^2 \\
& -0.015986p^2 - 0.0011936h_1^2 + 0.000269h_2^2 + 0.00062638h_3^2 - 0.01
\end{aligned}
\tag{4.32}
$$

where μ is the friction coefficient, p represents the brake pressure, h_1, h_2, and h_3 are respectively the thickness of friction material, disc thickness, and back plate thickness that these basic random variables as listed in Table 4.1 and (Fig. 4.22).

The reliability index using different methods of the iHL-RF enhanced by the inexact step size is presented in Fig. 4.23. As seen, all methods are converged to the same reliability index of 2.1444. The basic formulation of the iHL-RF with step size computed based on the merit, angle, and sufficient descent conditions provided a smooth iteration for computing the reliability index but the RHL-RF with maximum step size of more than 1 showed some noises in the iterative procedure for determining the MPP. The moderate nonlinearity type of the LSF given by the response surface with second-order polynomial function is given for this problem but the RHL-RF (AC-0.8) may produce a non-smooth iteration but it provided the efficient results among other methods. The merit function for this problem is most inefficient for

Table 4.1 Random variables for disc brake example

Random variable	h_1 (mm)	h_2 (mm)	h_3 (mm)	μ	p (MPa)
Mean	15	20	10	0.4	0.5
CoV	0.1	0.1	0.1	0.35	0.25
Distribution	Gumbel	Gumbel	Gumbel	Normal	Lognormal

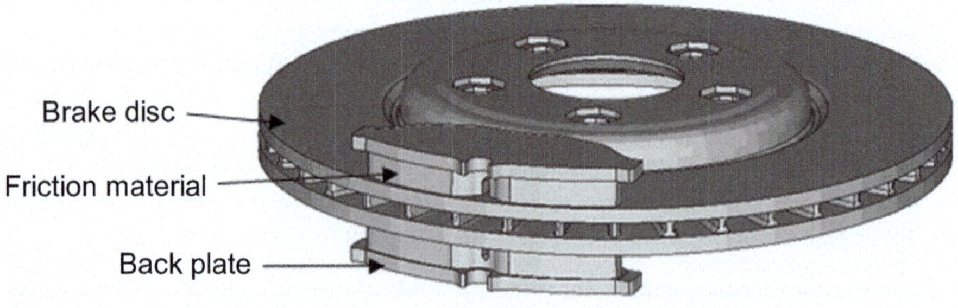

Fig. 4.22 The view of vehicle disc brake example. Reprinted from [14], with permission from Elsevier

both iterations and computing LSF. The robustness of the FORM formulation based on the proposed inexact search direction is satisfied for this example but the efficiency depends on the step size and methods for obtaining the stability.

Example 4.4 A two-degree dynamic system

A two-degree freedom primary-secondary dynamic system as shown in Fig. 4.24 is defined by the following LSF,

$$g = Fs - Ks \times P(E[x_s^2])^{1/2} \tag{4.33}$$

where, Fs denotes the force capacity. P is the peak factor and is set as 3. $E[x_s^2]$ is the mean-square relative displacement response of the secondary spring and is given by

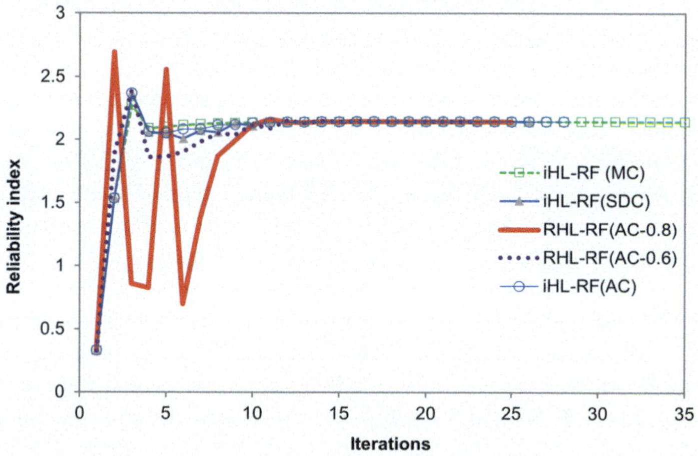

Fig. 4.23 Reliability indexes for different steepest descent methods improved by inexact line search for vehicle disc brake example

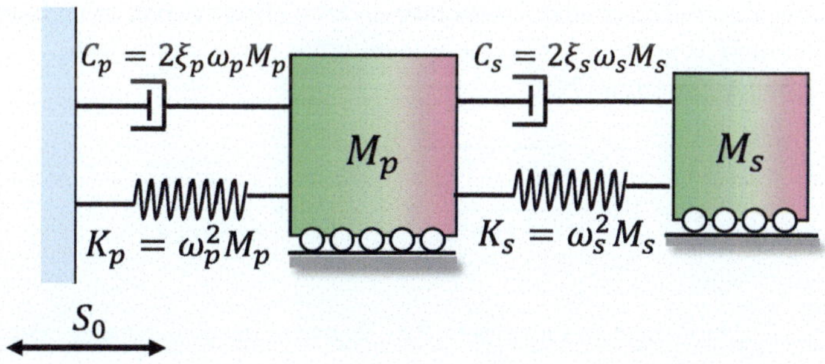

Fig. 4.24 Two-degree of freedom dynamic system. Reprinted from [15] with permission from Elsevier

$$E\left[x_s^2\right] = \frac{\pi S_0}{4\xi_s\omega_s}\left[\frac{\xi_a\xi_s}{\xi_p\xi_s\left(4\xi_a^2+\theta^2\right)+\gamma\xi_a^2} \times \frac{\left(\xi_p\omega_p^3+\xi_s\omega_s^3\right)\omega_p}{4\xi_a\omega_a^4}\right] \quad (4.34)$$

where $\gamma = \frac{M_s}{M_p}$ (mass ratio); $\omega_a = \frac{\omega_p+\omega_s}{2}$ (average frequency); $\xi_a = \frac{\xi_p+\xi_s}{2}$ (average damping ratio of two system); $\theta = \frac{\omega_p-\omega_s}{\omega_a}$ (a tuning parameter); S_0 is intensity of the white noise. This example includes eight lognormal random variables whose random parameters are listed in Table 4.2.

The converged reliability index of this nonlinear problem as the reliability index for studied methods is presented in Fig. 4.25 for the two-degree of freedom dynamic system example. As seen from the results of the reliability index, all methods are converged but several of the methods showed premature results and they converged to inaccurate MPP for this example such as the iHL-RF (SDC) and iHL-RF (MC) and RHL-RF (AC-0.8). The reliability index using the FORM iterative formula is given from literature [16] as $\beta = 2.0164$ using STM or finite step size method. By using the convergence criterion applied in the inexact line search, the robustness of the FORM formulation for this problem is always guaranteed. However, the fixed point is obtained after several iterations for several methods. The iHL-RF (AC) is accurately converged to global MPP with a reliability index of 2.0164 it is closely followed by the results of the literature. The efficiency of the iHL-RF (MC) is better than other methods but the iHL-RF (AC) is cited in the second position for comparing the results. By checking the LSF at converged MPP, the values for different conditions as iHL-RF are 0.1334 for merit condition, 0.5485 for sufficient descent condition and 8.37×10^{-7} for angle condition. It can be seen that the angle line search provides the lowest performance value at MPP. This example is one of the highly nonlinear reliability problems that is sensitive to the line search. By comparing the efficiency and globality, the iHL-RF (AC) method provides superior abilities among other methods while the sufficient and merit conditions should be adjusted based on the factors in the line search procedure for this problem.

Example 4.5 A Conical Structure

As shown in Fig. 4.26, a conical structure under a compressive axial load P and a bending moment M is given for this example which is given by the following relation for its performance function [7],

$$g = 1 - \frac{\sqrt{3\left(1-\nu^2\right)}}{\pi E t^2 \cos^2\alpha} \times \left(\frac{P}{0.66} + \frac{M}{0.41 r_1}\right) \quad (4.35)$$

Table 4.2 Statistical basic random variables for Two-degree of freedom dynamic system

Random variable	M_p	M_s	K_p	K_s	ξ_p	ξ_s	F_s	S_0
Mean	1	0.01	1	0.01	0.05	0.02	15	100
Standard deviation	0.1	0.001	0.2	0.001	0.02	0.01	1.5	10

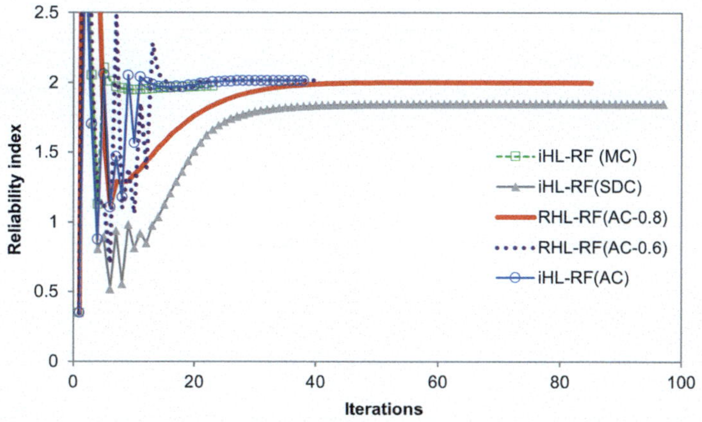

Fig. 4.25 Reliability indexes for different steepest descent methods improved by inexact line search for two-degree of freedom dynamic system example

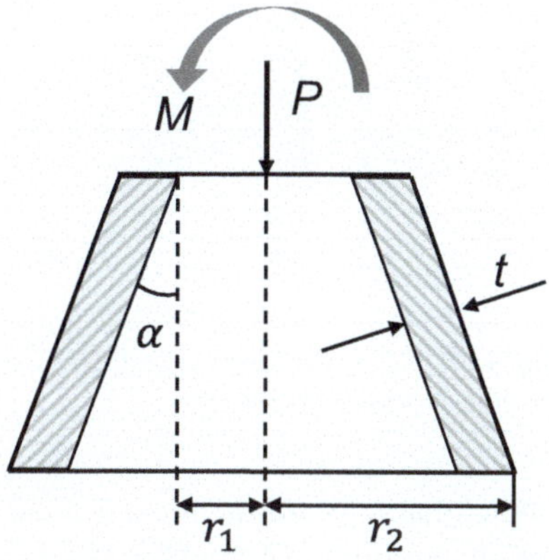

Fig. 4.26 The view of the conical structure. Reprinted from [15], with permission from Elsevier

The buckling failure performance mode is considered for LSF of conical structure, in which, v is the Poisson's ratio equal to 0.3. This problem includes six independent both normal and non-normal basic random variables whose statistical characteristics are listed in Table 4.3.

As presented results in Fig. 4.27 for the reliability index of different methods of this example, all studied methods are converged, robustly. The sufficient decent condition is strongly more efficient than other algorithms. This problem is a kind of nonlinear LSF that its nonlinearity follows as moderately. Thus, the SDC has a high convergence rate for this kind of nonlinear problem. The efficiency of other methods except iHL-RF (SDC) have shown the same efficiency and their iterations have shown no significant difference. Therefore, for this problem, the robustness of all methods is obtained, and the same efficiency is given based on the MC and AC.

Table 4.3 The basic random variables for conical structure example

Variables	Description	Distribution	CoV	Mean
E	Young's modulus (MPa)	Lognormal	0.05	70000
t	Thickness(m)	Normal	0.05	0.0025
α	Slop angle (rad)	Normal	0.02	0.524
r_1	Internal radius(m)	Normal	0.025	0.9
M	Bending moment(N-m)	Gumbel	0.08	80000
P	Axial load	Gumbel	0.08	70000

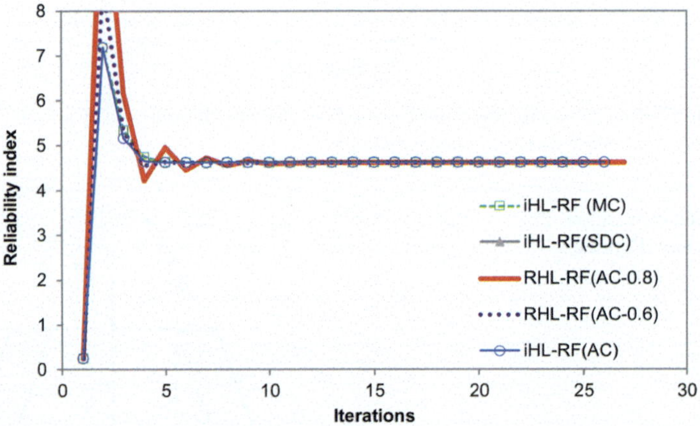

Fig. 4.27 Reliability indexes for different steepest descent methods improved by inexact line search for conical structure example

The globality of all methods for MPP search is given based on suitable line search for all algorithms of the FORM with inexact line search.

Example 4.6 A steel frame structure

A frame with span L and height H under a lateral load P, and distributed load W are plotted in Fig. 4.13. The LSF is defined as [9],

$$g = 0.05 - \Delta \tag{4.36}$$

The supports of this frame are clamped boundaries and simply support the connection with the earth at the left and right columns, respectively. The lateral displacement under external loads is given as LSF and its statistical properties of basic random variables are tabulated in Table 4.4 and (Fig. 4.28).

The reliability index for different methods of FORM-based inexact line searches and steepest descent formulation is given in Fig. 4.29. As seen, all methods are robustly converged to the global reliability index. However, their efficiency are different for this problem. The merit and angle conditions provide the results with

Table 4.4 Statistical properties of the random variables for steel frame example

Random variable	Mean	Standard deviation	Distribution	Random variable	Mean	Standard deviation	Distribution
A_1(m^2)	91×10^{-4}	9.1×10^{-4}	Lognormal	E_1 (kg/m^2)	2.38×10^{10}	1.9×10^9	Normal
I_1(m^4)	80.9×10^{-6}	8.09×10^{-6}	Lognormal	E_2 (kg/m^2)	2.17×10^{10}	1.74×10^9	Normal
A_2(m^2)	53.8×10^{-4}	5.38×10^{-4}	Lognormal	L (m)	6	0.48	Normal
I_2(m^4)	83.6×10^{-6}	8.36×10^{-6}	Lognormal	H (m)	4.5	0.36	Normal
A_3(m^2)	106×10^{-4}	10.6×10^{-4}	Lognormal	W (kg/m)	4500	630	Gumbel
I_3(m^4)	112.6×10^{-6}	11.26×10^{-6}	Lognormal	P (kg)	2000	350	Gumbel

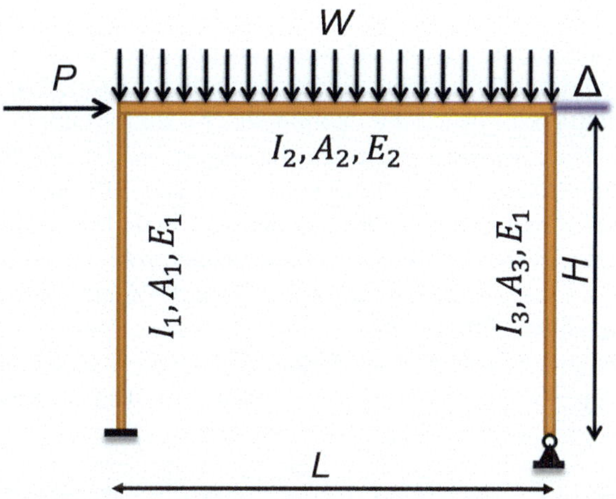

Fig. 4.28 The schematic view of the steel frame structure. Reprinted from [17], with permission from Elsevier

an efficient computational burden compared to the other methods. On the other hand, the application of the inexact search direction strongly affects the efficiency of the computational burden for this problem. It is a highly nonlinear problem that is strongly dependent on the step size of the steepest decent method. At the final iteration, this example may provide the chaotic solution for the step size with the same value at all iterations such as the STM with step size 0.1 and the HL-RF method with step size 1. Thus, a condition as an inexact line search to control the step size can improve the robustness and the efficacy of the nonlinear problems.

The simple computation of the FORM formulas for application in the engineering problem is one of the main efforts to develop the analytical approaches. The merit condition for satisfying the global results is required to give some information for the LSF at iteration thus its application is more complicated than other conditions

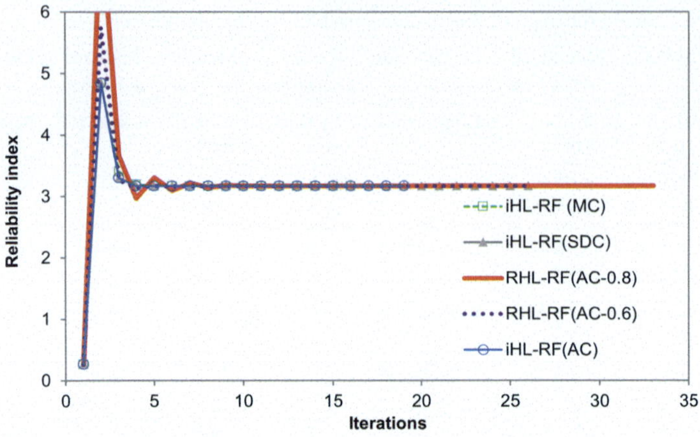

Fig. 4.29 Reliability indexes for different steepest descent methods improved by inexact line search for steel frame structure example

as sufficient descent and angle. As presented in this chapter, the descent and angle conditions have global properties for FORM formulas but to control their efficiency should be adjusted their parameters for each example.

The converged results as reliability index (β), iterations (ite) related to the number of the computation for gradient vector and Call Function (CF) related to the number of evaluating the performance function in the iterative process of FORM formula are summarized in Table 4.5 for all studied problems in this section.

As seen for Examples 4.3, 4.5, and 4.6 all methods are robustly converged to the global MPP by the same reliably index. However, the globality of these studied methods is different for Example 4.4. The iHL-RF with an inexact line search of angle property provides the global results. Thus, the globality of the angle condition

Table 4.5 Converged results of the various FORM formulas enhanced by inexact line search proposed in this chapter for Examples 4.3, 4.4, 4.5 and 4.6

Methods	Example 4.3		Example 4.4		Example 4.5		Example 4.6	
	β	ite(CF)	β	ite(CF)	β	ite(CF)	β	ite(CF)
IHL-RF (MC)	2.1444	34(444)	1.976	22(2484)	4.6277	23(851)	3.168	15(1754)
IHL-RF (SDC)	2.1444	25(300)	1.8502	96(9608)	4.6278	15(539)	3.168	25(2100)
IHL-RF (AC)	2.1444	27(324)	2.0164	37(3729)	4.6278	25(899)	3.1681	18(1512)
RHL-RF (AC-0.6)	2.1444	22(264)	2.0164	39(3902)	4.6278	24(867)	3.1681	25(2050)
RHL-RF (AC-0.8)	2.1444	24(288)	2.0035	84(8529)	4.6283	26(942)	3.1679	32(2628)

is more than other methods and the sufficient decent and merit condition can be improved based on their adjusting parameters.

All methods have shown the same robustness thus, the proposed method in this chapter using the inexact line search provides stable results, unlike the HL-RF method. The nonlinearity degree of the LSFs is strongly affected by the HL-RF method but by using three conditions in this chapter, the robustness of FORM-based steepest decent line search is enhanced. The robust convergence is given for every problem using the HL-RF method improved by the inexact step size which is computed by angle or sufficient decent conditions.

The efficiency as the computational burden of these studied FORM framework depends on their parameters and the nonlinearity degree of the engineering problems. For nonlinear problems such as Examples 4.1, 4.3, and 4.4, it is not shown which method is better for all problems. However, the efficiency of the AC for Examples 4.3 and 4.4 are similar to the MC and the SDC is more efficient than others for Example 4.3. For Example 4.2, the AC provided the efficient results. By this condition as efficiency, the angle condition may be given the first range, and the merit function is cited at the second level. Nevertheless, it cannot select a method that has superior efficiency than others for all problems.

Finally, by comparing robustness, all methods are good. The AC is suitable for high, moderate functions by comparing the globality and for efficiency, the AC and MC may generally provide the acceptable efficiency compared to the SDC. Consequently, based on this ranked-based method and their parameters applied in this chapter, the best to worst methods are ranked for FORM formulas enhanced by inexact lines search as angle, merit and sufficient decent conditions.

4.7 Summary

In this chapter, the FORM-based steepest descent search direction. The iterative formulation of FORM is established based on the steepest descent sensitivity vectors that almost all analytical methods of FORM are enhanced by a step size less than 1. In addition, all collected the FORM formulas are developed as an improved version of the HL-RF method named as the steepest descent method. The main improvement of these approaches is similar to the relaxed approach that is applied to improving the robustness of FORM-based HL-RF. The HL-RF method is the first method that we try to show its abilities and convergence feedback including chaotic and periodic results. By using the HL-RF method, we try to collect several approaches of FORM named (i) improve HL-RF, (ii) chaos control method, (iii) relaxed HL-RF, (iv) Non-negative constraint method that they are formulated by a refined HL-RF formulation for improving gradient line search direction based on a step size.

References

1. P.L. Liu, A. Der Kiureghian, Optimization algorithms for structural reliability. Struct. Saf. **9**(3), 161–177 (1991)
2. Y.X. Yuan, A new stepsize for the steepest descent method. J. Comput. Math. **24**(2), 149–156 (2006)
3. Y. Zhang, A. Der Kiureghian, Two improved algorithms for reliability analysis, in *Reliability and Optimization of Structural Systems: Proceedings of the sixth IFIP WG7. 5 working conference on reliability and optimization of structural systems* 1994, pp. 297–304 (1995)
4. T. Santosh, R. Saraf, A. Ghosh, H. Kushwaha, Optimum step length selection rule in modified HL-RF method for structural reliability. Int. J. Press. Vessel. Pip. **83**(10), 742–748 (2006)
5. B. Keshtegar, S. Chakraborty, An efficient-robust structural reliability method by adaptive finite-step length based on Armijo line search. Reliab. Eng. Syst. Saf. **172**, 195–206 (2018)
6. L. Bin, L. Gang, Armijo-based adaptive stability transformation method. Chin. J. Comput. Mech. (2018)
7. B. Keshtegar, M. Miri, An enhanced HL-RF Method for the computation of structural failure probability based on relaxed approach. Civ. Eng. Infrastruct.S J. **1**(1), 69–80 (2013)
8. P. Huang, H.Z. Huang, T. Huang, A novel algorithm for structural reliability analysis based on finite step length and Armijo line search. Appl. Sci. **9**(12), 2546 (2019)
9. B. Keshtegar, Stability iterative method for structural reliability analysis using a chaotic conjugate map. Nonlinear Dyn. **84**, 2161–2174 (2016)
10. M. Al-Baali, Descent property and global convergence of the Fletcher—Reeves method with inexact line search. IMA J. Numer. Anal. **5**(1), 121–124 (1985)
11. B. Keshtegar, Enriched FR conjugate search directions for robust and efficient structural reliability analysis. Eng. Comput. **34**, 117–128 (2018)
12. W.W. Hager, H. Zhang, A new conjugate gradient method with guaranteed descent and an efficient line search. SIAM J. Optim. **16**(1), 170–192 (2005)
13. Z. Meng, G. Li, B.P. Wang, P. Hao, A hybrid chaos control approach of the performance measure functions for reliability-based design optimization. Comput. Struct. **146**, 32–43 (2015)
14. B. Xia, H. Lü, D. Yu, C. Jiang, Reliability-based design optimization of structural systems under hybrid probabilistic and interval model. Comput. Struct. **160**, 126–134 (2015)
15. C. Luo, S.P. Zhu, B. Keshtegar, X. Niu, O. Taylan, An enhanced uniform simulation approach coupled with SVR for efficient structural reliability analysis. Reliab. Eng. Syst. Safet. **237**, 109377 (2023)
16. B. Keshtegar, Z. Meng, A hybrid relaxed first-order reliability method for efficient structural reliability analysis. Struct. Saf. **66**, 84–93 (2017)
17. B. Keshtegar, Chaotic conjugate stability transformation method for structural reliability analysis. Comput. Methods Appl. Mech. Eng. **310**, 866–885 (2016)

Chapter 5
Directional Steepest Descent Approaches for FORM

Abstract In this chapter, based on the directional steepest descent sensitivity vector named as direction FORM, several formulations computed by gradient method are presented. The main format of these approaches is developed by finite step length method including (i) directional stability transformation method, (ii) adaptive directional stability transformation method, (iii) enhanced finite step length method, (iv) dynamical relaxed directional method. The abilities for efficiency and robustness are discussed by several numerical/structural reliability problems.

Keywords Directional steepest decent sensitivity vector · Directional stability transformation method · Adaptive directional stability transformation method · Finite step length

5.1 Introduction

Engineering systems involve various uncertainties which are of vital importance for safe design under structural reliability analysis (SRA). Physical properties, dimensions, size, materials, and environmental effects as loads are uncertainties that they are inherently affect on engineering system, which are naturally produced in an engineering system. Uncertainties in existing structural/mechanical engineering systems can be estimated by reliability methods like simulation or analytical ones. To handle these uncertainties, the failure probability for a limited condition as the performance function of a structural system should be computed in the design phase to consider the safety level. The performance function is a probabilistic stepmodel involving various random variables. The main effort of this book is to introduce the analytical methods as one popular type of reliability method to find the most probable point (MPP) [1]. This type of MPP search approach can be approximated to the failure probability using MPP as a well-known approach named first-order reliability method (FORM).

Two abilities of efficiency and robustness are challenges of FORM-based SRA in finding the MPP process. The analytical methods-based SRA with low computational burden is a good method for evaluating uncertainties while it provides accurate MPP.

S.-P. Zhu and B. Keshtegar, *Structural Reliability Analysis*, Structural Integrity 31,
https://doi.org/10.1007/978-3-031-75197-4_5

As presented in Chap. 3, the Hasofer-Lind and Rackwitz-Flessler (HL-RF) method is commonly applied to approximate the MPP with the steepest descent formulation [2]. However, it might yield unstable outcomes for nonlinear limit state function (LSF) [3, 4]. The improved FORM-based HL-RF is formulated to compute MPP as presented in Chap. 2 using the steepest descent search direction (SDSD) [5] and also conjugate search direction can be applied for MPP search [6, 7]. The SDSD with a modified version of the normalized search direction vector (NSDV or sensitivity vector) can enhance the stability of FORM using HL-RF as results presented in Chap. 4. The line search direction involves the step size and normalized search direction vector in the FORM formulation. By using the steepest decent condition, we show the exact and inexact step size in Chaps. 3 and 4. The search direction vector is important as well as the step size. Therefore, several formulations of the modified SDSD method named directional stability transformation method (DSTM) [8], finite-step length (FSL) [9], and improve Hasofer-Lind and Rackwitz-Flessler (iHL-RF) formulated by Barzilai-Borwein rule [10] have been developed. The directional steepest decent (DSD) method is one of the new line search vectors, which is computed by gradient vector and previous point. In this approach, the line search is different from the steepest decent method.

In this chapter, analytical reliability methods-based FORMs are collected to evaluate the reliability of structures. In these methods, a new search direction vector is used for computing the line search of MPP. This normalized vector is followed by the directional vector which is determined by the gradient vector of LSF at the previous point. The abilities of these methods are compared together based on several examples.

5.2 Finite Step Length Method

It is supposed that the original random variables are followed by continuous probability distribution function. An LSF is computed by random variables that it can be detected the failure domain as $g < 0$. The main effort of FORM is to find MPP as a point on LSF with the lowest distance to origin in $\boldsymbol{U}$-(normal standard) space. These are the basic definitions in the FORM-based formula which is applied to search MPP. The MPP is searched based on an NSDV i.e. $\boldsymbol{\alpha}$ and a step size which are two main factors for computing line search. The β and related point in $\boldsymbol{U}$-space are related by NSDV as $\boldsymbol{U} = \beta\boldsymbol{\alpha}$, where, β is reliability index and $\boldsymbol{\alpha}$ = NSDV. Therefore, we have a description of the LSF at normal standard space as below,

$$g(\boldsymbol{U}_k) = g(\beta_k\alpha_{1k}, \beta_k\alpha_{2k}, \beta_k\alpha_{3k}, \ldots, \beta_k\alpha_{nk}) = 0 \tag{5.1}$$

where n represents the number of variables. Generally, we have a nonlinear LSF that by the numerical method, it computes the gradient vector in the normal standard space as $\nabla g(\boldsymbol{U}) = [\partial g/\partial U_1, \partial g/\partial U_2, \ldots, \partial g/\partial U_n]^T$. The new points are defined in the design domain as below,

$$\boldsymbol{u}_{k+1}^{\lambda} = \boldsymbol{U}_k - \lambda \nabla g(\boldsymbol{U}_k) \tag{5.2}$$

where $\lambda \gg 0$ defined as finite-step-length. This definition is applied for new NSDV which is used for MPP search. The NSDV for MPP search is defined as below,

$$\boldsymbol{\alpha}_{k+1} = \frac{\boldsymbol{u}_{k+1}^{\lambda}}{\left|\boldsymbol{u}_{k+1}^{\lambda}\right|} \tag{5.3}$$

And the reliability index using the new search direction is given as,

$$\beta_{k+1} = \frac{\nabla^T g(\boldsymbol{U}_k)\boldsymbol{U}_k - g(\boldsymbol{U}_k)}{\nabla^T g(\boldsymbol{U}_k)\boldsymbol{\alpha}_{k+1}} \tag{5.4}$$

Thus, the new point is determined as $\boldsymbol{U}_{k+1} = \beta_{k+1}\boldsymbol{\alpha}_{k+1}$. The search direction vector for a schematically LSF is presented in Fig. 5.1. It can be seen this line search is along the point $\boldsymbol{u}_{k+1}^{\lambda}$ for determining the new point. The mean factor of this line search is to approximate the NSDV by using the previous point and gradient vector.

As seen from Fig. 5.1 and Eq. 5.3, a fixed point is obtained when $\lambda = 0$. Note that, this method is related to HL-RF, when $\lambda \to \infty$ because $\lim\limits_{\lambda \to \infty} \frac{\boldsymbol{U}_k - \lambda \nabla g(\boldsymbol{U}_k)}{|\boldsymbol{U}_k - \lambda \nabla g(\boldsymbol{U}_k)|} = -\frac{\nabla g(\boldsymbol{U}_k)}{|\nabla g(\boldsymbol{U}_k)|}$. This is an iterative formulation of FORM that the line search is controlled based on the normalized finite point vector as presented in Eq. 5.3. In this method, the step size is not applied for giving the effect of the search direction vector similar to the steepest descent condition methods. Nonetheless, it is considered as finite step length that it may be selected more than 1. As seen when $\lambda \to \infty$, the FSL method is followed form the HL-RF method. Therefore, the HL-RF-based steepest descent method is a type of this formulation while it may give the step size as finite values. By this way, it is named as finite step length (FSL) method because in the framework of its iterative formulation, a finite step-length is applied for computing the NSDV.

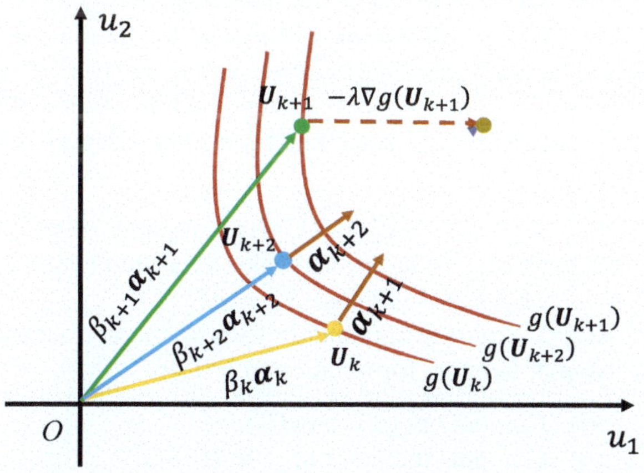

Fig. 5.1 Iterative process by using the finite-step length method

The convergence property of this method is given from the sufficient descent condition as $|\boldsymbol{U}_{k+1} - \boldsymbol{U}_k| < |\boldsymbol{U}_k - \boldsymbol{U}_{k-1}|$, thus we have,

$$\begin{aligned} |\boldsymbol{U}_{k+1} - \boldsymbol{U}_k| &= t_k|\boldsymbol{U}_k - \boldsymbol{U}_{k-1}| = t_k t_{k-1}|\boldsymbol{U}_{k-1} - \boldsymbol{U}_{k-2}| = \cdots \\ &= t_k t_{k-1} \ldots t_1 t_0 |\boldsymbol{U}_1 - \boldsymbol{U}_0| \end{aligned} \tag{5.5}$$

$$|\boldsymbol{U}_{k+1} - \boldsymbol{U}_k| = t_k t_{k-1} \ldots t_1 t_0 |\boldsymbol{U}_1 - \boldsymbol{U}_0| \tag{5.6}$$

where, $0 < t_{k,} t_{k-1}, \ldots, t_0 < 1$, thus $\lim_{k \to \infty} t_k t_{k-1} \ldots t_1 t_0 \approx 0$. This means as,

$$|\boldsymbol{U}_{k+1} - \boldsymbol{U}_k| \approx 0 \tag{5.7}$$

It is obtained a fixed point and the convergency is provided, when sufficient descent condition is applied in the FSL method. If the step size is given too large then the efficiency of this method may be improved, but the convergence iteration may be given the same condition with the HL-RF method. Consequently, this method may provide some issues for MPP search as well as the HL-RF iterative formulas that converged to unstable results.

For providing stability, the sufficient decent condition is a good rule for controlling its instability by adjusting the finite step length. Generally, λ should be adjusted based nonlinear degree of the LSF as a small value when analyzing an LSF with high-nonlinear degree. On the other hand, it has more computational effort when smaller the step length is selected. There is no rule for selecting the initial step length, thus it is given a large value at first, then it can be adjusted by sufficient decent conditions as below,

$$\lambda = \begin{cases} \lambda & if\ |\boldsymbol{U}_{k+1} - \boldsymbol{U}_k| < |\boldsymbol{U}_k - \boldsymbol{U}_{k-1}| \\ \lambda/c & if\ |\boldsymbol{U}_{k+1} - \boldsymbol{U}_k| \geq |\boldsymbol{U}_k - \boldsymbol{U}_{k-1} \end{cases} \tag{5.8}$$

The step length is adjusted based on the factor $1.2 < c < 1.5$. By using this factor, the step size is reduced based on the iteration process when the LSF is highly nonlinear thus at the final iteration, it may be given a fixed point and step length limits to zero. The FSL method as an iterative formulation of FORM with finite line search can be used for MPP search and computing the reliability index i.e. β by following steps:

Step 1: Define LSF $g(\boldsymbol{X})$, Give stopping criterion (ε), random variables with distributions, μ and σ, and also initial step length (λ) and adjusting factor $1.2 < c < 1$, Let $k = 1$, Choose $\boldsymbol{X}_1 = \boldsymbol{\mu}$;
Step 2: Transferring random variables to normal standard space;
Step 3: Iterative formulation of FSL by;

(i) Compute $g(\boldsymbol{U}_k)$, $\nabla g(\boldsymbol{U}_k)$;
(ii) Compute the point $u_{k+1}^{\lambda} = \boldsymbol{U}_k - \lambda \nabla g(\boldsymbol{U}_k)$;

(iii) Compute the normalized search direction vector as $\alpha_{k+1} = \frac{u_{k+1}^{\lambda}}{|u_{k+1}^{\lambda}|}$;

(iv) Determine the reliability index as $\beta_{k+1} = \frac{\nabla^T g(U_k)U_k - g(U_k)}{\nabla^T g(U_k)\alpha_{k+1}}$ and new point as $U_{k+1} = \beta_{k+1}\alpha_{k+1}$;

(v) For $k \geq 3$, adjust $\lambda = \begin{cases} \lambda & if \ |U_{k+1} - U_k| < |U_k - U_{k-1}| \\ \lambda/c & if \ |U_{k+1} - U_k| \geq |U_k - U_{k-1}| \end{cases}$;

Step 4: **If** $|\beta_{k+1} - \beta_k| < \varepsilon$ **then** stop iterative formulation and print $X^* = X_{k+1}$, $U^* = U_{k+1}$, $\beta = \beta_{k+1}$;
Else $k = k + 1$. Compute $X_{k+1} = T^{-1}(U_{k+1})$ go to Step 2.

The primary difference between this iterative formulation which is computed based on the finite step length and the improved version of the HL-RF that they are formulated based on the sufficient decent condition are summarized below:

(a) The line search direction is different in the FSL method and it is computed based on the previous point and gradient vector of the LSF while the sensitivity vector as line search is computed based on the gradient vector in the FORM-based steepest descent methods.
(b) The selecting step size less than one is removed in the FSL thus it is not a complicated formulation for MPP search compared to the modified version of the steepest decent methods.
(c) The adjusting step length is applied based on the sufficient decent condition for providing a fixed point in this method. The FSL is applied as a participation factor of the gradient vector at each iteration that it is adjusted as a finite value as a large at the first iteration to a small value at the final iterations for highly nonlinear problems.

As seen from the iterative procedure of this algorithm that they are presented in Steps 1–7. This method is as simple as the HL-RF method which can be introduced as a basic method for finite search direction vector in the FORM-based MPP search.

For showing the abilities of the convergence results of the FLS, two nonlinear examples are applied. One of them is described by non-normal variables [11] and the other is computed based on the normal variables.

Example 5.1 $g = x_1^4 + x_2^2 - 50$, with Lognormal variable x_1 defined by mean = 5 and standard deviation (STD) = 1, Gumbel variable x_2 defined by mean = 10 and STD = 10.

Example 5.2 $g = x_1^3 + x_1^2 x_2 + x_2^3 - 18$, where, normal variables of x_1 and x_2 have $\mu_1 = 10, \sigma_1 = 5$ and $\mu_2 = 9.9, \sigma_2 = 5$.

By using the parameters of this algorithm as $c = 1.25$ and $\lambda = 100$. The above two examples are solved based on the FSL method. The results of the reliability index for these two examples are presented in Fig. 5.2 as Fig. 5.2a for Example 5.1 and Fig. 5.2b for Example 5.2. As seen from FSL results, the stable iterations are obtained but the reliability index obtained by the FSL method is not converged as a smooth curve. This means that the instability of the FORM is not obtained at the first iteration. By applying the sufficient decent condition, the value of finite step length is reduced by increasing the number of iterations. The finite step length with respect to iteration is presented in Fig. 5.3. A smaller value of the finite step length is obtained for these two nonlinear problems at final iterations. If we neglect the adjusting process for step length then the unstable results are provided as well as the HL-RF method.

The iteration process for MPP search in the normal standard space is presented in Fig. 5.4 for Example 5.2 using the FSL method. As seen as well as the HL-RF, this method provides the instability as periodic solution at the first iterations. And then by increasing the iterations, the suitable line search is extracted by adjusting the step length at final iterations.

As presented results in Fig. 5.2, the reliability index for Example 5.1 is computed as $\beta = 3.2593$ after 124 iterations and for Example 5.2 is computed as $\beta = 2.2983$ after 137 iterations. As seen, this method of FORM is not formulated based on the efficient procedure for MPP search. The line search direction is changed as directional form but its efficiency strongly depends on the nonlinearity degree of the limit stat

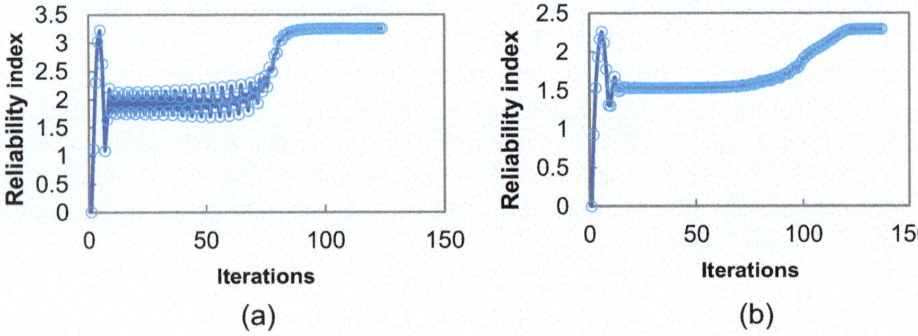

Fig. 5.2 Reliability index iterations obtained by FSL method **a** Example 5.1; **b** Example 5.2

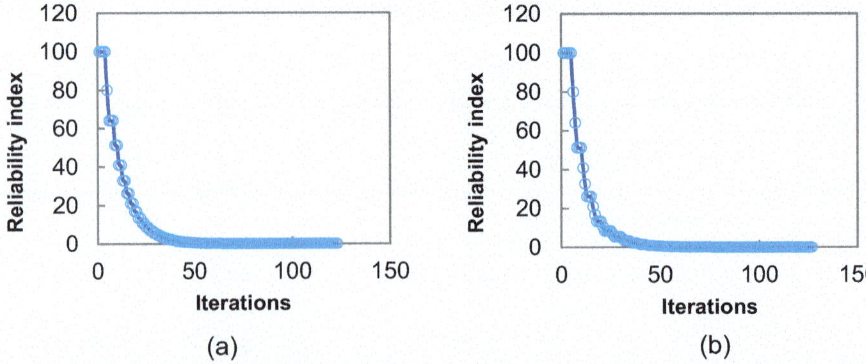

Fig. 5.3 Adjusting process for finite step length in FSL method using sufficient decent condition **a** Example 5.1 and **b** Example 5.2

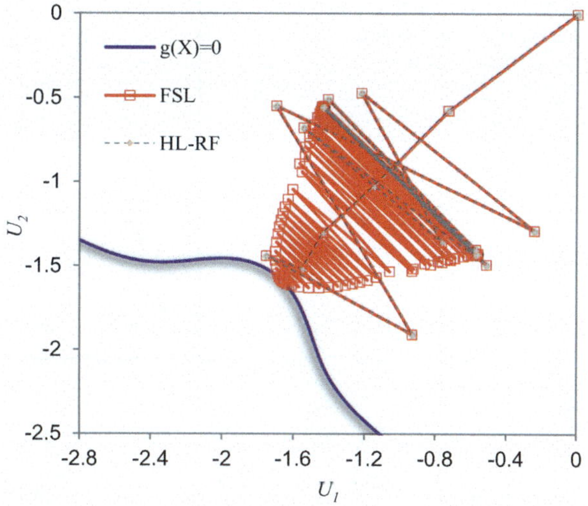

Fig. 5.4 Iteration for MPP search in normal standard space using FSL and HL-RF methods for Example 5.2

function. The finite step length is a strong force to yield stable results, while the efficiency of this method should be controlled based on the improvement of the finite step length. There is not any rules for computing the acceptable step length. In this chapter, several improved versions of the FSL methods are explored and discussed.

5.3 Directional Stability Transformation Method

The stability transformation method (STM) is a simple FORM method but it is more computationally. The efficiency of STM is enhanced by using the directional line search applied for computing NSDV. This line search can be formulated based on the STM-based chaos control method. Meng et al. [8] introduced the directional STM named DSTM for the FORM-based algorithm that the DSTM can provide a fixed point by using a nonlinear map more efficient than the STM. This method can be controlled the periodic and chaotic iterations of FORM-based SDSD as well as HL-RF method. Therefore, for complex LSFs computed by finite element or boundary element analyses, STM has limitation for application due to its inefficiency. Actually, the instability of HL-RF is important and it should be solved by a robust FORM algorithm.

The iterative formulation of the DSTM is established based on the STM. In the DSTM, the NSDV is given form STM as below,

$$\boldsymbol{U}_{k+1}^{STM} = \boldsymbol{U}_k^{DSTM} + \lambda\left[\boldsymbol{U}_{k+1}^{HL} - \boldsymbol{U}_k^{DSTM}\right] \tag{5.9}$$

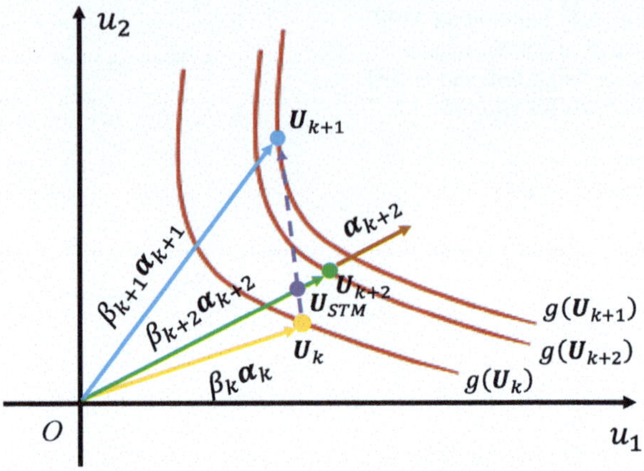

Fig. 5.5 Iterative process by using directional stability transformation method

in which, $0 < \lambda < 1$. The iterative procedure for DSTM is schematically plotted in Fig. 5.5 for the two-dimensional problem in the normal standard space. As presented in Fig. 5.5, the line search direction is determined based on $\boldsymbol{U}_{k+1}^{STM}$ that it is involved two points as previous point and point obtained by the HL-RF method as given from Eq. 5.9. Also, it can be seen that the new point using DSTM is computed based on the radial direction defined as the normalized search direction and reliability index as its scale thus we have,

$$\boldsymbol{U}_{k+1} = \beta_{k+1}\boldsymbol{\alpha}_{k+1} \tag{5.10}$$

where,

$$\beta_{k+1} = \frac{\nabla^T g(\boldsymbol{U}_k)\boldsymbol{U}_k - g(\boldsymbol{U}_k)}{\nabla^T g(\boldsymbol{U}_k)\boldsymbol{\alpha}_{k+1}} \tag{5.11}$$

And,

$$\boldsymbol{\alpha}_{k+1} = \frac{\boldsymbol{U}_k^{DSTM} + \lambda\left[\boldsymbol{U}_{k+1}^{HL} - \boldsymbol{U}_k^{DSTM}\right]}{\left|\boldsymbol{U}_k^{DSTM} + \lambda\left[\boldsymbol{U}_{k+1}^{HL} - \boldsymbol{U}_k^{DSTM}\right]\right|} \tag{5.12}$$

The chaos control factor with radial line search direction can be controlled by the convergence of DSTM.

DSTM provides the new search direction as well as FSL named as radial direction. However, the major difference between this method and FSL is to compute the normalized search direction. In the STM is used from the STM formulation for determining the normalized search direction vector in which applied the previous point and the HL-RF results are combined by the chaos control factor, while in the FSL, the normalized search direction vector is determined by the previous point and the gradient vector of LSF compared with a finite step length. The directional points u^{λ} in STM as shown in Fig. 5.1 and $\boldsymbol{U}_{k+1}^{STM}$ as presented in Fig. 5.5 are the major difference between these two FORM algorithms. On the other hand, the finite step length is adjusted based on the sufficient decent condition in the FSL method but for DSTM, the chaos control factor is given as constant value at all iterations. The DSTM -based iterative FORM formulation by using chaos control factor and STM point applied for line search can be used for MPP search that its computing the reliability index β is summarized by the following steps:

Step 1: Define LSF $g(\boldsymbol{X})$, Give stopping criterion (ε), random variables with distributions, μ and σ, and also set $0 < \lambda < 1$, $k = 1$, and $\boldsymbol{X}_1 = \boldsymbol{\mu}$;
Step 2: Transfer variables to normal standard space;
Step 3: Iterative procedure of DSTM by the following steps;

(i) Compute $g(U_k)$, $\nabla_u g(U_k)$;
(ii) Determine the new point by using the HL-RF method as below;

$$\boldsymbol{U}_{k+1}^{HL} = \frac{\nabla^T g(\boldsymbol{U}_k)\boldsymbol{U}_k - g(\boldsymbol{U}_k)}{\nabla^T g(\boldsymbol{U}_k)\nabla g(\boldsymbol{U}_k)}\nabla g(\boldsymbol{U}_k) \tag{5.13}$$

(iii) Compute the point $\boldsymbol{U}_{k+1}^{STM} = \boldsymbol{U}_k^{DSTM} + \lambda \boldsymbol{C}\left[\boldsymbol{U}_{k+1}^{HL} - \boldsymbol{U}_k^{DSTM}\right]$;
(vi) Compute normalized search direction vector as $\boldsymbol{\alpha}_{k+1} = \frac{\boldsymbol{U}_k^{DSTM} + \lambda \boldsymbol{C}\left[\boldsymbol{U}_{k+1}^{HL} - \boldsymbol{U}_k^{DSTM}\right]}{\left\|\boldsymbol{U}_k^{DSTM} + \lambda \boldsymbol{C}\left[\boldsymbol{U}_{k+1}^{HL} - \boldsymbol{U}_k^{DSTM}\right]\right\|}$;
(v) Determine $\beta_{k+1} = \frac{\nabla^T g(\boldsymbol{U}_k)\boldsymbol{U}_k - g(\boldsymbol{U}_k)}{\nabla^T g(\boldsymbol{U}_k)\boldsymbol{\alpha}_{k+1}}$ and new point as $\boldsymbol{U}_{k+1} = \beta_{k+1}\boldsymbol{\alpha}_{k+1}$;

Step 4: **If** $|\beta_{k+1} - \beta_k| < \varepsilon$ stop iterative formulation and print $\boldsymbol{X}^* = \boldsymbol{X}_{k+1}$, $\boldsymbol{U}^* = \boldsymbol{U}_{k+1}$, $\beta = \beta_{k+1}$;
Else $k = k + 1$. Compute $\boldsymbol{X}_{k+1} = T^{-1}(U_{k+1})$ go to Step 2.

By using the DSTM, Examples 5.1 and 5.2 as presented in Sect. 5.2 are solved. The reliability results of these two nonlinear examples are presented in Fig. 5.6. The stable results are obtained and the efficiency of the FSL method is strongly improved based on DSTM. Two factors of the directional line search and the chaos control formulation controlled the instability, and remarkably enhanced the computational burden compared to the HL-EF, STM and FSL for these two nonlinear examples.

The iterative process of the DSTM applied for MPP in the normal standard space is presented for Example 5.2 in Fig. 5.7. As seen, the STM is converged without chaos and a smooth line search is given by this method, that the MPP is $X* = [1.6839, 1.9696]$ and $U* = [-1.6630, -1.5863]$ related to reliability index of 2.2983 obtained after 12 iterations.

Example 5.3 three-span steel beam,

A three-span steel beam shown in Fig. 5.8 is defined by LSF to Eq. 5.14 which is computed based on uniform load W.

$$g = \frac{L}{360} - 0.0069\frac{WL^4}{EI} \tag{5.14}$$

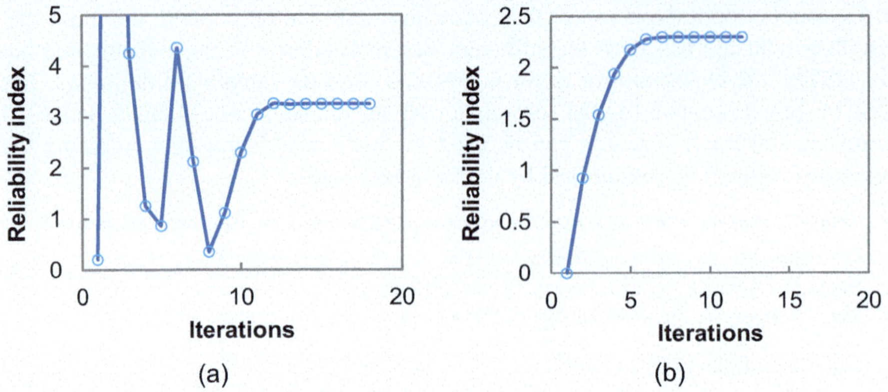

Fig. 5.6 Reliability index iterations obtained by FSL method **a** Example 5.1 **b** Example 5.2

Fig. 5.7 Iteration for MPP search in normal standard space using DSTM and HL-RF methods for Example 5.2

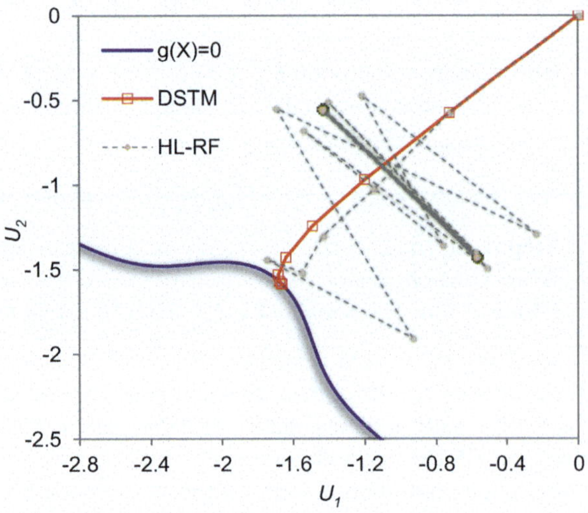

where $L = 5\,\text{m}$ defined as beam span. The variables W (the uniform load) $W \sim N(10, 0.4)\,\text{kN/m}$, E (modulus of elasticity) $E \sim N(2, 0.5) \times 10^7\,\text{kN/m}^2$ and I (moment of inertia of the cross-section) $I \sim N(8, 1.5) \times 10^{-4}\,\text{m}^4$ are followed the normal distribution.

The reliability index for DSTM is plotted in Fig. 5.9. As seen, the robust results are given DSTM with reliability index of 3.1805 after 75 iterations, but iterations for stabilization are more computationally inefficient. The selecting chaos control factor less than 1 in the DSTM provided convergence to achieve stable results, more slowly. For this problem with convex LSF, the DSTM provides inefficient results. Thus, we have a framework that it is good for concave function with highly nonlinear LSFs while it is high-computationally converged for concave problems.

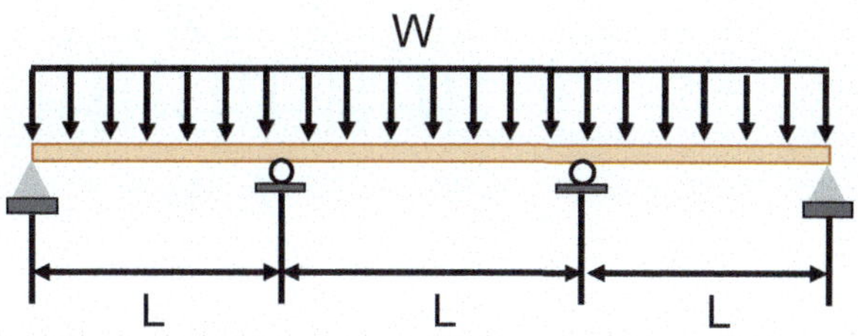

Fig. 5.8 Compact steel beam. Reprinted from [12], with permission from Elsevier

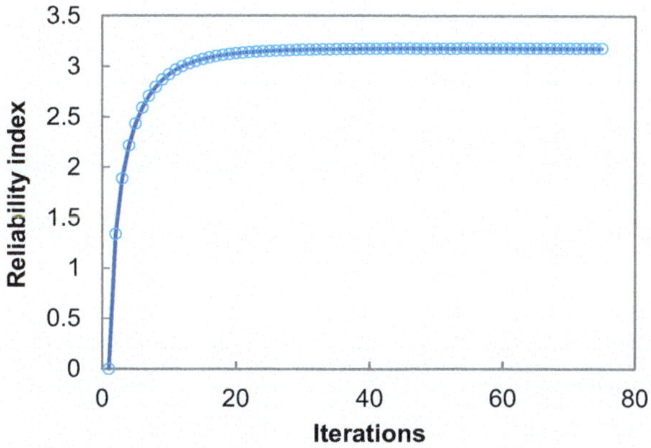

Fig. 5.9 Iterative of reliability index using DSTM for Example 5.3

Example 5.4 A steel frame,

As shown in Fig. 5.10, a steel frame with maximum lateral displacement for giving the LSF by the below relation [13],

$$g = 0.05 - \Delta \tag{5.15}$$

The finite element approach is utilized for computing Δ by using basic variables that listed in Table 5.1.

The results of reliability analysis based on the DSTM method are plotted in Fig. 5.11 for the steel frame structure. It can be found from the results of Fig. 5.11 that the DSTM-based FORM algorithm produced unstable results as a chaotic solution. The step size in this problem is given as 0.1, which provides stability for this problem. The STM and DSTM are two unstable algorithms for this case of the structural reliability analysis. The DSTM enhances the abilities of the FORM-based STM but for problems that is depended on the line search-based step size as this problem, the DSTM may be loosed robustness for giving stable results.

The DSTM provides a suitable line search direction for highly nonlinear performance function. It enhanced the instability of the FORM-based HL-RF algorithm and improved the efficiency of the STM method. However, its application for real engineering problems with convex performance functions is not suggested. It may provide unstable results when the convergence properties for an LSF are dependent

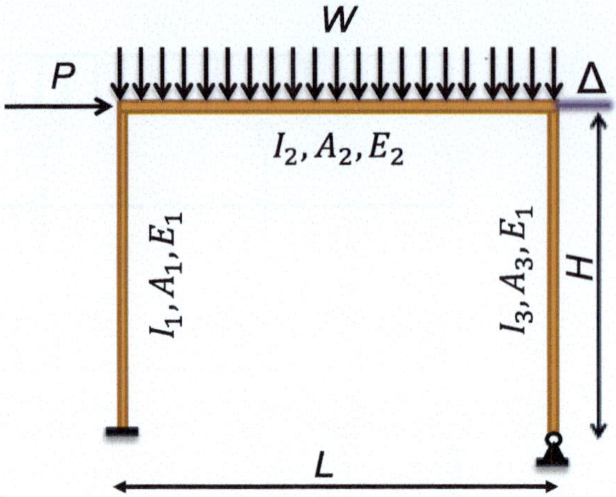

Fig. 5.10 Steel frame structure for Example 5.4. Reprinted from [13], with permission from Elsevier

Table 5.1 Variables of frame structure applied in Example 5.4

Variable	Mean	STD	PDF	Variable	Mean	STD	PDF
A_1 (m^2)	91×10^{-4}	9.1×10^{-4}	Lognormal	L (m)	6	0.48	Normal
A_2 (m^2)	53.8×10^{-4}	5.38×10^{-4}	Lognormal	E_1 (kg/m^2)	2.38×10^{10}	1.9×10^9	Normal
A_3 (m^2)	106×10^{-4}	10.6×10^{-4}	Lognormal	E_2 (kg/m^2)	2.17×10^{10}	1.74×10^9	Normal
I_1 (m^4)	80.9×10^{-6}	8.09×10^{-6}	Lognormal	H (m)	4.5	0.36	Normal
I_2 (m^4)	83.6×10^{-6}	8.36×10^{-6}	Lognormal	P (kg)	2000	350	Gumbel
I_3 (m^4)	112.6×10^{-6}	11.26×10^{-6}	Lognormal	W (kg/m)	4500	630	Gumbel

on the step size. On the other hand, the efficiency of the DSTM depends on the chaos control (CC) factor that may yield inefficient results for linear or convex engineering problems. Therefore, it is suggested that this method for further application in the structural reliable analysis should be improved based on the improving chaos control factor.

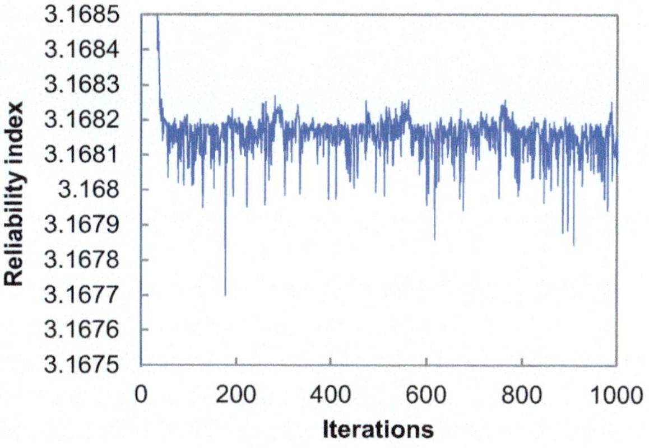

Fig. 5.11 The converged reliability index of DSTM for Example 5.4

5.4 Adaptive Directional Stability Transformation Method

The MPP search-based methods as DSTM can be used as a popular type of probabilistic model for SRA. As shown in Sect. 5.3, the DSTM provides efficient with stable iterations for highly LSFs but it may be loosed its robustness and efficiency for some cases of engineering problems. Meng et al. [4] enhanced the DSTM for efficiency and robustness when it is applied for MPP search by a rule as angle condition. They used the angle condition to enhance DSTM (EDSTM). A current and previous angle ratio is applied for adjusting the chaos control factor in EDSTM. The angle condition is schematically shown in Fig. 5.12 for applying this method. As shown in Fig. 5.12, the suitable line search is computed when the angles between two consecutive line search directions are generally decreased thus by decreasing angles iterative formulation by directional search direction, which provides the optimum MPP with stable converges.

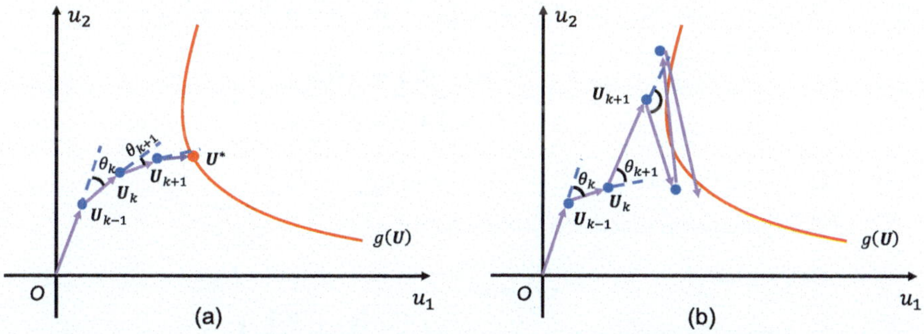

Fig. 5.12 Angle between several iterations of the EDSTM **a** stable results **b** unstable results as chaotic

The iterative process appears the instability when the angle between two line search directions is increased thus the DSTM formulation is required to enhance by adjusting the chaos control factor. The angle is computed as,

$$\theta_k = \arcos\left(\frac{\boldsymbol{U}_k[\boldsymbol{U}_k - \boldsymbol{U}_{k-1}]^T}{|\boldsymbol{U}_k|.|\boldsymbol{U}_k - \boldsymbol{U}_{k-1}|}\right) \tag{5.16}$$

It is conducted that the suitable chaos control factor is computed when $\theta_{k+1} \le \theta_k$, thus the chaos control factor is not adjusted and the previous value given for the new iteration. Otherwise i.e. $\theta_{k+1} > \theta_k$ for increasing angel by the iterative procedure as Fig. 5.12b, the chaos control factor should be updated for new iterations by using angle ratio θ_k/θ_{k+1}. It should be adjusted the chaos control factor with lowest value than the previous that the angle ratio less than 1. The adjusting formulation for the chaos control factor in EDSTM is given as,

$$\lambda = \begin{cases} \max\left[\frac{\lambda\theta_k}{\theta_{k+1}}, 0.4\lambda\right] & \theta_{k+1} > \theta_k \\ \lambda & \theta_{k+1} \le \theta_k \end{cases} \tag{5.17}$$

The minimum value for the angle ratio is given as 0.4 based on the above formulation, thus it does not provide a high-decreasing for chaos control factor and the angle ratio is selected as $0.4 \le \frac{\theta_k}{\theta_{k+1}} < 1$. The EDSTM is applied for FORM-based iterative scheme by the following steps:

Step 1: Define LSF $g(\boldsymbol{X})$, Give stopping criterion (ε) and random variables including distributions, μ and σ, and also give the step length and adjusting factor, $\lambda = 0.5$ Let $k = 1$, Choose $\boldsymbol{X}_1 = \boldsymbol{\mu}$;
Step 2: Transfer variables into normal standard space;
Step 3: Iterate to find MPP using EDSTM as below;

(i) Compute $g(U_k)$, $\nabla_u g(U_k)$;
(ii) Determine the new point by using the HL-RF method as below;

$$\boldsymbol{U}_{k+1}^{HL} = \frac{\nabla^T g(\boldsymbol{U}_k)\boldsymbol{U}_k - g(\boldsymbol{U}_k)}{\nabla^T g(\boldsymbol{U}_k)\nabla g(\boldsymbol{U}_k)}\nabla g(\boldsymbol{U}_k) \tag{5.18}$$

(iii) Compute the point $U_{k+1}^{STM} = U_k + \lambda C[U_{k+l}^{HL} - U_k]$;
(iv) Compute normalized sensitivity vector as $\boldsymbol{\alpha}_{k+1} = \frac{\boldsymbol{U}_{k+1}^{STM}}{|\boldsymbol{U}_{k+1}^{STM}|}$;
(v) Determine $\beta_{k+1} = \frac{\nabla^T g(\boldsymbol{U}_k)\boldsymbol{U}_k - g(\boldsymbol{U}_k)}{\nabla^T g(\boldsymbol{U}_k)\boldsymbol{\alpha}_{k+1}}$ and new point as $\boldsymbol{U}_{k+1} = \beta_{k+1}\boldsymbol{\alpha}_{k+1}$;
(vi) Compute the angle as $\theta_{k+1} = arcos\left(\frac{\boldsymbol{U}_{k+1}[\boldsymbol{U}_{k+1} - \boldsymbol{U}_k]^T}{|\boldsymbol{U}_{k+1}|.|\boldsymbol{U}_{k+1} - \boldsymbol{U}_k|}\right)$;
(vii) For $k \ge 2$, update the CC factor as below;

$$\lambda = \begin{cases} \max\left[\frac{\lambda\theta_k}{\theta_{k+1}}, 0.4\lambda\right] & \theta_{k+1} > \theta_k \\ \lambda & \theta_{k+1} \leq \theta_k \end{cases} \tag{5.19}$$

Step 4: **If** $|\beta_{k+1} - \beta_k| < \varepsilon$, iterate formulation is converged and then print, $\boldsymbol{X}^* = \boldsymbol{X}_{k+1}$, $\boldsymbol{U}^* = \boldsymbol{U}_{k+1}$, $\beta = \beta_{k+1}$;
Else $\boldsymbol{X}_{k+1} = T^{-1}(\boldsymbol{U}_{k+1})$ and set $k = k + 1$ and go to Step 2.

This method is as simple as the DSTM and FSL. This formulation is not complicated thus application of this method for structural reliability analysis is simple and user friendly. The main difference of this method with DSTM is to apply the adaptive chaos control factor that it is adjusted based on the angle condition. The robust convergence can be obtained based on this formulation by applying angle condition because it is given condition as $\theta_{k+1} \leq \theta_k$. Thus, at the final iteration, we have $\theta_{k+1} - \theta_k \approx 0$. This means that it is computed a fixed point as well as the FSL. Consequently, it can be conducted that the EDSTM is as robust as the FSL but it is more efficient. The efficiency of this method compared to the DSTM depends on the initial value of the chaos control factor. The initial value of the chaos control factor is selected as a middle value i.e. $\lambda = 0.5$.

Examples 5.1 and 5.2 given from the Sect. 5.2 is applied evaluating the abilities of the EDSTM method. the reliability index and the adjusting chaos control factor for Examples 5.1 and 5.2 are presented in Figs. 5.13 and 5.14, respectively. Note that, the EDSTM is converged robustly with efficient results. As seen, the chaos control factor is adjusted with the lowest values at the final iterations for these two nonlinear problems and it provides a suitable and stable search direction with smooth iterations for reliability index. By pointed out the results of DSTM, the EDSTM is generally more efficient than DSTM but for Example 5.2, it is inefficient. The adjusting process and the step size are the effective factors for obtaining efficient results.

The iterative process of MPP search for Example 5.2 using the EDSTM is presented in Fig. 5.15. As seen, the MPP is searched based on a line direction that

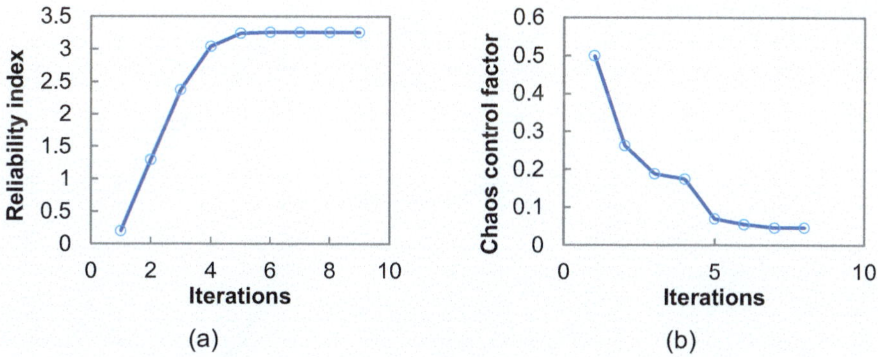

Fig. 5.13 Reliability results using EDSTM for Example 5.1 **a** reliability index **b** chaos control factor

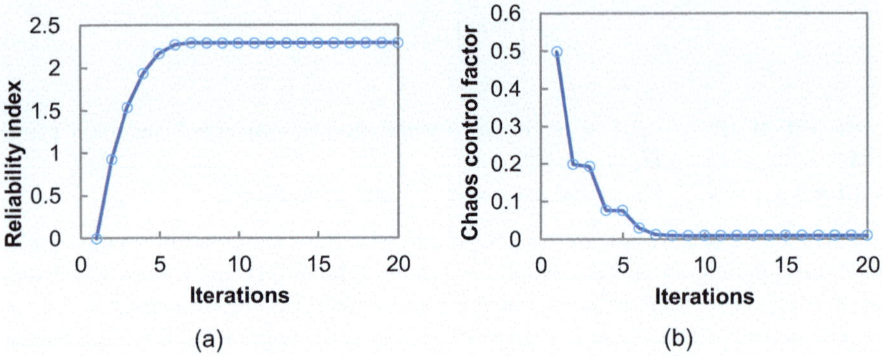

Fig. 5.14 Reliability results using EDSTM for Example 5.2 **a** reliability index **b** chaos control factor

has a smooth curve. At the first iterations, it shows a high-convergence rate but for controlling the instability at the final iterations, the EDSTM needs several iterations for providing a suitable step size. The angle condition can provide stability for this nonlinear problem.

As two extra engineering problems in Sect. 5.3, the three-span beam and steel frame structures are given for evaluating the abilities of the EDSTM. The beam problem is used for evaluating the efficiency of this method and the frame problem is for evaluating the robustness of the EDSTM. The results of the reliability index for beam and frame problems are plotted in Figs. 5.16 and 5.17, respectively. As seen from the results of Fig. 5.16, the EDSTM enhanced the efficiency of the DSTM but it is also converged, inefficiently. The reliability index of EDSTM is as same as the DSTM thus it has shown global results. The small value of the chaos control factor is

Fig. 5.15 Iteration for MPP search in normal standard space using EDSTM and HL-RF methods for Example 5.2

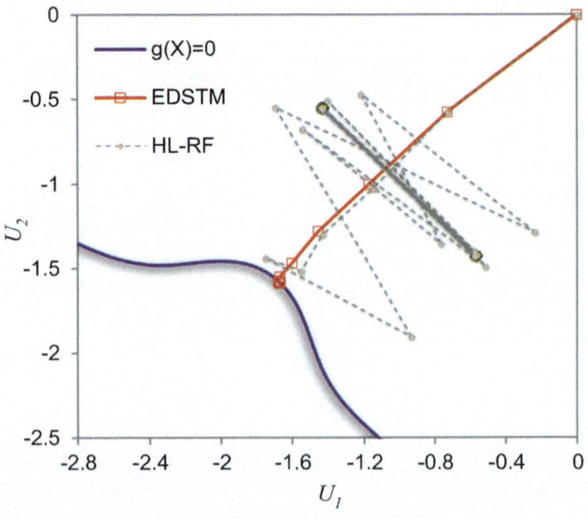

one of major issue of the EDSTM for convex problems. Thus, the EDSTM may yield an inefficient computational burden for determining the MPP for convex problems.

As results presented in Fig. 5.17, the robustness of the frame problem is given by the iterative formulation of EDSTM. It converges to reliability index of 3.1681 after 80 iterations. The angle condition provides the stabilization for nonlinear problems by adjusting the chaos control factor. Consequently, the EDSTM is a robust iterative formulation but its efficiency is strongly depended on the adjusting process of the chaos control factor. This method has acceptance abilities for solving the high-nonlinear problems but for moderate linear performance functions, it is an inefficient approach.

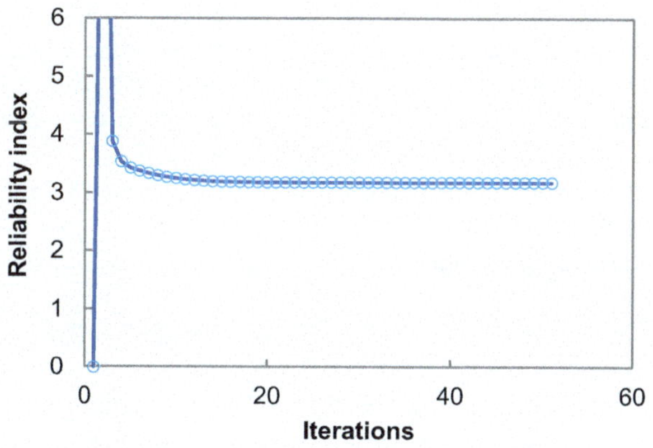

Fig. 5.16 The converged reliability index of EDSTM for Example 5.3

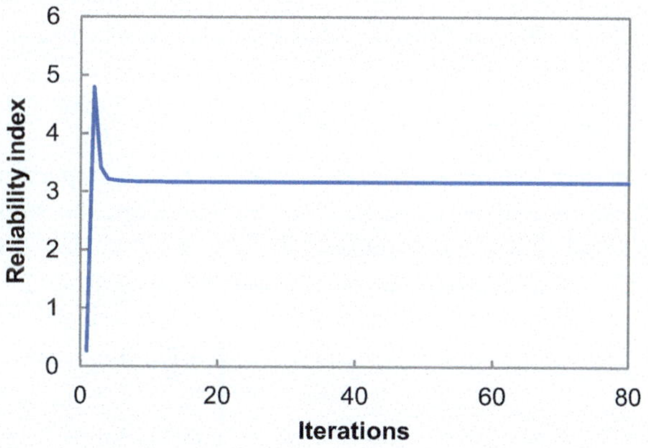

Fig. 5.17 The converged reliability index of EDSTM for Example 5.4

5.5 Enhanced FSL

The performance abilities of the FSL method is enhanced based on a new adjusting process for step length [14]. They used $|\boldsymbol{U}_{k+1} - \boldsymbol{U}_k| < |\boldsymbol{U}_k - \boldsymbol{U}_{k-1}|$ condition as well as the FSL method for adjusting step length. Zhang et al. [14] assumed that for nonlinear LSFs, the step length should be adjusted due to oscillation in each iteration. Therefore, the step length steadily decreases in the iterative process when $|\boldsymbol{U}_{k+1} - \boldsymbol{U}_k| \geq |\boldsymbol{U}_k - \boldsymbol{U}_{k-1}|$. They finding that the computational efforts and the global convergence as robust solutions of the FSL method are depended on finite-step length. For enhancing the convergence properties of the FSL method the novel adjusting procedure has been proposed named as enhance finite-step length (EFSL). The step length is not adjusted when $|\boldsymbol{U}_{k+1} - \boldsymbol{U}_k| < |\boldsymbol{U}_k - \boldsymbol{U}_{k-1}|$. The step length should be adjusted by the adjustment procedure when $|\boldsymbol{U}_{k+1} - \boldsymbol{U}_k| \geq |\boldsymbol{U}_k - \boldsymbol{U}_{k-1}|$. The step length is adaptively updated based on the nonlinearity degrees of LSFs. The step length is selected as a large value for weakly nonlinear functions but it is given a small value for highly nonlinear functions. The adjusting process of the EFSL method has two parts which is separated by the sufficient descent condition. In the part 1 when $|\boldsymbol{U}_{k+1} - \boldsymbol{U}_k| \geq |\boldsymbol{U}_k - \boldsymbol{U}_{k-1}|$, the step length is updated by below adjusting procedure,

$$\lambda_{k+1} = c\mathrm{P}(1 - \mathrm{P})\lambda_k \tag{5.20}$$

The logistic map i.e. $c\mathrm{P}(1 - \mathrm{P}) < 1$ (where $0 < \mathrm{P} < 1$ and $0 < \mathrm{c} < 4$) is applied for updating the finite step length. For giving the logistic map less than 1, the parameters c and P were suggested as $c = 2$ and $P = 0.05$ [14]. In the adjusting process, a new criterion as $\lambda_{k+1} \leq 0.15$ is defined for giving the smooth convergence, thus, we have a new adjusting formulation when $\lambda_{k+1} \leq 0.15$ as below,

$$\lambda_{k+1} = \exp[-0.125(k + 1)] \tag{5.21}$$

For part 2 when $|\boldsymbol{U}_{k+1} - \boldsymbol{U}_k| < |\boldsymbol{U}_k - \boldsymbol{U}_{k-1}|$, step length does not change for current iteration and we have $\lambda_{k+1} = \lambda_k$. However, the step length is adjusted when for second times, the condition $|\boldsymbol{U}_{k+1} - \boldsymbol{U}_k| < |\boldsymbol{U}_k - \boldsymbol{U}_{k-1}|$ is satisfied, thus when $\lambda_{k+1} = \lambda_k$, the new step size is adjusted as below,

$$\lambda_{k+1} = \begin{cases} 0.1\lambda_k & \lambda_k > 10 \\ \lambda_k & 5 \leq \lambda_k \leq 10 \\ 20\lambda_k & \lambda_k < 5 \end{cases} \tag{5.22}$$

Equation 5.22 provides a suitable balance for adaptive step length when we have a weakly or highly nonlinear LSF. Since it is not understood the nonlinear degree of limit state functions, thus the initial step length is given as $\lambda_1 = 2^{15}$. The steps for using the enhanced stability transformation method (ESTM) applied in a computer program are presented below:

Step 1: Define LSF $g(\boldsymbol{X})$, Give stopping criterion (ε), variables corresponding to distributions, μ and σ, and also give initial step length as $\lambda_1 = 2^{15}$ and adjusting factor $c = 2$ and $P = 0.05$. Let $k = 1, \boldsymbol{X}_1 = \boldsymbol{\mu}$;
Step 2: Transfer variables into normal standard space;
Step 3: Iterate ESTM for determine new point;

(i) Compute $g(U_k)$, $\nabla_u g(U_k)$;
(ii) Compute the point $\boldsymbol{u}_{k+1}^{\lambda} = \boldsymbol{U}_k - \lambda_k \nabla g(\boldsymbol{U}_k)$;
(iii) Compute NSDV as $\boldsymbol{\alpha}_{k+1} = \frac{\boldsymbol{u}_{k+1}^{\lambda}}{|\boldsymbol{u}_{k+1}^{\lambda}|}$;
(vi) Determine the reliability index as $\beta_{k+1} = \frac{\nabla^T g(\boldsymbol{U}_k)\boldsymbol{U}_k - g(\boldsymbol{U}_k)}{\nabla^T g(\boldsymbol{U}_k)\boldsymbol{\alpha}_{k+1}}$ and the new point as $\boldsymbol{U}_{k+1} = \beta_{k+1}\boldsymbol{\alpha}_{k+1}$;
(v) For $k \geq 2$, and if $|\boldsymbol{U}_{k+1} - \boldsymbol{U}_k| \geq |\boldsymbol{U}_k - \boldsymbol{U}_{k-1}|$, then $\lambda_{k+1} = c\mathrm{P}(1 - \mathrm{P})\lambda_k$, check for new step length if $\lambda_{k+1} \leq 0.15$, then $\lambda_{k+1} = \exp[-0.125(k + 1)]$;
(vi) For $k \geq 2$, and if $|\boldsymbol{U}_{k+1} - \boldsymbol{U}_k| < |\boldsymbol{U}_k - \boldsymbol{U}_{k-1}|$, then $\lambda_{k+1} = \lambda_k$, check for new step length if $\lambda_{k+1} = \lambda_{k-1}$, then update it as below;

$$\lambda_{k+1} = \begin{cases} 0.1\lambda_k \lambda_k > 10 \\ \lambda_k 5 \leq \lambda_k \leq 10 \\ 20\lambda_k \lambda_k < 5 \end{cases} \tag{5.23}$$

Step 4: **If** $|\beta_{k+1} - \beta_k| < \varepsilon$ the iterative formulation converged and print $\boldsymbol{X}^* = \boldsymbol{X}_{k+1}, \boldsymbol{U}^* = \boldsymbol{U}_{k+1}, \beta = \beta_{k+1}$;
Else $\boldsymbol{X}_{k+1} = T^{-1}(\boldsymbol{U}_{k+1})$ set $k = k + 1$ and go to Step 2.

The adjusting framework of the EFSL method for updating the step length is plotted as shown in Fig. 5.18. As seen, this method is similar to the FSL method but the adjusting process for determining the step length is different. In the EFSL, the step length is adjusted for both moderate linear and nonlinear LSF. The sufficient decent condition is used to provide stable results in this method but the step length is also given a value varied from 5 to 10 for linear problems as final iterations.

As complicated adjusting process is defined for updating the step length compared to DSTM and EDSTM. At the first iteration, the large value for step length as $\lambda_1 = 32768$ is given. This large value is good for moderate and convex problems while for highly nonlinear function, it should be computationally decreased. Consequently, the efficiency of the EFSL may be ignored for highly nonlinear functions.

The Examples 5.1 and 5.2 presented in Sect. 5.2 are used for comparing the ability of the EFSL method. The reliability results obtained by the EFSL method including the reliability index and the step length are proposed in Figs. 5.19 and 5.20 for Examples 5.1 and 5.2, respectively. As seen, a non-smooth reliability index is obtained at the first iterations but by increasing the iterations, the adaptive step length is decreased. It can be concluded that by decreasing the step length, the curve line of the reliability index is shown the smooth conditions. As seen, this method is as inefficient as the FSL method for highly nonlinear functions. Finally, it yields the

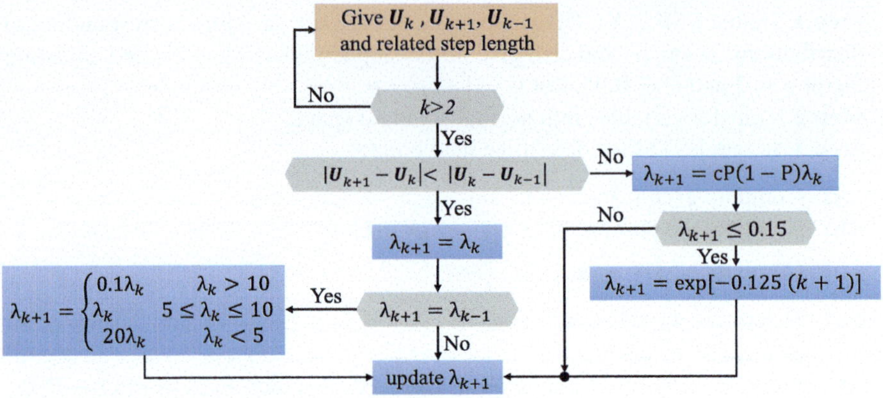

Fig. 5.18 Updating framework of step length in EFSL method

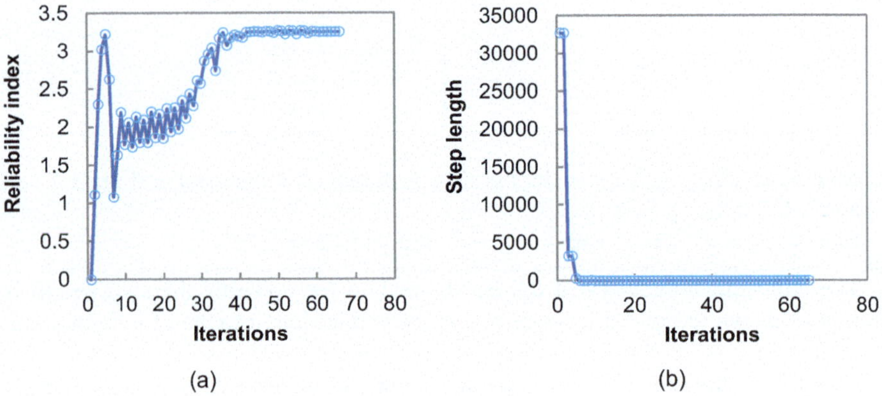

Fig. 5.19 Reliability results using EFSL for Example 5.1 **a** reliability index **b** step length

robust results, which may be enhanced the efficiency of the FSL method but it is not significantly.

The MPP search-based iterative process of the EFSL in normal standard space is plotted in Fig. 5.21 for Example 5.2. As seen, the oscillating results given by the EFSL as well as the FSL method. The adjusting procedure for nonlinear problems is computed very slowly when the step size is less than 0.15. For this example, the step size is computed less than 0.15. On the other hand, if the sufficient decent condition is satisfied, then the step size may be given a large value than the previous for step length less than 0.15. Consequently, it is needed to update the step length based on the several iterations, again. This is also given from the results of this method, it provides computational efforts to find the MPP.

Two engineering problems as presented in Sect. 5.3 are given for the abilities of EFSL for convex function as the three-span beam structure and for nonlinear

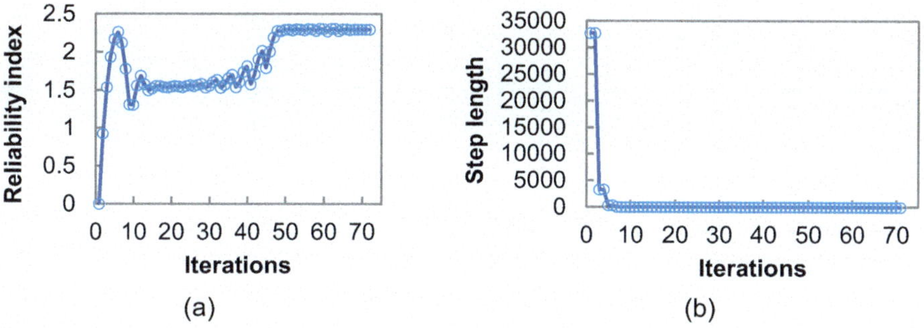

Fig. 5.20 Reliability results using EFSL for Example 5.2 **a** reliability index **b** step length

Fig. 5.21 Iteration for MPP search in normal standard space using EFSL and HL-RF methods for Example 5.2

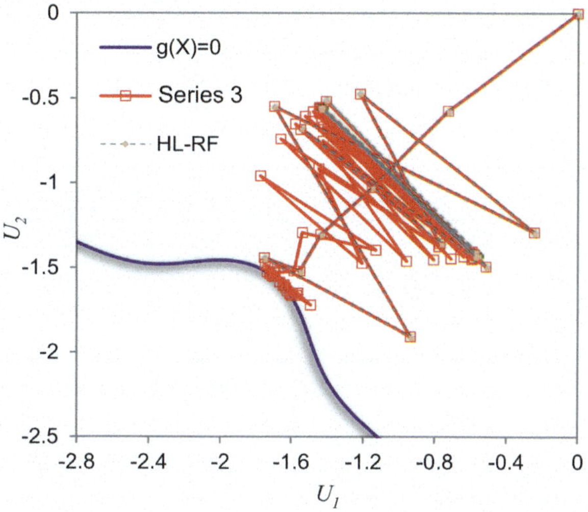

function as steel frame structure. The abilities for both robustness and efficiency of the EFLS are evaluated for these problems. The beam problem is used for evaluating the efficiency of the EFSL method and the frame problem is applied for evaluating the robustness of the EFSL.

The results of the reliability index for beam and frame problems are plotted in Figs. 5.22 and 5.23, respectively. As seen from the results of Fig. 5.22, the EFSL is an efficient method and it is converged, more efficiently. The efficiency of the DSTM is strongly improved based on FSL for convex problems because the line search at almost all iterations is similar to HL-RF-based SDSD.

The reliability index of EFSL is as same as the FSL and DSTM thus the globality of this method using the adjusting finite step length is obtained. The large finite-step length improved the efficiency of FORM-based EFSL for convex problems and it shows a superior performance for this problem obtained by EFSL. And also, it can be conducted that the efficiency of this method may be less than the FSL due to

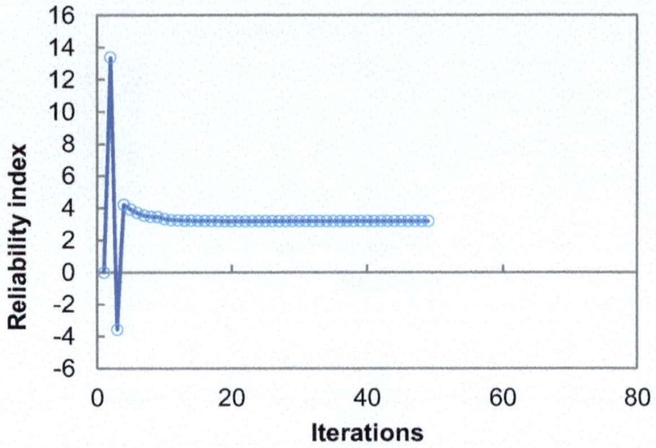

Fig. 5.22 The converged reliability index of EFSL for Example 5.3

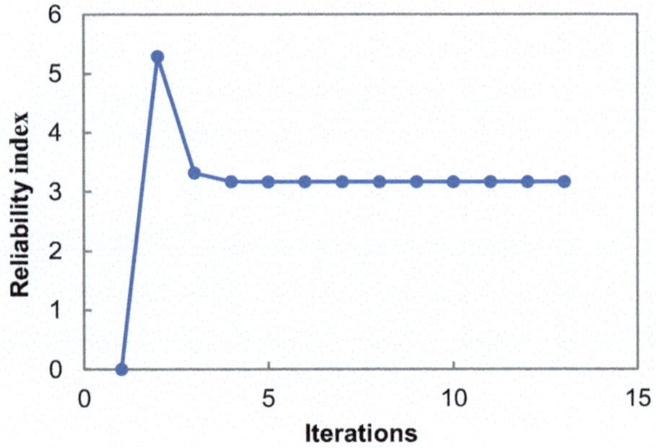

Fig. 5.23 The converged reliability index of EFSL for Example 5.4

adjusting step size by a small value between 5 and 10 for engineering reliability problems with convex type of LSF.

As results presented in Fig. 5.23, the robustness of the frame problem is given by the iterative formulation of EFSL as well as the EDSTM. The EFSL converges to reliability index of 3.1682 after 13 iterations. The adjusting process of the finite step length can provide the stabilization for this problem. Consequently, the EFSL is a robust iterative formulation. The efficiency of this method is strongly improved by comparing with the result of EDSTM. The adjusting process by using the sufficient descent conditions provides the robust and efficient results for this problem. By comparing the results presented in Fig. 5.22, the EFSL may be not remarkably suitable line search direction for convex problems. This method has shown acceptable abilities for solving moderate nonlinear problems but for nonlinear concave/convex functions, the EFSL is an inefficient method.

5.6 Dynamical Relaxed Directional Method

As an tool with balancing between efficient and accurate structural reliability methods for handling uncertainties, the FORM with a dynamical control factor is proposed for engineering problems [15]. This method named as dynamical relaxed directional method (DRD). The DRD-based iterative the relaxed approach is formulated that it is considered the sufficient descent condition by using an accelerated dynamical NSDV. The dynamical NSDV is formulated using sufficient decent conditions for achieving the robustness of FORM. In this section, the main formulation of FORM-based DRD is presented and then, the abilities of the DRD method are examined by four examples. In the comparing part for evaluating the performances, the DRD is compared with the obtained results of several methods of FORM.

5.6.1 Iterative Formulation DRD-Based FORM

Simultaneously SRA is very computationally expensive for engineering systems, thus an efficient technique must be developed to reduce the iteration burden of an SRA method applied in real engineering problems. The SDSD applied for NSDV as well as HL-RF is capable to run an efficient result. As presented in Chapte 2, the iterative HL-RF formula as $\boldsymbol{U}_{k+1}^{HL}$ provided unstable results for nonlinear LSFs [16]. Consequently, it is focused on enhancing the robustness of HL-RF based modification in NSDV such as STM or relaxed formulation applied in HL-RF [17]. Almost improved FORM given by directional-based NSDV are formulated by a controlling factor (step size) less than 1 as CC factor applied in STM [18], adaptive FSL [19], DSTM [8], or relaxed factor utilized in the relaxed HL-RF (RHL-RF) and hybrid RHL-RF [20]. The efficiency of improved FORM-based SDSD methods may be decreased by comparison with HL-RF for convex/moderate linear LSFs. Thus, it is a challenge to extract a robustness FORM formula with acceptable efficiency. The relaxed formulation which can be applied for determining directional sensitivity vector is presented to enhance the efficiency of directional FORM as below,

$$\tilde{\boldsymbol{U}}_{k+1}^{DR} = (1-\lambda)\boldsymbol{U}_k^{DR} + \lambda \boldsymbol{U}_k^{HL}, \boldsymbol{\alpha}_k) = \frac{\tilde{\boldsymbol{U}}_{k+1}^{DR}}{\left\| \tilde{\boldsymbol{U}}_{k+1}^{DR} \right\|} \tag{5.24}$$

where λ represents the relaxed factor, which has a dynamically computed for enhancing the efficiency of FORM-based directional SDSD vector. The relaxed factor is main parameter in DRD for increasing its efficiency which is given based on reducing the iterations for convergence. As seen for results of DSTM, a small value for step size as less than 1 (e.g. $\lambda = 0.1$) provided the acceptable efficiency for high LSF. By giving a larger λ (e.g. $\lambda > 0.75$), it may not robustly be converged for high-nonlinear LSF. For reducing the risk of instability for the method of FORM formulated by directional NSDV, a dynamically relaxed factor is introduced by self-adjusting formulation given by a merit function from 0 to 1.5 as below,

$$\lambda_k = \min\{1.5, 0.99M_k\} \tag{5.25}$$

where M_k denotes merit factor computed as below relation,

$$M_k = \frac{\left\| \boldsymbol{U}_k^{DR} - \boldsymbol{U}_{k-1}^{DR} \right\|}{\left\| \boldsymbol{U}_{k+1}^{HL} - \boldsymbol{U}_k^{DR} \right\|} \tag{5.26}$$

where, $\boldsymbol{U}_k^{DR}$ and $\boldsymbol{U}_{k-1}^{DR}$ are points computed by DRD approach at previous iterations and $\boldsymbol{U}_{k+1}^{HL}$ is point obtained by HL-RF in current iterations. As seen in Eq. 5.26, the merit factor is adjusted by $\boldsymbol{U}_{k+1}^{HL}$, $\boldsymbol{U}_k^{DR}$ and $\boldsymbol{U}_{k-1}^{DR}$ points. As first iteration, $\lambda_0 = 1$. Equation 5.25 can be performed by $\lambda_k < \frac{\left\| \boldsymbol{U}_k^{DR} - \boldsymbol{U}_{k-1}^{DR} \right\|}{\left\| \boldsymbol{U}_{k+1}^{HL} - \boldsymbol{U}_k^{DR} \right\|}$ as maximum relaxed factor given as 1.5. It provides a superior rate for convergence when relaxed factor is computed equal to 1.5. In relaxed factor Eq. 5.25, the first term i.e. $\lambda_k < \frac{\left\| \boldsymbol{U}_k^{DR} - \boldsymbol{U}_{k-1}^{DR} \right\|}{\left\| \boldsymbol{U}_{k+1}^{HL} - \boldsymbol{U}_k^{DR} \right\|}$ controls the instability, while the efficiency is enhanced by second term i.e. $\lambda_k = 1.5$. By computing $\lambda_k = 1.5$, the DRD formulation may be converged, more efficiently while the modified FORM-based SDSD as STM and DSTM are formulated using smaller step size. For nonlinear problems, it is applied the sufficient decent condition in merit function as $0.99M_k$ thus, we have $\left\| \boldsymbol{U}_{k+1}^{HL} - \boldsymbol{U}_k^{DR} \right\| < \left\| \boldsymbol{U}_k^{DR} - \boldsymbol{U}_{k-1}^{DR} \right\|$. By sufficient decent condition, it can be shown that,

$$\begin{aligned} \left\| \boldsymbol{U}_{k+1}^{HL} - \boldsymbol{U}_k^{DR} \right\| &= \kappa_k \left\| \boldsymbol{U}_k^{DR} - \boldsymbol{U}_{k-1}^{DR} \right\| = \kappa_k \kappa_{k-1} \left\| \boldsymbol{U}_{k-1}^{DR} - \boldsymbol{U}_{k-2}^{DR} \right\| \\ &= \cdots = \textstyle\prod_{i=1}^{k} \kappa_i \left\| \boldsymbol{U}_1^{DR} - \boldsymbol{U}_0 \right\| \end{aligned} \tag{5.27}$$

In which, $0 < \kappa < 1$ and thus for $k \to \infty$, $\lim\limits_{k \to \infty} \prod_{i=1}^{k} \kappa_i \approx 0$ thus $\left\| \boldsymbol{U}_k^{DR} - \boldsymbol{U}_{k-1}^{DR} \right\| \approx 0$. This means, stable results with fixed point are achieved. By applying the sufficient descent condition in merit function, the DRD method as well as the EDSTM and EFSL formulations provides stable results. The new line search is a directional line search that the step size in this method may be computed as more than 1 as a superior convergence rate for linear LSFs. Consequently, DRD can be introduced as a formulation with robustness and efficiency compared to other SDSD method of analytical approaches.

The proposed DRD approach can be coded for a computer program as presented by the following steps:

Step 1: Define $g(\boldsymbol{X})$, Give stopping criterion (ε) and variables with distributions μ and σ, and also set $\lambda_1 = 1$, $k = 1$, $\boldsymbol{X}_1 = \boldsymbol{\mu}$;
Step 2: Transfer the variables into normal standard space;
Step 3: Iterate the DRD method for finding new point and reliability index;

(i) Compute $g(\boldsymbol{U}_k)$, $\nabla_u g(\boldsymbol{U}_k)$;

(ii) Determine new point by HL-RF method as follows;

$$U_{k+1}^{HL} = \frac{\nabla^T g(U_k)U_k - g(U_k)}{\nabla^T g(U_k)\nabla g(U_k)}\nabla g(U_k) \tag{5.28}$$

(iii) Determine new directional point as $\tilde{U}_{k+1}^{DR} = (1-\lambda_k)U_k^{DR} + \lambda_k U_k^{HL}$;

(iv) Compute normalized sensitivity vector as $\boldsymbol{\alpha}_{k+1} = \frac{\tilde{U}_{k+1}^{DR}}{\left\|\tilde{U}_{k+1}^{DR}\right\|}$;

(v) Determine $\beta_{k+1} = \frac{\nabla^T g(U_k)U_k - g(U_k)}{\nabla^T g(U_k)\boldsymbol{\alpha}_{k+1}}$ and new point $U_{k+1} = \beta_{k+1}\boldsymbol{\alpha}_{k+1}$;

(vi) Update the merit factor for $k \geq 2$ as $M_k = \frac{\|U_k - U_{k-1}\|}{\left\|U_{k+1}^{HL} - U_k\right\|}$;

(vii) Determine new relaxed factor as $\lambda_k = \min\{1.5, 0.99M_k\}$;

Step 4: If $U_{k+1} - U_k < \varepsilon$ the iterative formulation is stopped and print $X^* = X_{k+1}$, $U^* = U_{k+1}$, $\beta = \beta_{k+1}$;
Else $X_{k+1} = T^{-1}(U_{k+1})$ and set $k = k + 1$ then go to Step 2.

The framework of the DRD method given by the above steps is plotted in Fig. 5.24 for searching the MPP. In this method, the relaxed factor is exactly computed as accelerated dynamical form given from previous results. This method is similar to the EDSTM but it has some differences with it as: (1) the relaxed step size is computed based on the sufficient descent condition with the limited value of 1.5, thus we may have a step size more than one for some linear reliability problems; (2) it is an exact line search and the step size is directly updated at each iteration but in the EDSTM, the chaos control factor is computed based on an inexact function. The formulation of the DRD method is as simple as the FSL and DSTM.

5.6.2 *Comparative Abilities of DRD*

The albites of the DRD method formulated by directional SDSD with a dynamically relaxed factor are compared with five FORMs formulated by SDSD methods using three examples. These FORM formulas-based SDSD are (1) HL-RF, (2) STM with $\lambda = 0.1$, (3) DSTM with $\lambda = 0.1$, (4) RHL-RF and (5) FSL with $c = 1.4$ and $\lambda = 100$. The DRD method is computed by self-adaptive parameters for computing the step size.

Example 5.5 a high-nonlinear function,

$$g = x_1^3 + x_1^2 x_2 + x_2^3 - 18 \tag{5.29}$$

where, normal x_1 and x_2 defined by $\mu_1 = 10$, $\mu_2 = 9.9$ and $\sigma_1 = \sigma_2 = 5$. Extracted results from Ref. [9] showed $\beta = 2.2983$ and $X^* = (1.6857, 1.9681)$. By results extended from Keshtegar [13], $\beta = 2.298222$, $U^* = (-1.66307, -1.58616)$,

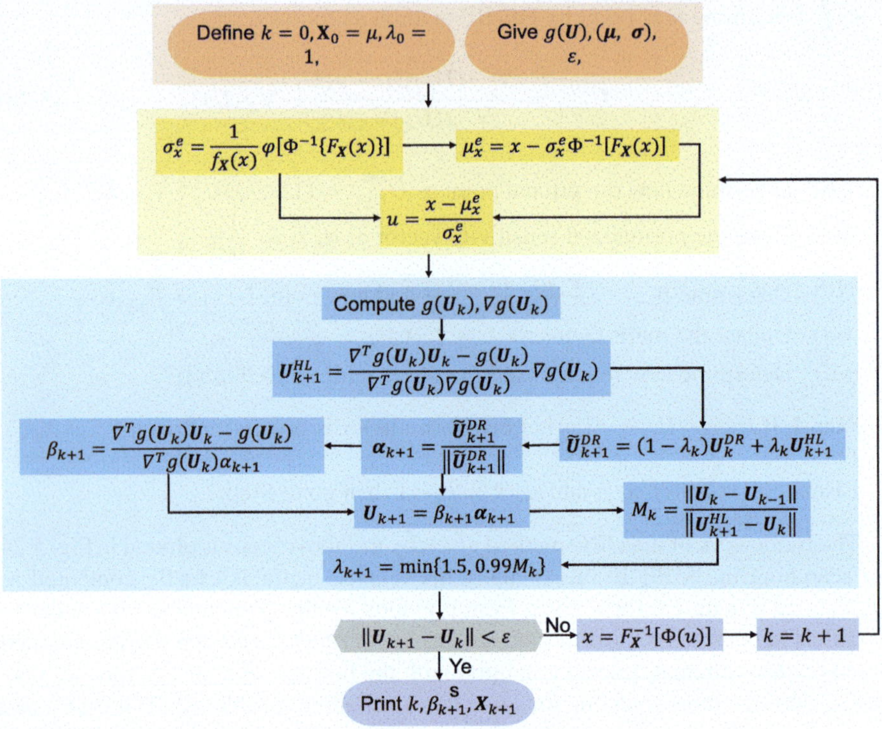

Fig. 5.24 Schematic diagram for DRD to determine MPP

and $\boldsymbol{X}^* = (1.68463, 1.96919)$. Using DRD method, the results are as $\beta = 2.298239$, $\boldsymbol{U}^* = (-1.66384, -1.58542)$, and $\boldsymbol{X}^* = (1.69381, 1.96113)$ which are captured after 10 iterations (182 call functions). It presented that DRD robustly converges and it follows the extracted results from literatures [9, 13].

The LSF, iterations for stabilization and MPP region computed by DRD, STM, RHL-RF, HL-RF, DSTM, and FSL are presented in Fig. 5.25 for Example 5.5. As seen, HL-RF does not converge but modified FORMs-based SDSD are converged, more robustly.

Example 5.6 three-span steel beam as plotted in Fig. 5.26 is defined by following LSF,

$$g = \frac{L}{360} - 0.0069\frac{WL^4}{EI} \tag{5.30}$$

where, W is uniform load, I is moment of inertia of the cross-section, E is modulus of elasticity and L is beam span given as 5 m. Three normal random variables are defined as $W \sim N(10, 0.4)$ kN/m, $E \sim N(2, 0.5) \times 10^7$ kN/m^2 and $I \sim N(8, 1.5) \times 10^{-4}$ m^4.

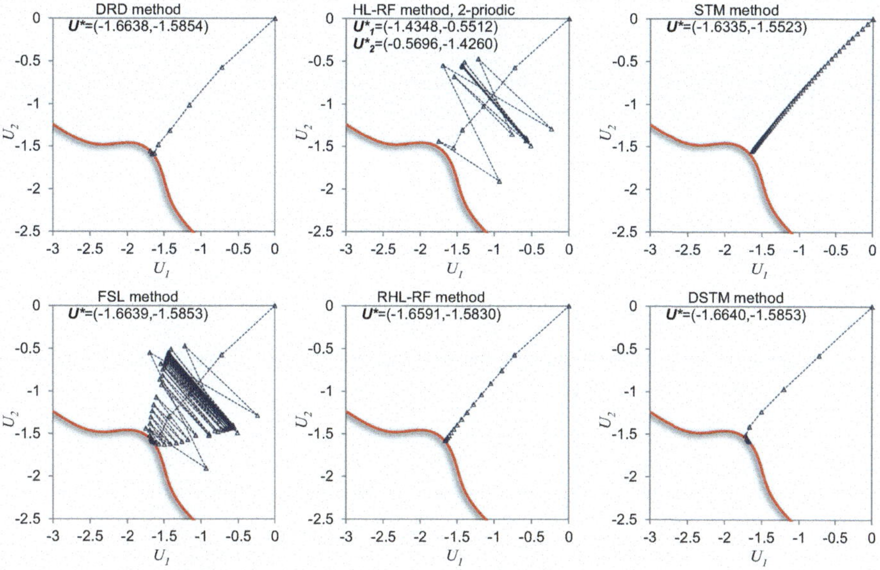

Fig. 5.25 Convergence for different FORM-based SDSD in Example 5.5

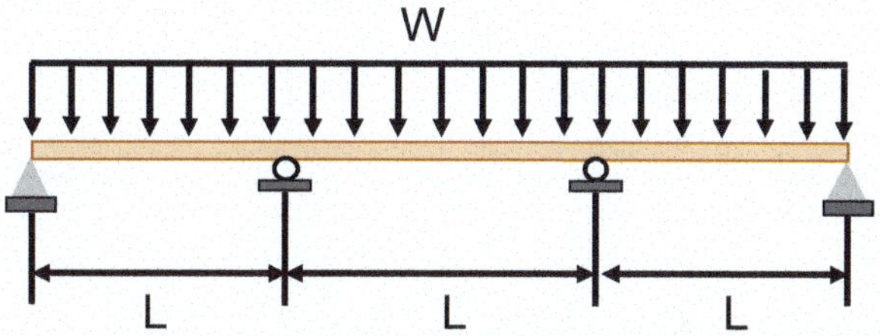

Fig. 5.26 Steel beam for Example 5.6. Reprinted from [12], with permission from Elsevier

The reliability results for different FORM-based SDSD are presented in Fig. 5.27. As seen, all methods are robust, but iterations for stabilization are differed. The applied step size with small value in STM, DSTM, RHL-RF and FSL does not provide efficient iterative formulation; thus, these methods are slowly converged while the DRD formulated by dynamically relaxed factor provides a step size more than 1 (see Fig. 5.27b) that it led to an efficient iteration.

Example 5.7 A dynamic system as presented in Fig. 5.28,

$$g = Fs - 3Ks \times \{\frac{\pi S_0}{4\xi_s\omega_s}[\frac{\xi_a\xi_s}{\xi_p\xi_s\left(4\xi_a^2+\theta^2\right)+\gamma\xi_a^2} \times \frac{\left(\xi_p\omega_p^3+\xi_s\omega_s^3\right)\omega_p}{4\xi_a\omega_a^4}]\}^{0.5} \quad (5.31)$$

where $\gamma = \frac{M_s}{M_p}$, $\omega_a = \frac{\omega_p+\omega_s}{2}$, $\xi_a = \frac{\xi_p+\xi_s}{2}$ and $\theta = \frac{\omega_p-\omega_s}{\omega_a}$. The Lognormal random variables are listed in Table 5.2.

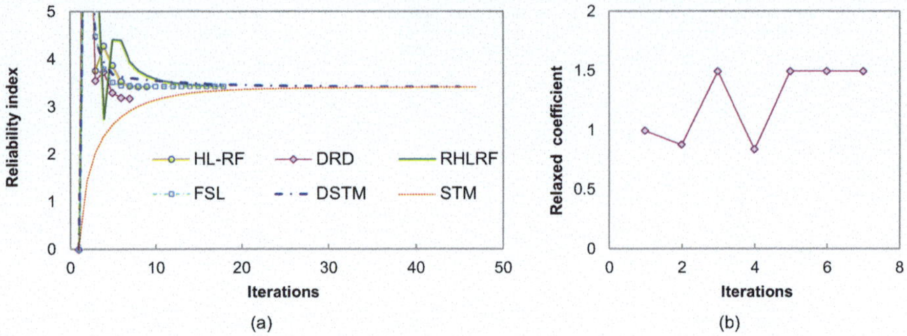

Fig. 5.27 **a** iteration for computing reliability index using different FORM formulas; **b** relaxed factor computed in DRD for Example 5.6

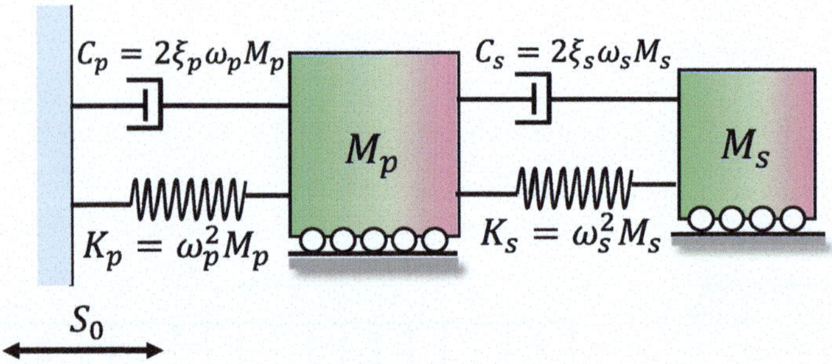

Fig. 5.28 Dynamic system applied in Example 5.7. Reprinted from [21], with permission from Elsevier

Table 5.2 Random variables for Example 5.7

Variable	Mean	STD	Variable	Mean	STD
M_P	1	0.1	ξ_P	0.05	0.02
M_S	0.01	0.001	ξ_s	0.02	0.01
K_p	1	0.2	S_0	100	10
K_S	0.01	0.002	F_s	15	0.15

Figure 5.29a demonstrates the iterations for reliability index and Fig. 5.29b shows the iterations for relaxed factor in DRD method. The DRD method is converged to $\beta = 2.0164$, by an iteration formula after 17 gradient evaluation that it related to 1100 evaluations of LSF. As seen from presented results in Fig. 5.29a, 2-periodic $\beta = [0.7733, 0.9396]$ is captured by HL-RF, while other methods are robust. The DRD is meaningfully faster than FSL and RHL-RF. The STM has last efficient among other, nevertheless, DRD performs highest albites among others. By computing an accelerated factor for adjusting SDSD which is determined in range from 1.5 to 0, the FORM-based SDSD can be enhanced for efficiency and robustness.

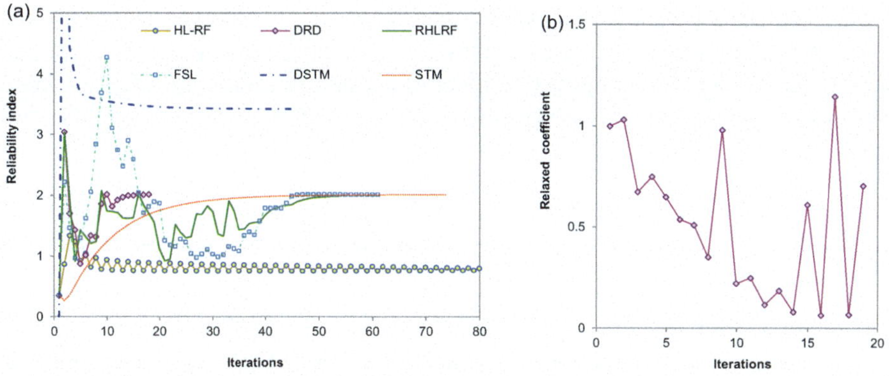

Fig. 5.29 **a** Iteration for computing reliability index using different FORM formulas; **b** relaxed factor computed in DRD for Example 5.7

Table 5.3 Reliability results of different methods for Examples 5.5, 5.6 and 5.7

Methods	Example 5.5		Example 5.6		Example 5.7	
	Iter/call	β	Iter/call	β	Iter/call	β
HL-RF	Not-converged		10/130	3.1805	Not-converged	
FSL	115/2258	2.2983	17/221	3.1807	60/3442	2.0169
STM	62/1253	2.2489	48/617	3.1785	74/5965	2.0156
DSTM	12/213	2.2983	45/558	3.1815	52/3343	2.0169
DRD	10/182	2.2982	7/90	3.1805	23/1443	2.0164
RHL-RF	17/321	2.2898	19/259	3.1805	55/3545	2.0165

The reliability results as iteration related to evaluating gradient vector, call function related to evaluating LSF and β for HL-RF, STM, FSL, RHL-RF, DSTM and DRD are listed in Table 5.3. Note that, DRD performed superior robustness and efficiency than other methods. The DSTM improved the efficiency of STM for Example 5.5 and 3 but by comparing iterations, it is inefficient in Examples 5.6 and 5.7. By comparing with the FSL, RHL-RF and STM, the RHL-RF shows better convergence ability. The DRD provides the stable reliability index which is computed by the lowest computational burden.

5.7 Discussion

The studied methods presented in this chapter named as FSL, DSTM, EDSTM, EFSL and DRD methods are compared for both efficiency and robustness through five structural/mechanical problems. In the studied FORM-based directional search direction, the different parameters are given for computing the line search as (1) FSL is computed MPP by initial step length equal to 100 i.e. $\lambda = 100$ and adjusted factor $c = 1.25$; (2) DSTM is searched the MPP based on the parameters of $\boldsymbol{C} = [\boldsymbol{I}]$ and $\lambda = 0.1$; (3) EDSTM is computed the reliability results as reliability index and MPP by using the parameters of $\boldsymbol{C} = [\boldsymbol{I}]$ and $\lambda_1 = 0.5$; the EFSL method is computed the reliability results by using the parameters of $\lambda_1 = 2^{15}$, $c = 2$ and $P = 0.05$; the new DRD method is computed the reliability results using a parameter of $\lambda_1 = 1$ for all studied examples in this section. A primary evaluating these methods is presented in Sects. 5.2–5.6. However, it is not given a suitable method for structural reliability analysis. Because, they have some advantages and disadvantages.

Example 5.8 Roof truss,

As shown in Fig. 5.30, a roof truss under uniform q load is given for the first example that top and bottom chords as compression and tension bars are respectively made by reinforced concrete and steel. The LSF is defined by displacement as presented in below relation,

$$g = 0.03 - \left(\frac{ql^2}{2}\right)\left(\frac{3.81}{AcEc} + \frac{1.13}{AsEs}\right) \tag{5.32}$$

Six normal variables are presented in Table 5.4 for roof truss in Example 5.8.

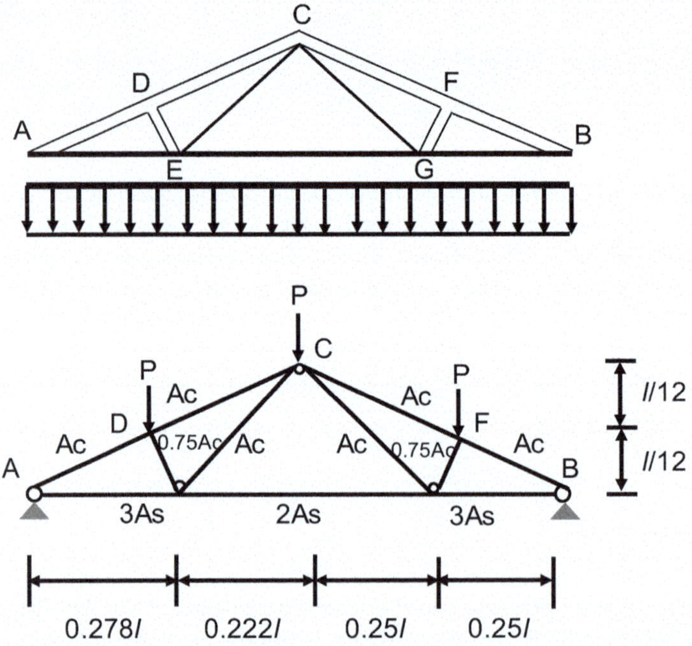

Fig. 5.30 A view for roof truss in Example 5.8. Reprinted from [22], with permission from Elsevier

Table 5.4 Variables for roof truss presented in Example 5.8

Variable	l(m)	q(N/m)	Ac(m^2)	As(m^2)	Ec(Pa)	Es(Pa)
Mean	12	20,000	0.04	9.82×10^{-4}	2×10^{10}	1×10^{11}
STD	0.12	1400	0.0048	5.9852×10^{-5}	1.2×10^{9}	6×10^{9}
Description	Length of truss	Uniform load	Cross-section of concrete	Cross-section of steel	Elastic modulus of concrete	Elastic modulus of steel

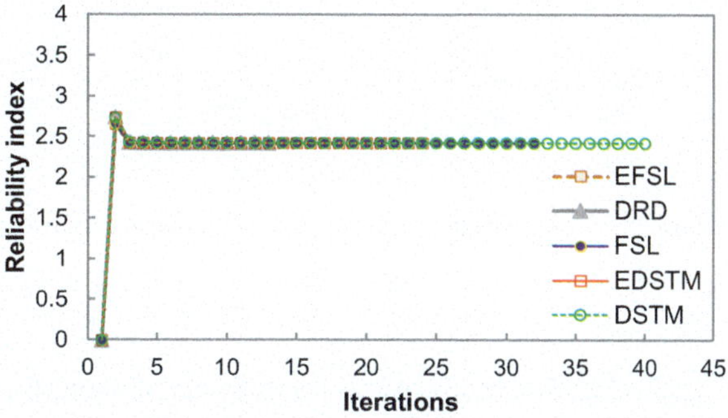

Fig. 5.31 Iterative of reliability index for different studied FORM-based directional line search for Example 5.8

The results of the reliability index obtained by different methods which are established based on the directional line search are presented in Fig. 5.31 for the Example 5.8. All methods are converged to the same reliability index of 2.4212. Therefore, the stability and globality of these methods are obtained for this problem, however, the efficiency of the studied methods depended on their line search as normalized search direction vector.

The convergence results include reliability index (β), failure probability (P_f), call function for determining the gradient vector of LSF computed by finite difference method (CF), number of iterations as an iterative procedure of methods (Iter) and the improved efficiency are given in Table 5.5 for roof truss problem. The improved efficiency factor (IEF) it determined as $IEF_i = \left(1 - \frac{\max(Iter)}{Iter_i}\right) \times 100$, where, $Iter_i$ is the number of iterations for i-the method and $\max(Iter)$ relates to maximum iteration for a method among other that they converged robustly. The IEF is used for a rank of these methods that the lowest gives the best rank as 1 and the highest value of IEF gives a rake of 5 for these five FORM methods. the Ranked number (RN) is also presented in Table 5.5. As can be seen, the robust results are obtained for this example using different methods. However, the computational burden for determining the MPP is different. The DSTM and FSL approaches are inefficient approaches among other methods. The DRD has remarkably improved the efficiency of reliability analysis for this problem. The DRD is converged about twice faster than EFSL and EDSTM. The EFSL and EDSTM have not significant difference in the iterations and it can be conducted that the EFSL is as efficient as the EDSTM but it is more inefficient than the DRD.

Example 5.9 A two-degree dynamic system,

As shown in Fig. 5.28, the LSF for a two-degree dynamic system is defined for force capacity (Fs) of system as below [13],

$$g = Fs - Ks \times 3(E[x_s^2])^{1/2} \tag{5.33}$$

where,

$$E[x_s^2] = \frac{\pi S_0}{4\xi_s\omega_s}\left[\frac{\xi_a\xi_s}{\xi_p\xi_s(4\xi_a^2 + \theta^2) + \gamma\xi_a^2} \times \frac{\left(\xi_p\omega_p^3 + \xi_s\omega_s^3\right)\omega_p}{4\xi_a\omega_a^4}\right] \tag{5.34}$$

where $\gamma = \frac{M_s}{M_p}$ (mass ratio); $\xi_a = \frac{\xi_p+\xi_s}{2}$ (average damping ratio of two system); $\omega_a = \frac{\omega_p+\omega_s}{2}$ (average frequency); $\theta = \frac{\omega_p-\omega_s}{\omega_a}$ (a tuning parameter). For this example, eight lognormal variables are considered with presented parameters in Table 5.6.

The reliability index results for different methods for a two-degree dynamic system are presented in Fig. 5.32. As seen, the EFSL is diverted for this example that it is related to computing step size when the sufficient decent condition is satisfied and a smaller value is computed for step size as $\lambda < 5$ thus the step size is increased about 20 times and at the next iteration it is loosed the suitable search direction vector given by the gradient vector that it is a property of this problem. Therefore, the EFSL method cannot find accurately the search direction for the MPP search at the final iterations. The FSL method is converged but it is inefficient. This problem has highly nonlinear properties thus fast adjusting to small step size (chaos control factor) is

Table 5.5 Reliability results for different studied FORM-based directional line search for Example 5.8

Methods	β	P_f	CF	Iter	IEF	RN
FSL	2.4217	0.007724	681	31	25.80	4
EFSL	2.4213	0.007733	505	23	69.56	3
DSTM	2.4212	0.007735	546	39	0	5
EDSTM	2.4212	0.007735	300	20	95	2
RDR	2.4212	0.007735	180	12	225	1

Table 5.6 Variables for Two-degree dynamic system in Example 5.9

Variable	M_s	M_p	K_s	K_p	ξ_s	ξ_p	F_s	S_0
Mean	0.01	1	0.01	1	0.02	0.05	15	100
STD	0.001	0.1	0.001	0.2	0.01	0.02	1.5	10

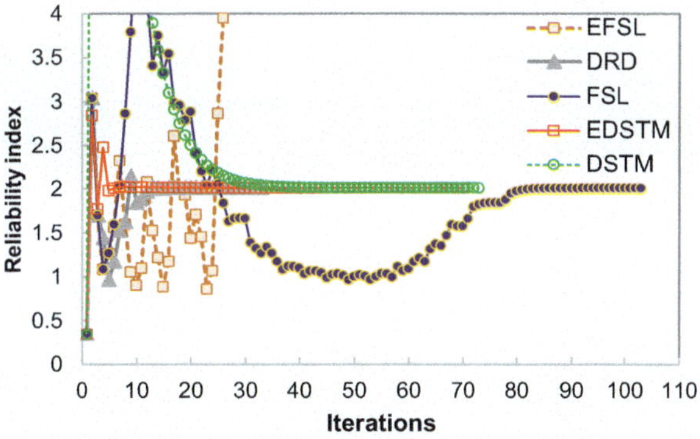

Fig. 5.32 Iterative of reliability index for different studied FORM-based directional line search for Example 5.9

an ability for the reliability method to provide stable results, more efficiently. The DSTM with constant step size also provided inefficient results compared to EDSTM. The DRD method is robust and efficient method that it is converged faster than others.

The reliability results for two-degree dynamic system problem are tabulated in Table 5.7. Note that, the FSL, DSTM, EDSTM, RDR provide stable results. The sufficient descent or angle conditions are applied in the FSL, EDSTM, DRD and EFSL but in the EFSL, the step length is increased for some conditions as 20 times when sufficient descent is satisfied and the $\lambda < 5$. For this nonlinear function, the finite step size should be selected a small value for controlling the stability of iteration. By increasing the step length as 20 times in EFSL, the EFSL method cannot find a suitable line search direction by increasing the step size and it is loosed stability for MPP search. The EDSTM method is efficient than the FSL and DSTM while it is converged twice faster than the FSL. The RDR is an efficient method and this method is remarkably improved the efficiency of the FORM-based directional line search for this example.

Example 5.10 A vehicle disc brake system,

The brake disc, friction materials and back plates is shown in Fig. 5.33, that as LSF based on the response surface function for friction loads of the disc system is

Table 5.7 Reliability results for different studied FORM-based directional line search for Example 5.9

Methods	β	P_f	CF	Iter	IEF	RN
FSL	2.0164	0.021879	5408	102	0	4
EFSL	Diverged		NON	NON	–	5
DSTM	2.0164	0.021879	4663	72	41.66667	3
EDSTM	2.0165	0.021874	4552	69	47.82609	2
RDR	2.0165	0.021874	2064	32	218.75	1

Fig. 5.33 A view for disc brake system applied in Example 5.10. Reprinted from [23], with permission from Elsevier

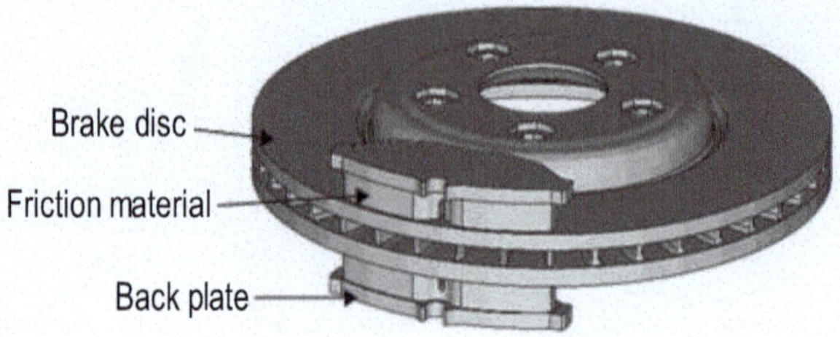

expressed as follow,

$$\begin{aligned} g = {} & 0.046287 + 0.20458\mu - 0.059821p - 0.00036549h_1 \\ & - 0.010037h_2 + 0.013836h_3 + 0.24308\mu p - 0.0037884\mu h_1 \\ & + 0.0023358\mu h_2 - 0.016981\mu h_3 + 0.029287ph_1 - 0.015872ph_2 \\ & - 0.0028333ph_3 + 0.0007175h_1h_2 - 0.00046158h_1h_3 \\ & - 0.00036486h_2h_3 - 0.39076\mu^2 - 0.015986p^2 - 0.0011936h_1^2 \\ & + 0.000269h_2^2 + 0.00062638h_3^2 - 0.01 \end{aligned} \quad (5.35)$$

The basic variables applied for computing the LSF of Eq. 5.35 are listed in Table 5.8.

The converged results of the different reliability methods are tabulated in Table 5.9. The reliability index for different methods-based directional line search approaches using the FORM formula is plotted in Fig. 5.34. As seen from the results presented in Table 5.9 and Fig. 5.34, all methods are robust but their efficiency is strongly depended on the line search formula for both the step length (chaos control factor) and the gradient vector of LSF (in FSL and EFSL) or steepest descent gradient vector (in DSTM, EDSTM, DRD). The step size is plied an effective rule for the efficiency of these methods. The EFSL is more robust than the DSTM and EDSTM but it is inefficient than the FSL. The dynamical adjusting procedure for DSTM is strongly improved the efficiency of FORM formulation in DRD method. This example shows that the moderate nonlinear function and convex problem provide some issues for convergence of the FORM which are established based on DSTM formulation as seen

Table 5.8 Variables for disc brake system applied in Example 5.10

Variable	p (MPa)	μ	h_1 (mm)	h_2 (mm)	h_3 (mm)
Mean	0.5	0.4	15	20	10
CoV	0.25	0.35	0.1	0.1	0.1
PDF	Lognormal	Normal	Gumbel	Gumbel	Gumbel
Description	Brake pressure	Friction coefficient	Friction material thickness	Disc thickness	Back plate thickness

Table 5.9 Reliability results for different studied FORM-based directional line search for Example 5.10

Methods	β	P_f	CF	Iter	IEF	RN
FSL	2.1444	0.016	209	19	1063.15	2
EFSL	2.1444	0.016	286	26	750	3
DSTM	2.1444	0.016	2658	221	0	5
EDSTM	2.1444	0.016	948	79	179.74	4
RDR	2.1444	0.016	168	14	1478.57	1

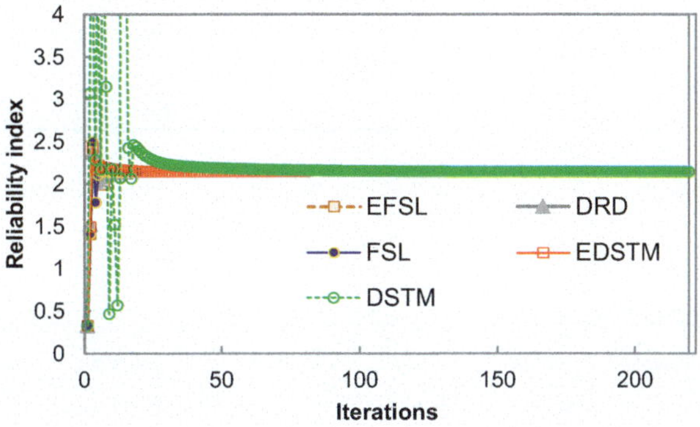

Fig. 5.34 Iterative of reliability index for different studied FORM-based directional line search for Example 5.10

in DSTM and EDSTM. The FSL and EFSL respectively improved the efficiency of the DSTM about 10 and 8 times while the EDSTM improved the efficiency about twice. The dynamical search direction method by using the relaxed factor more than 1 provided a high-convergence rate compared to other methods. By using DRD, a step size more than 1 is computed for this example thus we have a superior convergency.

Example 5.11 A steel frame structure,

A frame as shown in Fig. 5.35 with span L and height H under a lateral load P, and distributed load W is given for computing the LSF for the maximum lateral load that LSF-based displacement as shown in Fig. 5.35 is defined as [11],

$$g = 0.05 - \Delta \tag{5.36}$$

The lateral displacement under external loads is given as LSF and its statistical properties of basic random variables are tabulated in Table 5.10.

The results of different methods extended by the directional line search are plotted for reliability index in Fig. 5.36 and are tabulated in Table 5.11. Note that, the reliability method-based DSTM provides unstable results as chaotic solutions but other methods are robust. The EDSTM provides a global convergence with stable results but the computational burden of this method is remarkably more than the DRD and EFSL. The EFSL has superior convergence performance for this problem.

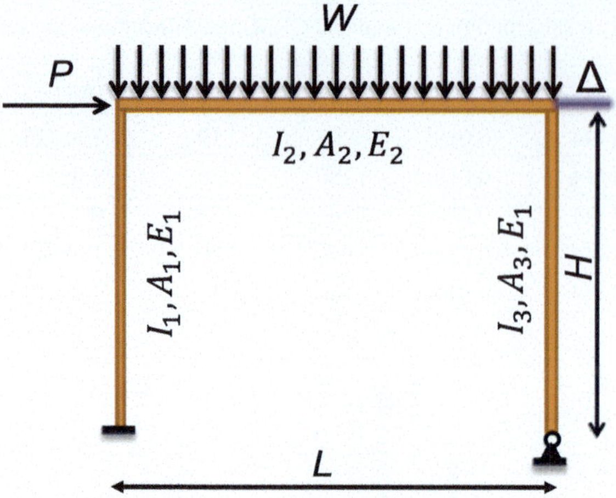

Fig. 5.35 A view of steel frame structure applied in Example 5.11. Reprinted from [13], with permission from Elsevier

Table 5.10 Variables for steel frame example applied in Example 5.11

Variable	Mean	STD	PDF	Variable	Mean	STD	PDF
A_1(m^2)	91×10^{-4}	9.1×10^{-4}	Lognormal	E_1 (kg/m^2)	2.38×10^{10}	1.9×10^9	Normal
I_1(m^4)	80.9×10^{-6}	8.09×10^{-6}	Lognormal	E_2 (kg/m^2)	2.17×10^{10}	1.74×10^9	Normal
A_2(m^2)	53.8×10^{-4}	5.38×10^{-4}	Lognormal	L (m)	6	0.48	Normal
I_2(m^4)	83.6×10^{-6}	8.36×10^{-6}	Lognormal	H (m)	4.5	0.36	Normal
A_3(m^2)	106×10^{-4}	10.6×10^{-4}	Lognormal	W (kg/m)	4500	630	Gumbel
I_3(m^4)	112.6×10^{-6}	11.26×10^{-6}	Lognormal	P (kg)	2000	350	Gumbel

It can be concluded that the EFSL method provides the suitable line search which is provided robust results with efficient computational burden iterations for moderately nonlinear performance functions.

The DRD and EFSL improved the computational burden of EDSTM about three times and five times, respectively. The EDSTM has also shown some unstable iterations but it is converged at the final iteration for checking the angle condition. The application of DSTM is simpler than the EFSL method for computing the step size but it cannot provide low compactional burden results for convex problems or moderate problems as well as this problem.

Example 5.12 A space dame truss.

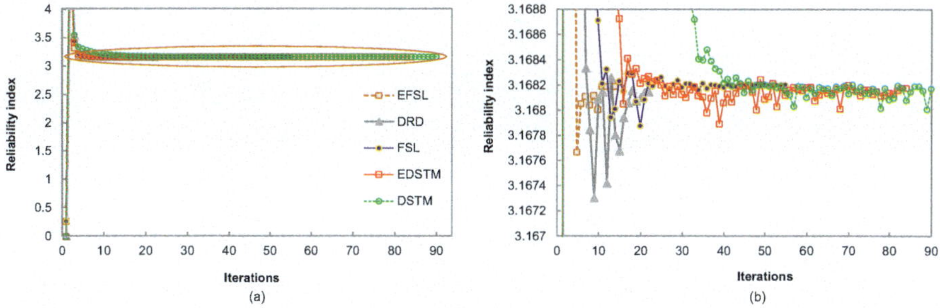

Fig. 5.36 Iterative of reliability index for different studied FORM-based directional line search for Example 5.11 **a** methods comparison; **b** details

Table 5.11 Reliability results for different studied FORM-based directional line search for Example 5.11

Methods	β	P_f	CF	Iter	IEF	RN
FSL	3.1682	0.000767	4430	54	53.70	3
EFSL	3.1682	0.000767	1066	13	538.46	1
DSTM	Chaotic		Not-converged		NON	5
EDSTM	3.1682	0.000767	6812	83	0	4
RDR	3.1681	0.000767	1721	21	295.23	2

As space 24-truss element presented in Fig. 5.37 is taken for this example. The implicit LSF is defined in Fig. 5.37 which is considered for evaluating maximum z-direction displacement (Δ_{P1}^{z}),

$$g = 0.01 - \left|\Delta_{P1}^{z}\right| \tag{5.37}$$

As seen from Fig. 5.37, the top point is subjected to load P_1 while second level points have constrained load P_2-P_7. The LSF is computed by a linear finite element model by using 32 variables presented in Table 5.12.

The reliability results for the space truss problem are listed in Table 5.12 and the reliability index given by the different directional reliability methods is shown in Fig. 5.38. The reliability index extracted from Ref. [16] is obtained as $\beta = 1.66955$. as seen from the results presented in Table 5.12, the reliability index for robust method is closely agreement with the results given form the literature.

By comparing the results in Table 5.13 and Fig. 5.38, the DSTM provided unstable results but the stability of the DSTM is enhanced by using angle conditions for adjusting the chaos control factor in the EDSTM. Inefficient results are given by the EDSTM and the FSL approaches but the EFSL strongly improved the efficiency of the FSL for this problem. As seen, the DRD and EFSL yield efficient results compared to EDSTM and FSL.

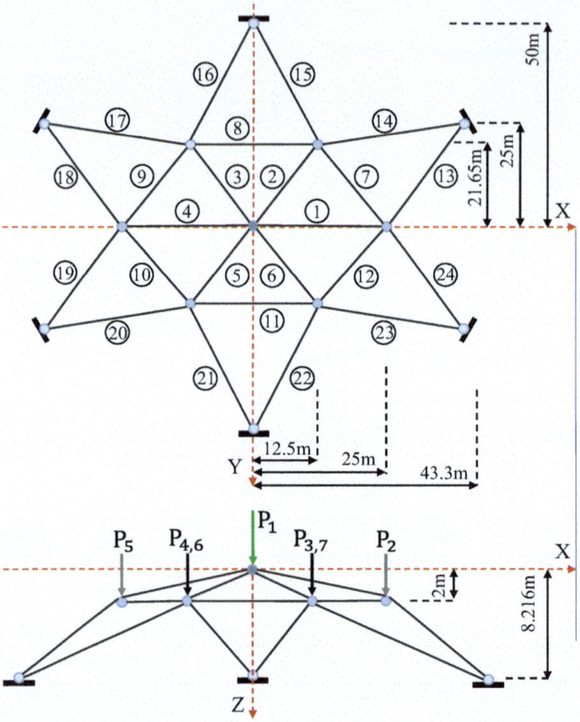

Fig. 5.37 A view of space truss structure for Example 5.12. Reprinted from [24], with permission from Elsevier

Table 5.12 Variables for space truss structure applied in Example 5.12

Variable	$A_1 - A_6$ (m^2)	$A_7 - A_{12}$ (m^2)	$A_{13} - A_{24}$ (m^2)	E(Gpa)	P_1(kN)	$P_2 - P_7$ (kN)
Mean	0.013	0.01	0.016	205	20	10
CoV	0.1	0.1	0.1	0.12	0.15	0.12
PDF	Normal	Normal	Normal	Normal	Gumbel	Gumbel
Description	Cross-sectional bars			Young's modulus	Load	Load

Table 5.13 Reliability results for different studied FORM-based directional line search for Example 5.12

Methods	β	P_f	CF	Iter	IEF	RN
FSL	1.6492	0.049553	5344	50	0	4
EFSL	1.649	0.049574	2027	19	163.15	1
DSTM	Chaotic		Not-converged		NON	5
EDSTM	1.6656	0.047897	2210	34	47.05	3
RDR	1.6492	0.049553	1560	24	108.33	2

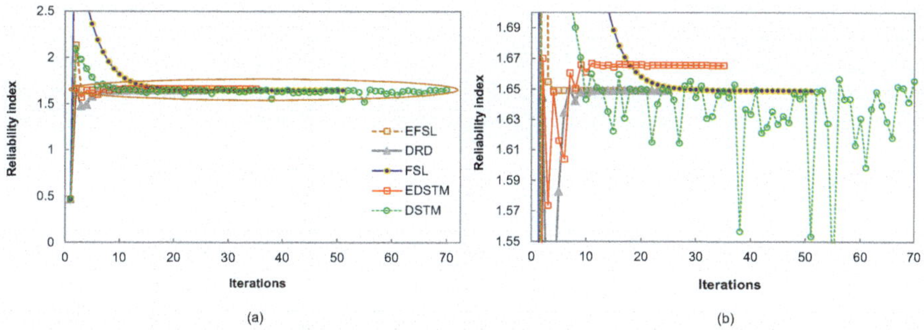

Fig. 5.38 Iterative of reliability index for different studied FORM-based directional line search for Example 5.12 **a** methods comparison; **b** details

This problem is also a moderately nonlinear problem with convex property thus the EFSL can adjust the step length with acceptable formulation. A large value of step length at the first iteration is pushed to iteration process to move as well as the steepest descent search direction in EFSL while at the final iteration, it is computed a small step size by adjusting formulation of EFSL and the convergence is obtained by robust conditions and efficient computational burden for EFSL method.

Finally, we introduce several good methods by presenting results in this section. It can be given the robust results by using the EDSTM, DRD and FSL approaches. Thus, for highly nonlinear problems, the EDSTM, DRD and FSL are suitable for computing the reliability index and by using the angle and sufficient descent conditions, these methods can provide stable results for nonlinear and high-nonlinear problems.

By comparing the efficiency, the EDSTM and DSTM are the acceptable methods for highly nonlinear problems but for nonlinear problems they are more inefficient than the FSL method. The DRD method by using a dynamical step size which is computed between 1.5 and 0 can be strongly improved the efficiency of the DSTM and EDSTM. For linear problems, the FSL and EFSL are the acceptable methods as well as the DRD.

For convex problems, the EDST and DSTM provide inefficient results but the robustness of the method is depended on the nonlinearity degree of the limit sate function. The FSL and EFSL as well as the DRD by following the steepest descent search direction provided efficient results for convex problems compared to EDSTM and DSTM which are computed using the small chaos control factors.

The ranked levels of different models given by average value of RN for five examples are presented in Fig. 5.39. This figure can be helped to the best view about selecting a method. As seen by comparing the efficiency, the methods are ranked from best to lowest as DRD, EFSL, EDSTM, FSL and DSTM. The EDSTM is followed to rank of the FSL while the DRD and EFSL are better than EDSTM and FSL. Lowest performers are presented by DSTM. By comparing the efficiency and the robustness, the methods are ranked as RDR, EDSTM, EFSL, FSL and DSTM. Thus, the DRD showed the superior abilities than others.

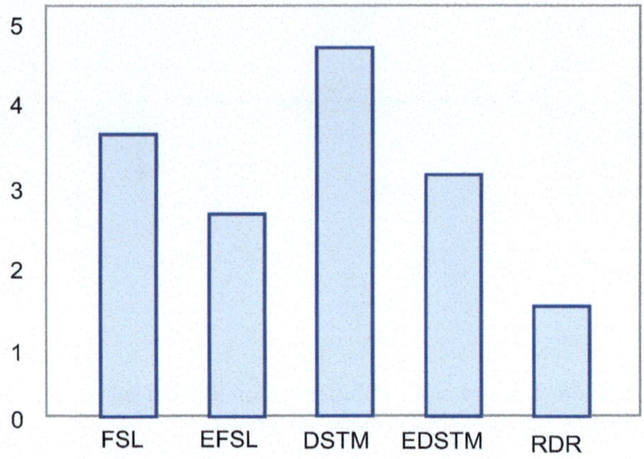

Fig. 5.39 Ranked method by efficiency for different FORM (directional steepest descent methods)

5.8 Summary

In this chapter, another type of FORM-based steepest descent condition is developed as FORM formulated by the directional steepest decent sensitivity vector. In addition, several novel formulations computed by gradient method are presented that the lines search is computed based on a normalized selectivity vector which is computed by a direction FORM. The directional steepest decent sensitivity vector is determined by using gradient vector and the previous point. The main format of this approach is developed by the finite step length method. And then, several approaches including (i) directional stability transformation method, (ii) adaptive directional stability transformation method, (iii) enhanced finite step length method, (iv) dynamical relaxed directional method have been introduced for determining the MPP. Finally, several examples are utilized for comparing the abilities of the directional steepest decent methods.

References

1. C. Jiang, S. Han, M. Ji, X. Han, A new method to solve the structural reliability index based on homotopy analysis. Acta Mech. **226**(4), 1067–1083 (2015)
2. B. Keshtegar, Enriched FR conjugate search directions for robust and efficient structural reliability analysis. Eng. Comput. **34**, 117–128 (2018)
3. B. Keshtegar, D. Meng, M.E.A. Ben Seghier, M. Xiao, N.T. Trung, D.T. Bui, A hybrid sufficient performance measure approach to improve robustness and efficiency of reliability-based design optimization. Eng. Comput. **37**, 1695–1708 (2021)
4. Z. Meng, Y. Pu, H. Zhou, Adaptive stability transformation method of chaos control for first order reliability method. Eng. Comput. **34**(4), 671–683 (2018)
5. M. Bagheri, M. Miri, N. Shabakhty, Fuzzy reliability analysis using a new alpha level set optimization approach based on particle swarm optimization. J. Intell. Fuzzy Syst. **30**(1), 235–244 (2016)

6. M.E.A.B. Seghier, B. Keshtegar, B. Elahmoune, Reliability analysis of low, mid and high-grade strength corroded pipes based on plastic flow theory using adaptive nonlinear conjugate map. Eng. Fail. Anal. **90**, 245–261 (2018)
7. B. Keshtegar, P. Hao, A hybrid self-adjusted mean value method for reliability-based design optimization using sufficient descent condition. Appl. Math. Model. **41**, 257–270 (2017)
8. Z. Meng, G. Li, D. Yang, L. Zhan, A new directional stability transformation method of chaos control for first order reliability analysis. Struct. Multidiscip. Optim. **55**, 601–612 (2017)
9. J.X. Gong, P. Yi, A robust iterative algorithm for structural reliability analysis. Struct. Multidiscip. Optim. **43**, 519–527 (2011)
10. W. Zhao, Y. Chen, J. Liu, An effective first order reliability method based on Barzilai-Borwein step. Appl. Math. Model. **77**, 1545–1563 (2020)
11. B. Keshtegar, Stability iterative method for structural reliability analysis using a chaotic conjugate map. Nonlinear Dyn. **84**, 2161–2174 (2016)
12. H. Zhao, Z. Yue, Y. Liu, Z. Gao, Y. Zhang, An efficient reliability method combining adaptive importance sampling and Kriging metamodel. Appl. Math. Model. **39**(7), 1853–1866 (2015)
13. B. Keshtegar, Chaotic conjugate stability transformation method for structural reliability analysis. Comput. Methods Appl. Mech. Eng. **310**, 866–885 (2016)
14. D. Zhang, J. Zhang, M. Yang, R. Wang, Z. Wu, An enhanced finite step length method for structural reliability analysis and reliability-based design optimization. Struct. Multidiscip. Optim. **65**(8), 231 (2022)
15. M. Bagheri, S.A. Hosseini, B. Keshtegar, Dynamical relaxed directional method for fuzzy reliability analysis. Structures **34**, 169–179 (2021)
16. B. Keshtegar, S.P. Zhu, Three-term conjugate approach for structural reliability analysis. Appl. Math. Model. **76**, 428–442 (2019)
17. B. Keshtegar, M. Bagheri, Fuzzy relaxed-finite step size method to enhance the instability of the fuzzy first-order reliability method using conjugate discrete map. Nonlinear Dyn. **91**, 1443–1459 (2018)
18. D. Yang, Chaos control for numerical instability of first order reliability method. Commun. Nonlinear Sci. Numer. Simul. **15**(10), 3131–3141 (2010)
19. B. Keshtegar, S. Chakraborty, An efficient-robust structural reliability method by adaptive finite-step length based on Armijo line search. Reliab. Eng. Syst. Saf. **172**, 195–206 (2018)
20. B. Keshtegar, Z. Meng, A hybrid relaxed first-order reliability method for efficient structural reliability analysis. Struct. Saf. **66**, 84–93 (2017)
21. C. Luo, S.P. Zhu, B. Keshtegar, X. Niu, O. Taylan, An enhanced uniform simulation approach coupled with SVR for efficient structural reliability analysis. Reliab. Eng. Syst. Saf. **237**, 109377 (2023)
22. C. Luo, S.P. Zhu, B. Keshtegar, W. Macek, R. Branco, D. Meng, Active Kriging-based conjugate first-order reliability method for highly efficient structural reliability analysis using resample strategy. Comput. Methods Appl. Mech. Eng. **423**, 116863 (2024)
23. B. Xia, H. Lü, D. Yu, C. Jiang, Reliability-based design optimization of structural systems under hybrid probabilistic and interval model. Comput. Struct. **160**, 126–134 (2015)
24. B. Keshtegar, A hybrid conjugate finite-step length method for robust and efficient reliability analysis. Appl. Math. Model. **45**, 226–237 (2017)

Chapter 6
Directional Conjugate Gradient Approaches for FORM

Abstract In this chapter, based on the basic formulation of conjugate FORM-based FR, several improving first order reliability methods formulated by conjugate search direction vectors are established including (i) limited FR, (ii) Enhanced FR-based variant type in gradient vector, (iii) Chaotic conjugate chaos control method, (iv) chaotic conjugate stability transformation method. The step size is adaptively adjusted based on the sufficient decent condition and the Armijo-type condition in these formulations. These formulations are compared with step size and without step size for several engineering problem.

Keywords Conjugate FORM · Enhanced FR-based variant type · Conjugate search direction vector · Armijo-type condition

6.1 Introduction

The basic first order reliability method (FORM) formulas have been extended by steepest descent search direction (SDSD) as presented in Chap. 1 in which, the line search direction is determined using gradient vector. The conjugate gradient (CG) is an optimization method that can be used for unconstrained optimization problems. The CG method has shown the best characteristics as low memory requirements, strong to find local optimum and fast global convergence. In the CG method, the search direction vector (SDV) is computed based on information of the new gradient vector and the conjugate SDV obtained at previous iterations. Firstly, the CG method was established by Hestenes et al. step solving linear algebraic systems and then this method is applied to find the optimum quadratic finite dimensional functions and non-quadratic functions. The CG methods can be applied to the nonlinear unconstrained optimization problem given as below relation,

$$\min_{U^* \in R^n} f\left(\boldsymbol{U}^*\right) \tag{6.1}$$

S.-P. Zhu and B. Keshtegar, *Structural Reliability Analysis*, Structural Integrity 31,
https://doi.org/10.1007/978-3-031-75197-4_6

where, $f{:}R^n \rightarrow R$ is a real-continuous differentiable function thus we have a gradient vector of function f as $\nabla f(\boldsymbol{U}^*)$. The main effort of the above relation is to find X^* as the optimum point which provides a minimum result for function $f(x)$. The CG method as a continuous optimization method is started by an initial point i.e. x_0 for searching the optimum based on the generation of points by following relation,

$$\boldsymbol{U}_{k+1}^* = \boldsymbol{U}_k^* + \alpha_k \boldsymbol{d}_k \tag{6.2}$$

where α_k denotes the step size which is computed by a line search condition, and d_k is the CG vector which is updated as follows,

$$\boldsymbol{d}_k = \begin{cases} -\nabla f\left(\boldsymbol{U}_k^*\right) & k = 0 \\ -\nabla f\left(\boldsymbol{U}_k^*\right) + \theta_k \boldsymbol{d}_{k-1} & k \geq 1 \end{cases} \tag{6.3}$$

where, θ_k relates to the CG factor that it is shown the effects of the previous conjugate line search and different methods of CG are established based on definition of this factor. The search-based iteration of gradient and CG methods for finding the optimum point is schematically presented in Fig. 6.1 As seen from this figure, the CG method applied the previous line search for computing the new line search based on the factor θ_k. It can be used as a novel method in the FORM-based iterations. However, the formulation of this method for FORM is very difficult because we have a constrained probabilistic model which is used as an optimization problem of the FORM formulation.

Keshtegar and Miri [1] formulated the CG method given by Conjugate-Descent (CD) factors of CD and Liu–Storey (LS) applied on structural reliability analysis (SRA) which is determined by below unconstraint probabilistic optimization model,

$$f(\boldsymbol{U}) = \frac{\boldsymbol{U}^T \cdot \boldsymbol{U}}{2} + \lambda g(\boldsymbol{U}) \tag{6.4}$$

where $f(\boldsymbol{U})$ is the unconstrained function applied for optimization in reliability analysis and λ is the penalty coefficient. The penalty coefficient should be given $\boldsymbol{U} + \lambda g(\boldsymbol{U}) = 0$ to provide $g(\boldsymbol{U}) = 0$. Therefore, penalty coefficient is determined as below,

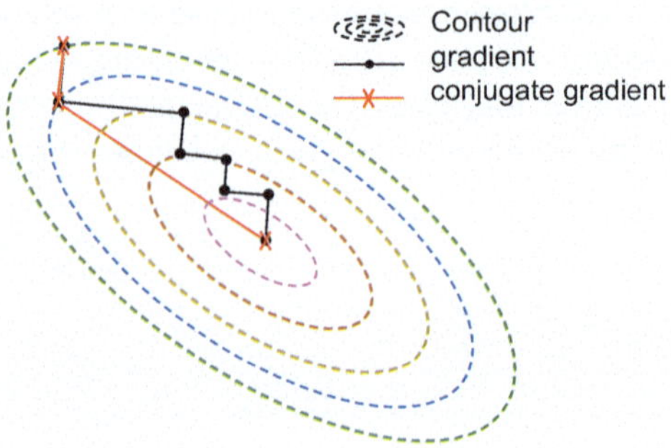

Fig. 6.1 Comparing the gradient and CG search procedure for funding the optimum results

$$\lambda = -\frac{\nabla^T g(\boldsymbol{U}) \cdot \boldsymbol{U}}{\nabla^T g(\boldsymbol{U}) \nabla g(\boldsymbol{U})} \tag{6.5}$$

Thus, the gradient vector for the optimization probabilistic model of FORM is computed as below,

$$\nabla f(\boldsymbol{U}_k^*) = \boldsymbol{U}_k^* - \frac{\nabla^T g(\boldsymbol{U}_k^*)\boldsymbol{U}_k^* - g(\boldsymbol{U}_k^*)}{\nabla^T g(\boldsymbol{U}_k^*)\nabla g(\boldsymbol{U}_k^*)} \nabla g(\boldsymbol{U}_k^*) \tag{6.6}$$

The step size in the CG is an effective factor to provide stable results. It is impossible to develop a practical computational framework to find the suitable values for step size. However, the step size can be approximated based on line search rules including Armijo, Strong Wolfe, and Wolfe Rules. In the SRA method, the CG search direction is computed based on an exact step size given by the Wolfe line search [1]. In Wolfe line search conditions, the step size as a positive value should be satisfied in the below relations,

$$f(\boldsymbol{U}_k^* + \alpha_k \boldsymbol{d}_k) - f(\boldsymbol{U}_k^*) \leq \delta \alpha_k \nabla^T f(\boldsymbol{U}_k^*) \boldsymbol{d}_k \tag{6.7}$$

$$\nabla^T f(\boldsymbol{U}_{k+1}^*) \boldsymbol{d}_k \geq \sigma \nabla^T f(\boldsymbol{U}_k^*) \boldsymbol{d}_k \tag{6.8}$$

The giving step size based on the above relation is time consuming and the direct application of the CG line search may be difficult for the iterative procedure of FORM. On the other hand, for the convex reliability problem, the CG-based FORM named (CFORM) may provide very slowly convergent results. On the other hand, the exact step size is difficult for these conditions but it can be checked the angle or sufficient descent conditions for providing stable results using the CG method. This is a challenge for future research works in the numerical methods developed for structural reliability analysis. Consequently, due to direct considering the gradient vector of limit state function (LSF), the direction finite step size is an acceptable line search for application of the CG families.

In this chapter, several CG methods that are formulated in the structural reliability analysis are presented. All the CFORM methods are formulated based on the directional conjugate line search computed by finite step size. Therefore, the computing process for determining step size in CG line search is removed from the formulation procedure. This means that a simple way is found to apply the CG line search by transferring to directional line search. Finally, applications of the conjugate finite step length methods are presented for fuzzy reliability analysis and reliability analysis of corroded pipeline.

6.2 Conjugate HL-RF Method

In FORM, the main effort is to find the most probable point (MPP. $\boldsymbol{U}^*$) which is defined as related to a point in normal standard space on LSF with the minimum distance to origin. Thus, we have $\beta = \|\boldsymbol{U}^*\|$ and $\mathrm{P}_f \approx \Phi(-\beta) = \Phi(-\|\boldsymbol{U}^*\|)$. The MPP search is vital important in SRA-based FORM. As known, the Hasofer-Lind and Rackwitz-Flessler (HL-RF) method can be used to obtain the MPP by the below main steps:

Step 1: Define LSF, basic variables and their information and set initial parameters of FORM;

Step 2: Transform variables into $\boldsymbol{U}$-space as,

$$u = \frac{x - \mu_x^e}{\sigma_x^e} \tag{6.9}$$

where, μ_x^e and σ_x^e are equivalent mean and standard deviation, respectively. μ_x^e and σ_x^e is obtained as [2],

$$\sigma_x^e = \frac{1}{f_X(x)}\varphi\left[\Phi^{-1}\{F_X(x)\}\right] \tag{6.10}$$

$$\mu_x^e = x - \sigma_x^e \Phi^{-1}[F_X(x)] \tag{6.11}$$

where, $F_X(x)$ is the cumulative distribution function (CDF), $f_X(x)$ is the probability density function (PDF), Φ^{-1} denotes inverse standard normal CDF and φ denotes standard normal PDF;

Step 3: Iterate the HL-RF for determining the reliability index as,

$$\mathbf{U}_{k+1} = \beta_{k+1}\boldsymbol{\alpha}_{k+1} \tag{6.12}$$

where,

$$\boldsymbol{\alpha}_{k+1} = -\frac{\nabla^T g(\boldsymbol{U}_k)}{\|\nabla^T g(\boldsymbol{U}_k)\|} \tag{6.13}$$

$$\beta_{k+1} = \frac{g(\boldsymbol{U}_k) - \nabla^T g(\boldsymbol{U}_k)\boldsymbol{U}_k}{\|\nabla^T g(\boldsymbol{U}_k)\|} \tag{6.14}$$

where the gradient vector of LSF is shown as $\nabla g(\boldsymbol{U}) = [\partial g/\partial u_1, \partial g/\partial u_2, \ldots, \partial g/\partial u_n]^T$;

Step 4: Check the convergence if given it, then extract the reliability index and compute $\mathrm{P}_f \approx \Phi(-\beta)$.

The basic iterative formulation was parameters for FORM using the directional of the CG line search direction named conjugate HL-RF (CHL-RF) algorithm. They estimated the MPP by a conjugated normalize search direction as below,

$$\boldsymbol{U}_{k+1} = \beta_{k+1}\boldsymbol{\alpha}_{k+1} \tag{6.15}$$

where, $\boldsymbol{\alpha}_{k+1}^{\lambda}$ is the conjugate normalized search direction or conjugate sensitivity vector of random variables computed by following relation,

$$\boldsymbol{\alpha}_{k+1} = \frac{\boldsymbol{U}_{k+1}^{c\lambda}}{\left\|\boldsymbol{U}_{k+1}^{c\lambda}\right\|} \tag{6.16}$$

in which, point $\boldsymbol{U}_{k+1}^{C\lambda}$ is a point alone the conjugate line search direction as demonstrated in Fig. 6.2. $\boldsymbol{U}_{k+1}^{C\lambda}$ is determined as below,

$$\boldsymbol{U}_{k+1}^{C\lambda} = \boldsymbol{U}_k + \lambda \boldsymbol{d}_k \tag{6.17}$$

where $\lambda > 0$ is the step length and d_k is CG vector of the LSF at k_{th} iteration, that d_k is given by following relation,

$$\boldsymbol{d}_k = -\nabla g(\boldsymbol{U}_k) - \frac{||\nabla g(\boldsymbol{U}_k)||^2}{\|\nabla g(\boldsymbol{U}_{k-1})\|^2}\boldsymbol{d}_{k-1} \tag{6.18}$$

The reliability index in Eq. (6.15) is given as follows,

$$\beta_{k+1} = \frac{g(\boldsymbol{U}_k) - \nabla^T g(\boldsymbol{U}_k)\boldsymbol{U}_k}{\nabla^T g(\boldsymbol{U}_k)\boldsymbol{\alpha}_{k+1}} \tag{6.19}$$

As seen from Fig. 6.2, the finite step length is applied as a scale factor for the conjugate line search which is the scalar of gradient and previous CG vectors. The new CG vector is not along the gradient vector direction; thus, the conjugated normalized

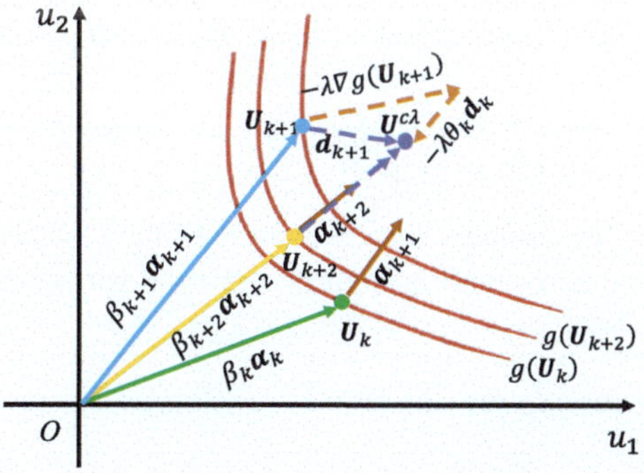

Fig. 6.2 CHL-RF scheme with conjugate normalized sensitivity vector

sensitivity vector (NSV) is different with the previous NSV which is obtained by with the HL-RF normalized search direction when steepest descent for $\lambda \to \infty$ and $\boldsymbol{d}_k = 0$. This means that $\boldsymbol{\alpha}_{k+1}^{\lambda} \neq \boldsymbol{\alpha}_{k+1}$ as schematically shown in Fig. 6.2.

It can be concluded that, $\boldsymbol{U}_{k+1}^{C\lambda} = \boldsymbol{U}_k$ when $\lambda = 0$. This means that the point $\boldsymbol{U}_{k+1}^{C\lambda}$ provides a fixed vector related to previous thus the iterative CHL-RF formulation is stopped or may be provided results with slowly convergence. For overcome to computational burden issue, it is given larger step length as $\lambda > 0$. On the other hand, for controlling the instability of the CHL-RF algorithm, it is supposed that λ satisfies sufficient descent conditions, i.e. $\nabla^T g(\boldsymbol{U}_k)\boldsymbol{d}_k < -\|\nabla g(\boldsymbol{U}_k)\|^2$, thus we have an adjustment for step length as below,

$$\lambda = \begin{cases} \lambda & \|\boldsymbol{U}_{k+2} - \boldsymbol{U}_{k+1}\| > \|\boldsymbol{U}_{k+1} - \boldsymbol{U}_k\| \\ c_1\lambda & \text{otherwise} \end{cases} \tag{6.20}$$

By using the above relation, the steps for computing the reliability index using CHL-RF applied in a computer program are presented as below:

Step 1: Define $g(\boldsymbol{U}) = 0$, give random variables as information of μ and σ and set $c_1 \in [0.75, 0.95]$ and ε (stopping criterion) < *0; Let* $k = 0$ and initial values as $\boldsymbol{X}_0 = \mu$, $d_0 = 0$ and $\lambda_0 \in [5, 500]$;

Step 2: Transform variables into $\boldsymbol{U}$-space as well as HL-RF method;

Step 3: Iterative process of CHL-RF by the following relations;

(i) Determine the gradient vector $\nabla g(\boldsymbol{U})$ and $g(\boldsymbol{U})$;
(ii) Compute $\boldsymbol{d}_k = -\nabla g(\boldsymbol{U}_k) - \frac{\|\nabla g(\boldsymbol{U}_k)\|^2}{\|\nabla g(\boldsymbol{U}_{k-1})\|^2}\boldsymbol{d}_{k-1}$;
(iii) Determine $\boldsymbol{U}_{k+1}^{C\lambda} = \boldsymbol{U}_k + \lambda\boldsymbol{d}_k$ and then $\boldsymbol{\alpha}_{k+1}^{\lambda} = \frac{\boldsymbol{U}_{k+1}^{C\lambda}}{\|\boldsymbol{U}_{k+1}^{C\lambda}\|}$;
(iv) Determine reliability index $\beta_{k+1} = \frac{g(\boldsymbol{U}_k) - \nabla^T g(\boldsymbol{U}_k)\boldsymbol{U}_k}{\nabla^T g(\boldsymbol{U}_k)\boldsymbol{\alpha}_{k+1}^{\lambda}}$;
(v) Compute the new point by $\boldsymbol{U}_{k+1} = \beta_{k+1}\boldsymbol{\alpha}_{k+1}^{\lambda}$;
(vi) Adjust step length as $\lambda = \begin{cases} \lambda & \|\boldsymbol{U}_{k+2} - \boldsymbol{U}_{k+1}\| > \|\boldsymbol{U}_{k+1} - \boldsymbol{U}_k\| \\ c_1\lambda & \text{otherwise} \end{cases}$;

Let the next iterate using $X_{k+1} = \mu_X^e + \sigma_X^e U_{k+1}$ and $k = k + 1$;

Step 4: If $\frac{\|\boldsymbol{U}_k - \boldsymbol{U}_{k+1}\|}{\|\boldsymbol{U}_k\|} < \varepsilon$, then presents results as $P_f \approx \Phi(-\beta))$ and $\boldsymbol{\alpha} = \boldsymbol{\alpha}_{k+1}^{\lambda}$, else Go to Step 2.

To compare the robustness of the CHL-RF algorithm, three nonlinear mathematical LSFs with normal/non-normal variables are given as Examples 6.1, 6.2, and 6.3 where $\varepsilon=10^{-6}$, $c = 0.9$ and $\lambda = 100$.

Example 6.1 the LSF is considered as,

$$g(\mathrm{X}) = x_1^3 + x_1^2 x_2 + x_2^3 - 18 \tag{6.21}$$

where, normal x_1 and x_2 variables are defined by $\mu_1 = 10, \mu_2 = 9.9$ and $\sigma_1 = \sigma_2 = 5$. As extracted results from the literature, reliability index obtained by Keshtegar and Miri [1] is 2.29825 with related MPP of $\boldsymbol{X}^* = (1.6857, 1.9681)$. this problem is solved by HL-RF and CHL-RF that the reliability index of these methods is presented in Fig. 6.3. Note that the CHL-RF is robustly converged with results of $\beta = 2.2983$ obtained after 12 iterations with related MPP in original space as $\boldsymbol{X}^* = [1.6855, 1.9679]$ and MPP in normal standard space as $\boldsymbol{U}^* = [-1.6629, -1.5864]$. However, the HL-RF provided the unstable results as 2-periodic solution as $\beta = [1.5355, 1.5370]$. The CHL-RF is a robust formulation method but its efficiency is strongly improved for this nonlinear problem. The CHL-RF with directional formulation can provide efficient and robust results for concave problems.

The iterations for MPP search of HL-RF and CHL-RF methods are presented in Fig. 6.4 for Example 6.1 in normal standard space, As seen, the CHL-RF provides the smooth curve for this problem on a linear form without any chaos. The CHL-RF holds sufficient descent conditions for this problem but almost iterations is near the MPP to provide stabilization as globality.

Example 6.2 A non-normal LSF as follows,

$$g(\mathbf{X}) = x_1^4 + x_2^2 - 50 \tag{6.22}$$

where, variables Lognormal x_1 and Gumbel x_2 variables are defined as means of $\mu_1 = 5$, $\mu_2 = 10$ and the standard deviation (STD) of $\sigma_1 = 1$, $\sigma_2 = 10$. Using FORM, the reliability index was extracted from Ref. [2] as $\beta = 3.259315$ while the MPP and reliability index using CHL-RF is computed as $\beta = 3.2593$ and $X^* = [2.6475, 0.9314]$, respectively. The HL-RF is not converged to stable results and its solutions are obtained, periodically as 4-reliability index of [2.5117, 1.8550, 2.3857, 1.5966].

The reliability index obtained by the HL-RF and CHL-RF methods are presented in Fig. 6.5 for Example 6.2, respectively. The stable results are obtained by using the CHL-RF method while the HL-RF provides the instability as 4-priodic solution. The smooth curve for convergence of the reliability index is given by the CHL-RF method thus this approach provides the suitable line search for finding the MPP

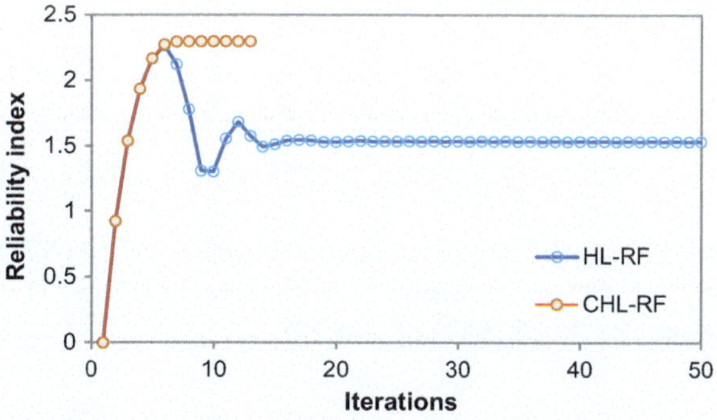

Fig. 6.3 Reliability index obtained by CHL-RF and HL-RF for Example 6.1

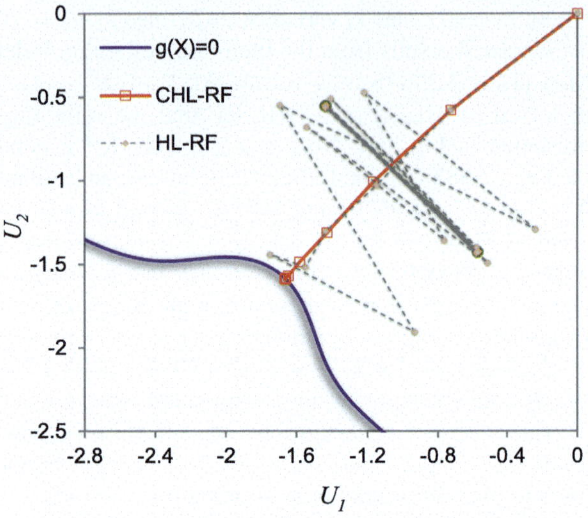

Fig. 6.4 MPP search of HL-RF and CHL-RF methods in normal standard space for Example 6.1

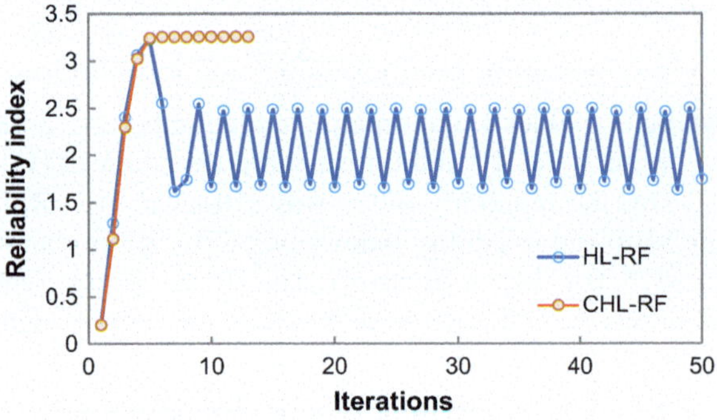

Fig. 6.5 Reliability index using CHL-RF and HL-RF for Example 6.2

by directional conjugate line search. The efficiency of the reliability results is also significantly improved for this problem.

Example 6.3 A nonlinear/non-normal LSF,

$$
\begin{aligned}
g(\boldsymbol{X}) &= 1.1 - 0.00115x_1x_2 + 0.00157x_2^2 + 0.00117x_1^2 + 0.0135x_2x_3 \\
&-0.0705x_2 - 0.00534x_1 - 0.0149x_1x_3 - 0.0611x_2x_4 + 0.0717x_1x_4 \\
&-0.226x_3 + 0.0333x_3^2 - 0.558x_3x_4 + 0.998x_4 - 1.339x_4^2
\end{aligned}
\tag{6.23}
$$

where, type-II largest value x_1 with mean of 10 (STD of 5), Normal x_2 with mean of 25 (STD of 5), Normal x_3 with mean of 0.8 (STD of 0.2) and Lognormal x_4 with mean of 0.0625 (STD of 0.0625).

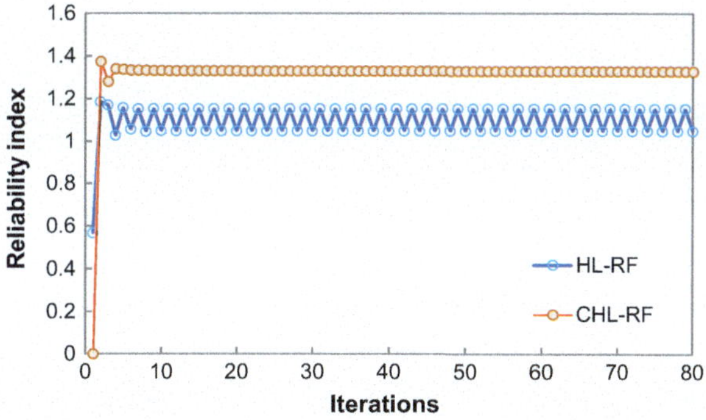

Fig. 6.6 Reliability index computed by HL-RF and CHL-RF for Example 6.3

Using extracted results from Yang [3] compared this example that extracted reliability index and MPP are 1.3304 and $\boldsymbol{X}^* = [14.906, 25.067, 0.8995, 0.04606]$, respectively.

This problem is applied to compare the ability of the CHL-RF method that converged results of this method are the $\beta = 1.3306$, $\boldsymbol{U}^* = [1.2950, 0.0147, 0.2989, 0.0621]$ and $\boldsymbol{X}^* = [14.8992, 25.0733, 0.8598, 0.0465]$, which closely agree well with the reliability results extracted from literature. The globality of the CHL-RF is guaranteed for this problem but it needed more computational burden for giving the MPP by the CHL-RF. The reliability index computed by CHL-RF and HL-RF methods is illustrated in Fig. 6.6. As seen, the CHL-RF provided stable results after 79 iterations but the HL-RF method has unstable solutions as 2-prdiodic for reliability index. This method is slowly converged for some reliability problems to control the stability. The line search vector as finite CG vector satisfies the sufficient descent conditions thus, the formulation of the CG is not changed by the finite step size (FSS). A large value more than 1 can provide stable results without adjusting the procedure. Consequently, the type of the LSF is directly affected on the computational burden of the CHL-RF method.

To illustrate satisficing the sufficient descent conditions by the CG SDV using CHL-RF formulation, five different step lengths are given as same value at all iterations without adjusting factors as $\lambda = 1, 10, 100, 1000, 10{,}000$. Then, the number of iterations for these three examples are tabulated in Table 6.1 for Examples 6.1, 6.2, and 6.3 at different step lengths. Note from Table 6.1, that the CHL-RF is converged with a similar iteration for Examples 6.1 and 6.2. The stable results are obtained based on all constant step size in all problems. Thus, the CHL-RF method is a framework of FORM free-step size and by holding the sufficient descent condition, it is always a robust method but its efficiency is depended on the nonlinearity of LSF. A constant step length more than 100 is acceptable for this approach.

Table 6.1 Iterations of CHL-RF for different step length without adjusting

Step length	$\lambda_k = 1$	10	100	1000	10,000
6.1	12	12	12	12	12
6.2	12	11	12	12	12
6.3	266	110	79	77	76

6.3 Limited FR Conjugate Search Direction

The improved FORM formula applied by the steepest descent search direction presented in Chap. 2 including improved HL-RF, relaxed HL-RF [4] and the stability transformation method (STM) [3] may be produced unstable results. However, it is shown that the CFORM formulated by directional conjugate line search is improved the robustness of FORM by a free step size. In the CHL-RF, the conjugate search direction method to computed the normalized sensitivity vector is formulated based Fletcher-Reeves (FR) CG method. As seen, this is robust but may provide inefficient results.

The CG is particularly simple with low storage, and acceptable numerical abilities as robust formulation and global convergency. However, it is essential to develop a FORM framework that provides global convergence while it is simple with a fast convergence rate by applying in structural reliability analysis. As seen in Sect. 6.2, two factors of conjugate factor and finite step size are important for developing a CFORM formulation but it should be guaranteed stable results with low computational burden. Generally, the FORM-based conjugate search direction is more robust than the FORM-based steepest descent search direction, but it provided the efficient results. The efficient property of the CG-based FORM is free to a given step size but the conjugate factor can be a rule for enhancing the computational burden.

A simple iterative CG-based FORM algorithm was established to enhance the stability of FORM. The conjugate FR scale factor is limited in the iterative process to search the MPP named as limited FR (LFR)-based CFORM. In LFR, a FSS is established by Armijo condition. Unlike the FORM-based SDSD formula, the LFR by applying the finite step length reduces the parallel risk of new and previous normalized SDVs. The CG vector $\boldsymbol{d}_k$ is calculated based on Eq. (6.24),

$$\boldsymbol{d}_k = \begin{cases} -\nabla g(\boldsymbol{U}_k) & k = 0 \\ -\nabla g(\boldsymbol{U}_k) + \theta_k \boldsymbol{d}_{k-1} & k \geq 1 \end{cases} \tag{6.24}$$

where θ_k is a scalar. the Fletcher-Reeves (FR) method is a well-known CG method with below θ_k,

$$\theta_k^{FR} = \frac{||\nabla g(\boldsymbol{U}_k)||^2}{||\nabla g(\boldsymbol{U}_{k-1})||^2} \tag{6.25}$$

In CHL-RF, the FR formulation is applied for computing the CG vector. In this section, a limited FR (LFR) scalar factor is used for structural reliability to improve the efficiency of the CFORM by FR (CHL-RF) using the below limited CG factor,

$$\theta_k^{LFR} = \min\{\delta, \theta_k^{FR}\} \tag{6.26}$$

in which $0.5 \leq \delta \leq 1.0$ is defined as an adjusted factor. Using Eq. (6.26), the limited FR by this scalar factor holds,

$$0 \leq \theta_k^{LFR} \leq \theta_k^{FR} \leq \delta \tag{6.27}$$

This means that θ_k^{LFR} has the same global convergence with the FR method. It is supposed that θ_k^{LFR} should satisfy sufficient descent condition (SDC) thus, it provides results without chaotic or periodic solutions. The globally is checked by the following rule,

$$\nabla^T g(U_k)\boldsymbol{d}_k < -c_1 \|\nabla g(\boldsymbol{U}_k)\|^2 \tag{6.28}$$

where, $0 < c_1 < 1$. To show the above condition, a cycle of LFR is presented in Fig. 6.7, it is shown angle between negative $\nabla g(U)$ and CG vector d is γ_k i.e. $0 \leq \gamma_k < \frac{\pi}{2}$ which is determined as below,

$$\tan \gamma_k = \frac{\theta_k^{LFR} \|\boldsymbol{d}_{k-1}\|}{\|\nabla g(\boldsymbol{U}_k)\|} \tag{6.29}$$

If $\gamma_k \rightarrow \frac{\pi}{2}$ then, $\frac{\theta_k^{LFR} \|\boldsymbol{d}_{k-1}\|}{\|\nabla g(\boldsymbol{U}_k)\|} \rightarrow \infty \Rightarrow \theta_k^{LFR} \rightarrow \infty$, thus, the LFR algorithm failed. For controlling the instability of scalar factor, it is given a limitation as $\tan \gamma_k = \delta$ and then we have $0.5 \leq \delta \leq 1.0$ as a maximum value for $\tan \gamma_k$. It provides a limit value for angle thus $\delta = \frac{\theta_k^{LFR} \|\boldsymbol{d}_{k-1}\|}{\|\nabla g(\boldsymbol{U}_k)\|} \leq 1$ is applied for maximum value of $\tan \gamma_k$. On the other hand, by applying SDC i.e. $\|\boldsymbol{d}_{k-1}\| \leq \|\nabla g(\boldsymbol{U}_{k-1})\|$, the CG factor of FR is given in range from 0 to δ. Consequently, the iterative LFR formula is computed by a limited CG scalar factor as below,

$$\theta_k^{LFR} = \min\{\delta, \theta_k^{FR}\} \tag{6.30}$$

Step size is significant factor as well as the limited CG factor in LFR algorithm, the stable convergence of LFR iterative FORM method is obtained when λ is selected as a suitable value. The finite step size λ satisfies Armijo-type condition, thus, we have,

$$\lambda_{\max} = M \frac{\|\boldsymbol{d}_k^{LFR}\|^2}{\|\nabla g(\boldsymbol{U}_k)\|^2} \tag{6.31}$$

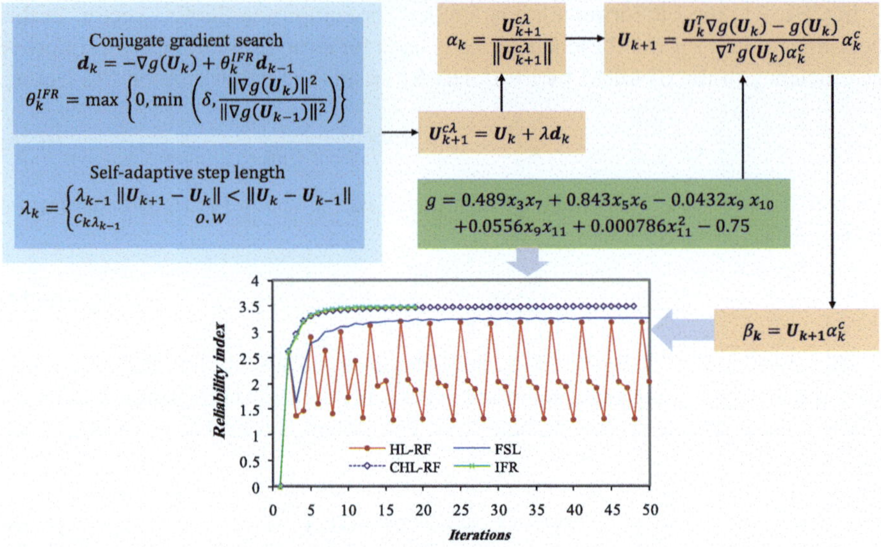

Fig. 6.7 An iterative schematic view for LFR method

where $M >> 1$. Based on $\left\|d_k^{LFR}\right\| < \|\nabla g(U_k)\|$, we give a $\lambda \le c\lambda_{\max}$ for $0 < c < 1$. Thus, we have,

$$\lambda_k = \begin{cases} \lambda_{k-1} & \|U_{k+2} - U_{k+1}\| < \|U_{k+1} - U_k\| \\ c_k\lambda_{k-1} & \text{otherwise} \end{cases} \tag{6.32}$$

In which $c_k = \frac{3}{10}\frac{\|\nabla g(U_k)\|^2}{\left|\nabla^T g(U_k) d_k^{LFR}\right|}$ and $\lambda_{int} = M$. The factor c_k is applied for adjusting the finite step size. The dynamic adaptive factor c_k is computed less than 1 by applying $\left\|d_k^{LFR}\right\| < \|\nabla g(U_k)\|$. The dynamical adjusting factor i.e. c_k is a major difference between CHL-RF and LFR methods. The steps for computing the reliability index and approximating failure probability for coding a computer program are presented below for the LFR method:

Step 1: Define ε (stopping criterion) and $g(U) = 0$, give variables as distribution, μ and σ and set $M \in [5, 500]$, $\delta \in [0.5, 1.0]$, $k = 0$ with initial values as $X_0 = \mu$, $d_0 = 0$ and $\lambda_0 = M$;

Step 2: Transform variables into U-space as $u_k = \frac{x_k - \mu_x^e}{\sigma_x^e}$, where $\sigma_x^e = \frac{1}{f_X(x_k)}\varphi\left[\Phi^{-1}\{F_X(x_k)\}\right]$ and $\mu_x^e = x_k - \sigma_x^e\Phi^{-1}[F_X(x_k)]$;

Step 3: Iterative process of LFR by the following relations;

(i) Determine the gradient vector $\nabla g(U)$ and $g(U)$;
(ii) Compute $\theta_k^{LFR} = \min\{\delta, \theta_k^{FR}\}$ with, $\theta_k^{FR} = \frac{\|\nabla g(U_k)\|^2}{\|\nabla g(U_{k-1})\|^2}$;

(iii) Compute $\boldsymbol{d}_k = -\nabla g(\boldsymbol{U}_k) - \theta_k^{LFR}\boldsymbol{d}_{k-1}$;

(iv) Determine $\boldsymbol{U}_{k+1}^{C\lambda} = \boldsymbol{U}_k + \lambda\boldsymbol{d}_k$ and then $\alpha_{k+1}^{\lambda} = \frac{\boldsymbol{U}_{k+1}^{C\lambda}}{\|\boldsymbol{U}_{k+1}^{C\lambda}\|}$;

(v) Determine reliability index $\beta_{k+1} = \frac{g(\boldsymbol{U}_k) - \nabla^T g(\boldsymbol{U}_k)\boldsymbol{U}_k}{\nabla^T g(\boldsymbol{U}_k)\alpha_{k+1}^{\lambda}}$;

(vi) Compute the new point by $\boldsymbol{U}_{k+1} = \beta_{k+1}\boldsymbol{\alpha}_{k+1}^{\lambda}$;

(vii) Adjust step length as $\lambda_k = \begin{cases} \lambda_{k-1} & \|\boldsymbol{U}_{k+2} - \boldsymbol{U}_{k+1}\| < \|\boldsymbol{U}_{k+1} - \boldsymbol{U}_k\| \\ c_k\lambda_{k-1} & \text{otherwise} \end{cases}$

where $c_k = \frac{3}{10}\frac{\|\nabla g(\boldsymbol{U}_k)\|^2}{|\nabla^T g(\boldsymbol{U}_k)\boldsymbol{d}_k^{LFR}|}$;

Let the next iterate using $X_{k+1} = \mu_X^e + \sigma_X^e U_{k+1}$ and $k = k + 1$;

Step 4: If $\frac{\|\boldsymbol{U}_k - \boldsymbol{U}_{k+1}\|}{\|\boldsymbol{U}_k\|} < \varepsilon$, then presents results as $P_f \approx \Phi(-\beta))$ and $\boldsymbol{\alpha} = \boldsymbol{\alpha}_{k+1}^{\lambda}$, else go to Step 2.

In the LFR algorithm, an adjusted factor is proposed for dynamically adapting the step size and the CG scalar factor is determined using SDC that these two factors are utilized for determining conjugate line search. To illustrate the performances of LFR as robustness and efficiency, this method is compared with and HL-RF, STM [3] (with $\lambda = 0.1$) and FR [5] (with $M = 50$ and $\delta = 0.75$) by four mathematical/mechanical examples as presented in Table 6.2. For all Examples, the same stopping criterion is given as $\varepsilon = 10^{-9}$.

The convergence results of CFORM given by iterative formulation of FR and LFR are recorded in Table 6.3. As seen from this Table, the LFR as well as FR is converged. The LFR is strongly more efficient than FR and LFR showed the same robustness with the FR method. In addition, the LFR by satisfying the SDC can guarantee robust convergence for FORM iterative scheme applied in highly non-linear problems (Fig. 6.8).

It is obvious, the LFR has significantly enhanced the efficiency of FR method for moderately nonlinearity problems as Examples 6.5, 6.6, and 6.7. The results presented in Table 6.3 are confirmed the capability of LFR for reliability analysis. The reliability index obtained by LFR and FR methods is closely followed to reliability index extracted in Table 6.2 given from literatures.

Based on results of Example 6.4 as presented in Fig. 6.9, 2-periodic solutions (β = [0.9267, 0.9863]) as unstable results are extracted by using HL-RF formulation while other methods are robust and RHL-RF provides the reliability index 2.365373 after 67 iterations. However, The LFR and FR (CHL-RF) provided stable reliability index of 2.3654 with related MPP of X^* = [1.8048, 1.4720] which captured after 22 iterations.

For Example 6.5, extracted results given from Ref. [6] for MPP is $\boldsymbol{X}^*$ = [4.01, 24,193, 0.822, 19.58, 509.02, 157.01]. The converged reliability index for Example 6.5 is presented in Fig. 6.10. In this example, FR and LFR methods have same converged results with similar accuracy and robustness, but LFR is more efficient. The safety index and MPP by LFR are respectively extracted after 6 iterations as β = 2.66984 and $\boldsymbol{X}^*$ = [4.0057, 24,204.1004, 0.8227, 19.5816, 514.3549, 155.8608]

Table 6.2 Examples for comparing LFR method

Example	Performance function	Random variables	Descriptions	β (Refs.)
6.4	$g(X) = x_1^4 + 2x_2^4 - 20$	× *1,2*: (10,5)	Normal non-linear LSF	2.365373 [5]
6.5	$g(X) = x_1x_2x_3x_4 - x_5x_6^2/8$	Weibull × *1*: (4,0.1) Lognormal × 2:(25,000,2000), Gumbel × *3*: (0.875,0.1) Uniform × *4*: (20,1) Exponential × *5*: (100,100) Normal × *6*: (150,10)	Non-normal LSF	2.6698 [6]
6.6	$g = 3r - \left\lvert \frac{2F}{mW^2}\sin(\frac{Wt_1}{2}) \right\rvert$ $W^2 = \frac{C_1+C_2}{m}$	*m*: (1, 0.05), *r*: (0.5,0.05) C_1: (1, 0.1), C_2: (0.1, 0.01), t_1: (1, 0.2), *F*: (1, 0.2)	Engineering LSF with normal variables as shown in Fig. 6.8	1.8652 [7]
6.7	$g(X) = 1.1 - 0.00115x_1x_2 + 0.00157x_2^2 + 0.00117x_1^2 - 0.0705x_2 + 0.0135x_2x_3 - 0.00534x_1 - 0.0149x_1x_3 - 0.0611x_2x_4 + 0.0717x_1x_4 - 0.226x_3 + 0.0333x_3^2 - 0.558x_3x_4 + 0.998x_4 - 1.339x_4^2$	Extreme value II × *1*: (10,5) Normal × *2*: N(25,5), Normal × *3*: N(0.8,0.2) Lognormal × *4*: (0.0625,0.0625)	A pipeline with response surface function applied for non-normal LSF	1.330537 [5] 1.3304 [3]

Table 6.3 Reliability results for Examples 6.1, 6.2, 6.3, and 6.4 using FR and LFR

Examples	FR			LFR		
	Iter	β	P_f	Iter	β	P_f
6.4	22	2.36549	0.009	22	2.36549	0.009
6.5	42	2.67018	0.0038	42	2.67018	0.0038
6.6	28	1.86513	0.0311	28	1.86513	0.0311
6.7	31	1.33053	0.0916	31	1.33053	0.0916

which are confirmed by agreement with results of Ref. [6]. The STM is an inefficient method and HL-RF and LFR are the efficient methods with iterations of 7 and 6 iterations, respectively.

A mechanical example shown in Fig. 6.8 with rectangular pulse load is analyzed using CFORM-based FR and LFR that reliability indexes are respectively as $\beta =$

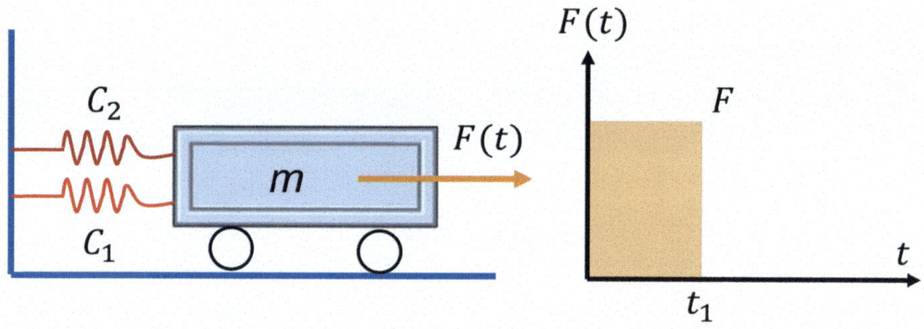

Fig. 6.8 Engineering oscillator applied in Example 6.6. Reprinted from [7], with permission from Elsevier

Fig. 6.9 Comparing LFR for reliability index in Example 6.4

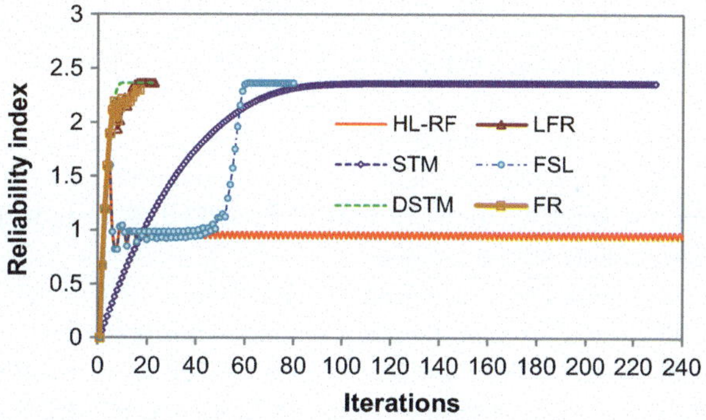

Fig. 6.10 Comparing LFR for reliability index in Example 6.5

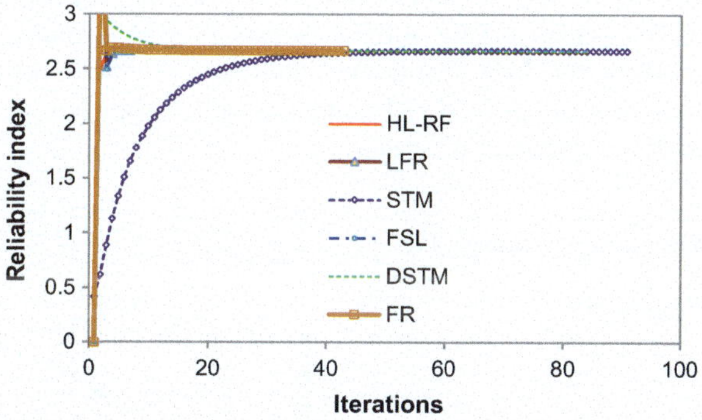

1.86513 and $\beta = 1.86511$ as related iterations of 28 and 6. It is shown that the LFR is fast converged compared to FR and it improved efficiency of FR about five-times. The converged results of the reliability index for different methods are plotted in Fig. 6.11 for Example 6.6. As results of Fig. 6.11, the LFR, and HL-RF provide the stable solution with similar efficiency but STM is strangely inefficient.

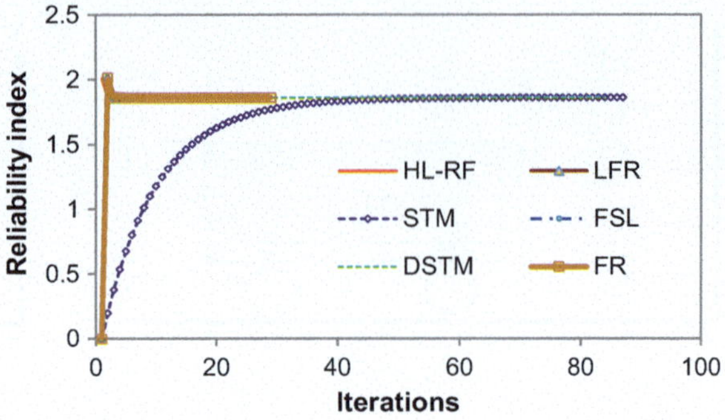

Fig. 6.11 Comparing LFR for reliability index in Example 6.6

In Example 6.7, given the results from Yang [3] showed that $\beta = 1.3482$ and $\boldsymbol{X}^* = [14.906, 25.067, 0.8995, 0.04606]$. The converged results obtained after 11 iterations using LFR are $\beta = 1.330538$, $\boldsymbol{X}^* = [14.9036, 25.0675, 0.85965, 0.04614]$, and $\boldsymbol{U}^* = [1.2956, 0.01349, 0.29824, 0.05162]$ which are agreement with results of literature. The reliability index using different models applied for comparing the LFR is presented in Fig. 6.12. It is shown that the LFR is improved the efficiency of the FR and robustness of FORM-based HL-RF algorithm. Unlike HL-RF, STM and FR provide stable results.

Iterations by evaluating $\nabla g(\boldsymbol{U})$ showed as *Iter* and converged β are respectively utilized to demonstrate the efficiency and robustness of different reliability methods that these two comparative factors given from results of FORMs are arranged in Table 6.4. As presented results in Table 6.4, HL-RF has an unstable iterative formulation for Examples 6.4 and 6.7 where it periodically converged to safter factors of $\beta = [0.9267, 0.9863]$ and $\beta = [1.0496, 1.1536]$, respectively. However, the STM-based SDSD as well as LFR and FR-based CFORM formulas showed stable suctions, more robustly but, LFR is more efficient among others. As results for Examples 6.5 and 6.6, the LFR and HL-RF have the same efficiency but STM provided results with high-computational burden.

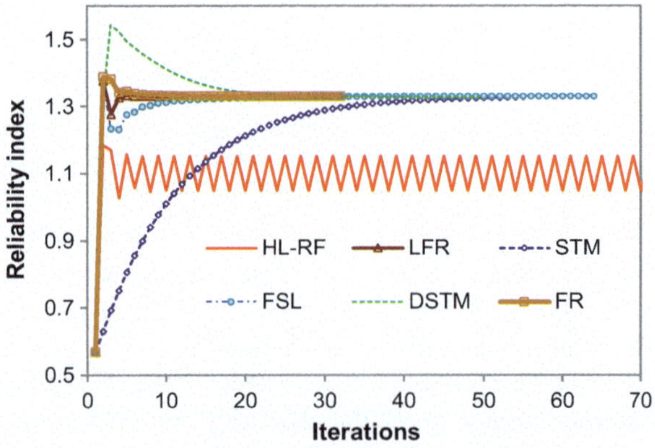

Fig. 6.12 Comparing LFR for reliability index in Example 6.7

Table 6.4 Comparative results for different FORM methods in Examples 6.4, 6.5, 6.6, and 6.7

Examples	HL-RF		STM		LFR		FR	
	Iter	β	Iter	β	Iter	β	Iter	β
6.4	Periodic-2		229	2.36546	22	2.36546	22	2.36549
6.5	7	2.66986	91	2.66974	6	2.66984	42	2.67018
6.6	7	1.86511	87	1.86491	6	1.86511	28	1.86513
6.7	Periodic-2		55	1.32734	11	1.33054	31	1.33053

An iterative conjugate framework is introduced for SRA that it formulated by CFORM MPP search. The LFR scheme is compared by four functions with CFORM-based FR, HL-RF and STM algorithms. The LFR method enhances the efficiency of FORM and CFORM-based FR. Therefore, it is suggested that the LFR be implemented for applicable SRA reliability analyses and applied in reliability-based design optimization problems, in future.

6.4 Enhanced CG Search Direction

Enhanced conjugate FR algorithms are presented including the spectral-variant FR (SVFR), improved FR (IFR), and modified FR (MFR) in this section. An adaptive finite-step length as well as LFR is applied to control the instabilities. These conjugate maps are applied for computing the search direction which is utilized to MPP search.

In the previous chapter, we showed that the instabilities of FORM iterative formulas were controlled in STM [3] with a small step size while the robustness of FORM can be extracted with FSL with an adjusting step size given from SDC [8]. The relaxed approaches of FORM [9] have been established using a step size given by SDCs [9] and a direct relation. Although, the adaptive CC as well as STM is established using a step size driven as 1 for linear function or 0.1 for nonlinear functions. The STM and FSL are converged with slow rate, when it is applied a smaller step size and evaluated as LSF with high- nonlinearity degree, respectively [10–12]. The almost modification if FORM formulas are established based on a step size less than 1 thus their efficiency is less than HL-RF for convex or low-nonlinear LSFs [3, 13].

The CHL-RF using FSS [5], and LFR [14] as CFORM can be applied for enhancing the ability of MPP search method due to their robustness. For developing the CFORM in SRA, two vital factors are robustness and efficiency. As presented by CHL-RF and LFR methods, the conjugate SDV can be improved robustness of the FORM. The CFORM is particularly formulated with simple relations which provided a good numerical performance as robustness in searching MPP. Nevertheless, the efficiency of CFORM approaches is a major challenge. In this section, three CFORM given from the FR method including limited FR (LFR), modified FR (MFR) and spectral-variant FR (SVFR) are discussed that they are formulated to approximate the reliability index. Finally, the performances of SVFR and MFR are evaluated and compared with HL-RF, FR [5] and LFR [14] by five LSFs.

6.4.1 Basic Formulation of Enhanced FR

In the CG method, it is simply formulated for eliminating the Hessian matrix in its formulation in reliability analyses. In CG methods, the CG vector is determined as follows,

$$\boldsymbol{d}_k = \begin{cases} -\nabla g(\boldsymbol{U}_k) & k = 0 \\ -\nabla g(\boldsymbol{U}_k) + \theta_k \boldsymbol{d}_{k-1} & k \geq 1 \end{cases} \tag{6.33}$$

where θ_k is the CG factor that it computed by Fletcher-Reeves (FR) CG factor as below,

$$\theta_k^{FR} = \frac{||\nabla g(\mathbf{U}_k)||^2}{\|\nabla g(\mathbf{U}_{k-1})\|^2} \tag{6.34}$$

As mentioned, the CHL-RF is formulated by FR method [5]. Keshtegar [14] applied a simple framework for computing the LFR CG factor. LFR, the CG factor of FR is limited to a maximum value as below,

$$\theta_k^{LFR} = \min\{\delta, \theta_k^{FR}\} \tag{6.35}$$

In which, δ is limited CG factor i.e. $0.5 < \delta \leq 1.0$.

It is presented two CG vectors computed by variant-type of FR which are computed as follows,

$$\mathbf{d}_k^{MFR} = -\eta_k \nabla g(\mathbf{U}_k) + \theta_k^{FR} \mathbf{d}_{k-1} \tag{6.36}$$

$$\mathbf{d}_k^{VSFR} = -\eta_k \nabla g(\mathbf{U}_k) + \theta_k^{VFR} \mathbf{d}_{k-1} \tag{6.37}$$

Equation 6.37 is an SVFR method CG factor is computed as below,

$$\theta_k^{VFR} = \frac{||\nabla g(\mathbf{U}_k)|| \cdot \left|\nabla^T g(\mathbf{U}_{k-1}) \nabla g(\mathbf{U}_k)\right|}{\|\nabla g(\mathbf{U}_{k-1})\|^3} \tag{6.38}$$

The participation factor for cooperating with the gradient vector in CG vector (η_k) is determined by the following relation,

$$\eta_k = \frac{\mathbf{d}_{k-1}^T \left[\nabla g(\mathbf{U}_k) - \nabla g(\mathbf{U}_{k-1})\right]}{\|\nabla g(\mathbf{U}_{k-1})\|^2} \tag{6.39}$$

It presents a variant type of FR method that the gradient vector is also controlled by a factor η_k as well as the CG scalar factor. Figure 6.13 plots the reliability index, LSF, and point in new iteration in two-dimensional standard normal space for different improved FR methods. It is shown that the d_k at point U_k is not paralleled to the direction of the negative vector of gradient i.e. $-\nabla g(\boldsymbol{U}_k)$. Therefore, $\boldsymbol{\alpha}_k$ and $\boldsymbol{\alpha}_k^c$ vectors are not the same. Consequently, the CFORM provides the results without periodically solutions as well as HL-RF for high-nonlinear problems. In variant-type FR methods, two scalar factors θ_k and η_k are applied for hybrid two vectors of $\nabla g(\boldsymbol{U}_k)$ and d_{k-1}.

The formulation of CFORM is given based on the type-variant modified FR as follows,

$$\boldsymbol{U}_{k+1} = \left(\frac{\boldsymbol{U}_k^T \nabla g(\boldsymbol{U}_k) - g(\boldsymbol{U}_k)}{\boldsymbol{\alpha}_k^{c\lambda T} \nabla g(\boldsymbol{U}_k)} \right) \boldsymbol{\alpha}_k^{c\lambda} \tag{6.40}$$

where, α_k^c is normalized conjugate SDV, which is given as,

$$\boldsymbol{\alpha}_k^{c\lambda} = \frac{\boldsymbol{U}_{k+1}^{c\lambda}}{\left\| \boldsymbol{U}_{k+1}^{c\lambda} \right\|} \tag{6.41}$$

where,

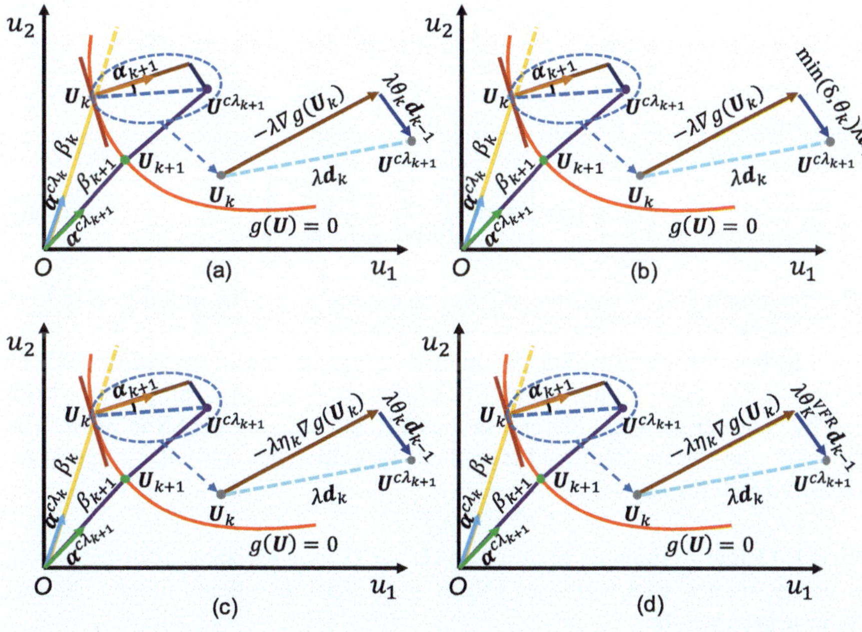

Fig. 6.13 An iteration of different CFORM **a** FR, **b** LFR, **c** MFR and **d** SVFR. Reprinted from [15], with permission from Springer Nature

$$U_{k+1}^{c\lambda} = U_k + \lambda_k d_k \tag{6.42}$$

FSS (λ) and CG vector ($\boldsymbol{d}$) are two vital factors in FORM formulation extended by- conjugate SDV. If λ is selected a suitable value then, the CFORM-based FR algorithms formulated by CG factors in Eqs. (6.330), (6.35), and (6.36) can provide a robust and efficient scheme. By adaptive FSS λ, it can be controlled the efficiency of CFORM. The FSS can be computed as large value in first iteration using Armijo rule as follows,

$$\lambda_{\max} = M \frac{d_k^2}{\|\nabla g(U_k)\|^2} \tag{6.43}$$

The FSS should be satisfied the SDC as $\|U_{k+1} - U_k\| > \|U_k - U_{k-1}\|$ for $\lambda \leq c_1\lambda_{\max}$ and $M >> 1$. If SDC is not considered then, $\lambda_k = c_k\lambda_{k-1}$ with adjusting coefficient c_k. This FSS holds SDC as follows,

$$\lambda_k = \begin{cases} \lambda_{k-1} & \|U_{k+1} - U_k\| < \|U_k - U_{k-1}\| \\ c_k\lambda_{k-1} & \|U_{k+1} - U_k\| \geq \|U_k - U_{k-1}\| \end{cases} \tag{6.44}$$

The SDC is applied based on a maximum FSS computed by Armijo rule and adjusting coefficient in Eq. (6.44) where, maximum FSS is selected by Eq. (6.45):

$$\lambda_0 = \frac{100}{\left\|\nabla g\left(U|_{U=\mu}\right)\right\|} \tag{6.45}$$

in which, $\nabla g\left(U|_{U=\mu}\right)$ is the mean point gradient vector, and λ_o is the initial FSS. The adjusting coefficient is computed as below,

$$c_k = \min\left\{0.95, \frac{3}{10}\frac{\|\nabla g(U_k)\|^2}{\left|\nabla^T g(U_k) d_k\right|}\right\} \tag{6.46}$$

The maximum FSS is adjusted at every intention by c_k. The initial FSS in LFR is set as a constant value at the first iteration between 5 to 100, while in this current FSS, the LFR method is provided a normalized conjugate sensitivity vector (NCSV) computed by Eq. (6.44). Consequently, the LFR is provided stabilization for the MPP search by considering a CG factor less than FR and dynamical FSS in MFR. The steps of the type-variant modified FR methods for coding in a computer program are summarized as follows:

Step 1: Define ε (stopping criterion), $g(U) = 0$, give variables as distributions, μ and σ, set $\delta \in [0.5, 1.0]$ for LFR, $M \in [5, 500]$, $k = 0$ and initial values as $X_0 = \mu$ and $d_0 = 0$;

Step 2: Transfer variables into $\boldsymbol{U}$-space as $u_k = \frac{x_k - \mu_x^e}{\sigma_x^e}$, where $\sigma_x^e = \frac{1}{f_X(x_k)}\varphi\left[\Phi^{-1}\{F_X(x_k)\}\right]$ and $\mu_x^e = x_k - \sigma_x^e \Phi^{-1}[F_X(x_k)]$;

Step 3: Iterative process of modified version of FR methods by the following relations;

(i) Determine the gradient vector $\nabla g(\boldsymbol{U})$ and $g(\boldsymbol{U})$;

(ii) if $k = 0$ then compute $\lambda_0 = \frac{100}{\|\nabla g(\boldsymbol{U}|_{U=\mu})\|}$;

(iii) Compute (i) $\theta_k^{FR} = \frac{\|\nabla g(\boldsymbol{U}_k)\|^2}{\|\nabla g(\boldsymbol{U}_{k-1})\|^2}$; (ii) $\eta_k = \frac{\boldsymbol{d}_{k-1}^T[\nabla g(\boldsymbol{U}_k) - \nabla g(\boldsymbol{U}_{k-1})]}{\|\nabla g(\boldsymbol{U}_{k-1})\|^2}$; $\theta_k^{LFR} = \min\{\delta, \theta_k^{FR}\}$ (iv) $\theta_k^{VFR} = \frac{\|\nabla g(\boldsymbol{U}_k)\| \cdot |\nabla^T g(\boldsymbol{U}_{k-1})\nabla g(\boldsymbol{U}_k)|}{\|\nabla g(\boldsymbol{U}_{k-1})\|^3}$;

(iv) Determine the CG vectors as;
  (a) FR $\boldsymbol{d}_k^{FR} = -\nabla g(\boldsymbol{U}_k) + \theta_k^{FR}\boldsymbol{d}_{k-1}$
  (b) LFR $\boldsymbol{d}_k^{LFR} = -\nabla g(\boldsymbol{U}_k) + \theta_k^{LFR}\boldsymbol{d}_{k-1}$
  (c) MFR $\boldsymbol{d}_k^{MFR} = -\eta_k \nabla g(\boldsymbol{U}_k) + \theta_k^{FR}\boldsymbol{d}_{k-1}$
  (d) SVFR $\boldsymbol{d}_k^{VSFR} = -\eta_k \nabla g(\boldsymbol{U}_k) + \theta_k^{VSFR}\boldsymbol{d}_{k-1}$

(v) Determine $\boldsymbol{U}_{k+1}^{c\lambda} = \boldsymbol{U}_k + \lambda \boldsymbol{d}_k$ and then $\boldsymbol{\alpha}_{k+1}^{c\lambda} = \frac{\boldsymbol{U}_{k+1}^{c\lambda}}{\|\boldsymbol{U}_{k+1}^{c\lambda}\|}$;

(vi) Determine reliability index $\beta_{k+1} = \frac{g(\boldsymbol{U}_k) - \nabla^T g(\boldsymbol{U}_k)\boldsymbol{U}_k}{\nabla^T g(\boldsymbol{U}_k)\boldsymbol{\alpha}_{k+1}^{\lambda}}$;

(vii) Compute the new point by $\boldsymbol{U}_{k+1} = \beta_{k+1}\boldsymbol{\alpha}_{k+1}^{c\lambda}$;

(viii) Adjust step length as $\lambda_k = \begin{cases} \lambda_{k-1} & \|\boldsymbol{U}_{k+1} - \boldsymbol{U}_k\| < \|\boldsymbol{U}_k - \boldsymbol{U}_{k-1}\| \\ c_k\lambda_{k-1} & \|\boldsymbol{U}_{k+1} - \boldsymbol{U}_k\| \geq \|\boldsymbol{U}_k - \boldsymbol{U}_{k-1}\| \end{cases}$

where $c_k = \min\left\{0.95, \frac{3}{10}\frac{\|\nabla g(\boldsymbol{U}_k)\|^2}{|\nabla^T g(\boldsymbol{U}_k)\boldsymbol{d}_k|}\right\}$;

Let the next iterate using $\boldsymbol{X}_{k+1} = \mu_{\boldsymbol{X}}^e + \sigma_{\boldsymbol{X}}^e \boldsymbol{U}_{k+1}$ and $k = k + 1$;

Step 4: If $\frac{\|\boldsymbol{U}_k - \boldsymbol{U}_{k+1}\|}{\|\boldsymbol{U}_k\|} < \varepsilon$, then presents results as $P_f \approx \Phi(-\beta))$ and $\boldsymbol{\alpha} = \boldsymbol{\alpha}_{k+1}^{\lambda}$, else go to Step 2.

The formwork for giving an iterative formulation of results in modified versions of FR based on Step 3 is presented in Fig. 6.14. It provides a simple framework with direct relations that are computed by results of new point for updating line search components as CG vectors and step size.

The above algorithms presented in Fig. 6.14, LFR is as simple as the FR but two variant-type FR methods have a complicated formulation compared to FR. In MFR and SVFR, the NCSV is computed by the participation factor η_k for the better hybrid of the gradient vector. The major difference of MFR and SVFR with FR are controlling factor for gradient vector participation and CG. On the other hand, a self-adaptive FSS is presented based on SDC with a dynamical adjusting factor computed by the nonlinearity of LSF.

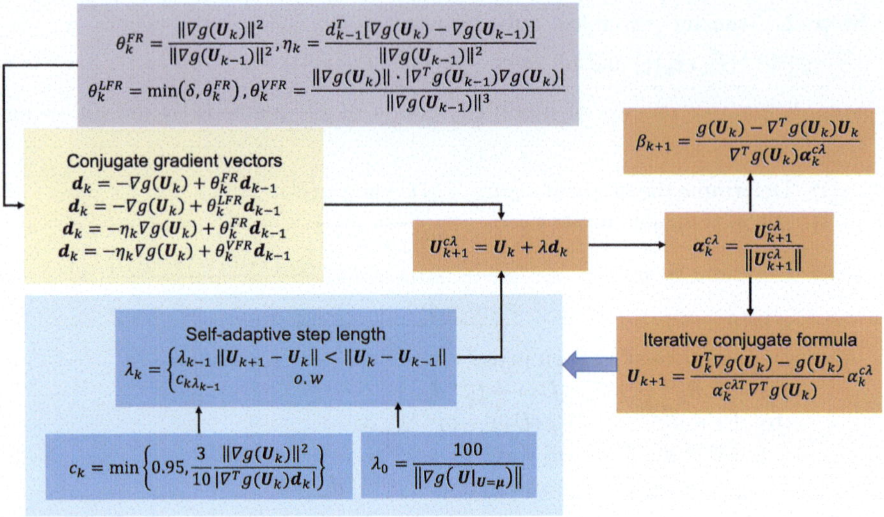

Fig. 6.14 Iterative step for modified FR method applied in CFORM

6.4.2 *Results of Numerical Examples*

Five reliability problems are applied to validate and discuss the SVFR, MFR, and LFR methods which are compared with the results of FR and HL-RF. The same stopping criterion ($\varepsilon = 10^{-6}$) is used, and the initial FSS-based Armijo are computed in all CFORM including FR, LFR, MFR, and SVFR. The limited CG factor in the LFR method is selected as $\delta = 0.75$. The numbers of evaluating the $\nabla g(U)$ named Iter and β are applied to demonstrate the efficiency and robustness of these sachems.

Example 6.8 An LSF with below relation,

$$g(\boldsymbol{X}) = x_1^3 + x_1^2 x_2 + x_2^3 - 18 \tag{6.47}$$

where, normal x_1 and x_2 variables have $\mu_1 = 10$, $\mu_2 = 9.9$ and $\sigma_1 = \sigma_2 = 5$. Extracted reference reliability index is 2.2983. Table 6.5 demonstrates reliability results for FR, LFR, MFR, and SVFR algorithms. The results show that the LFR converges with similar iterations with SVFR and MFR that they are converted faster than FR. The CFORMs established by enhanced FR as well as the FR method have robust iterative formulation. However, they are strongly efficient methods compared to FR for this example.

Example 6.9 A vehicle side-impact with response surface surrogate performance function, is given as follows,

$$\begin{aligned} g(\boldsymbol{X}) &= 0.489 x_3 x_7 + 0.843 x_5 x_6 - 0.0432 x_9 x_{10} \\ &+ 0.0556 x_9 x_{11} + 0.000786 x_{11}^2 - 0.75 \end{aligned} \tag{6.48}$$

Table 6.5 Reliability results of modified version conjugate FR for Example 6.8

Method	FR	LFR	SVFR	MFR
β (Iter)	2.283895 (57)	2.298247 (13)	2.297892 (11)	2.298009 (14)
MPP	(−1.6535, −1.5755)	(−1.6629, −1.5863)	(−1.6627, −1.5861)	(−1.6628, −1.5862)

where, $i = 1 \sim 7 x_i \sim N(1, 0.005), i = 8 \sim 9 x_i \sim N(0.3, 0.006), i = 10 \sim 11 x_i \sim N(0, 10.0)$ Table 6.6 gives reliability results as iterations, reliability index, and MPP ($\boldsymbol{X}^*$) for FR, LFR, SVFR, and MFR. It is demonstrated, that two CFORM methods of SVFR and LFR are efficient with twice convergent rate faster than FR. By comparing MFR and FR, MFR shows better performance.

Example 6.10 A automobile front axle with schematically view in Fig. 6.15 is studied by the following LSF which is computed by maximum stress of joint loads of bending moment and torque,

$$g = \sigma_s - \sqrt{\sigma_m^2 + 3\tau^2} \tag{6.49}$$

Table 6.6 Reliability results of modified version conjugate FR algorithms for Example 6.9

Method	FR	LFR	SVFR	MFR
β (Iter)	3.472649 (31)	3.497464 (17)	3.497487 (16)	3.511264 (21)
MPP ($\boldsymbol{X}^*$)	(0.9726, 0.9536, 0.9536, 0.9726, 0.3014, 30.0172, −8.3662)	(0.9654, 0.9434, 0.9434, 0.9654, 0.3021, 30.7313, −8.4658)	(0.9724, 0.9533, 0.9533, 0.9724, 0.3014, 30.2012, −8.3653)	(0.9723, 0.9532, 0.9532, 0.9723, 0.3015, 30.3430, −8.3685)

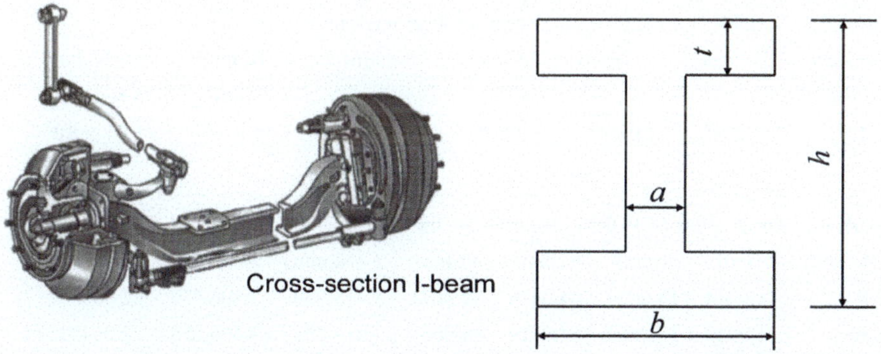

Fig. 6.15 A view for the front axle of automobile for Example 6.10. Reprinted from [16], with permission from Elsevier

where σ_s denotes limit stress given as 460 Mpa related to yielding material. σ_m and τ are respectively the maximum normal and shear stresses as expressed in Eq. (6.50) are computed by bending (M) and torque (T) moments,

$$\sigma_m = \frac{M}{W_x}$$
$$\tau = \frac{T}{W_\rho} \tag{6.50}$$

in which, W_x and W_ρ are the section factors given as below,

$$W_x = \frac{a(h-2t)^3}{6h} + \frac{b}{6h}[h^3 - (h-2t)^3]$$
$$W_\rho = 0.8bt^2 + 0.4\frac{a^3(h-2t)}{t} \tag{6.51}$$

where, a, b, t, and h are the dimensional sizes of I-beam (as plotted in Fig. 6.15). Normal variables are cited in Table 6.7 for computing the LSF in Eq. (6.49).

Table 6.8 summarizes the convergence results obtained by FR, LFR, SVFR, and MFR schemes. As seen in Table 6.8, the LFR, SVFR, and MFR procedures are robust methods as well as FR but LFR is efficient with reliability index of $\beta = 2.064536$. Whereas, SVFR and has the same efficiency as MFR and FR methods.

Table 6.7 Variables for the front axle of automobile in Example 6.10

Variable	a (mm)	t (mm)	b (mm)	h (mm)	T (N. mm)	M (N. mm)
Mean	12	14	65	85	3.1×10^6	3.5×10^6
STD	0.060	0.070	0.325	0.425	1.55×10^5	1.75×10^5

Table 6.8 The results of CFORM-based modified FR for Example 6.10

Methods	a (mm)	t (mm)	b (mm)	h (mm)	T (N. mm)	M (N. mm)	β (Iter)
FR	11.9914	13.9806	64.9426	84.9671	3,415,029	3,506,759	2.065667 (7)
LFR	11.9913	13.9804	64.9419	84.9668	3,414,731	3,506,658	2.064536 (4)
SVFR	11.9914	13.9806	64.9426	84.9672	3,415,028	3,506,759	2.065660 (7)
MFR	11.9914	13.9806	64.9426	84.9672	3,415,028	3,506,759	2.065660 (7)

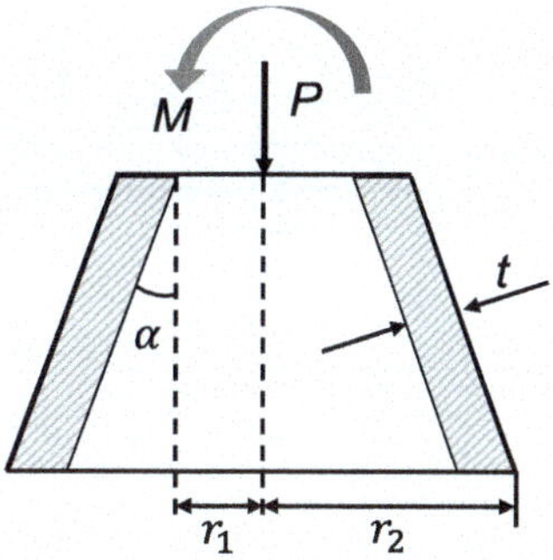

Fig. 6.16 The conical structure applied in Example 6.11. Reprinted from [16], with permission from Elsevier

Example 6.11 A structure as plotted in Fig. 6.16 under buckling failure mode is considered by the following LSF [11],

$$g = 1 - \frac{\sqrt{3(1-\nu^2)}}{\pi E t^2 \cos^2 \alpha} \times \left(\frac{P}{0.66} + \frac{M}{0.41 r_1} \right) \tag{6.52}$$

in which $\nu = 0.3$. Table 6.9 presents the variables as normal/non-normal distributions for Example 6.11.

After 17 iterations, the LFR converged to $\beta = 4.627765$ with related MPP of $\boldsymbol{X}^*$ = [65378.078, 0.002096, 0.52747, 0.88874, 108,411.745, 72,707.579]. By CFORM using FR, MFR, and SVFR, the reliability indexes are respectively as 4.635483 (obtained after 30 iterations), 4.723984 (obtained after 101 iterations), and 4.635342 (obtained after 30 iterations). It concluded that LFR is a superior framework by comparing efficiency. Nonetheless, the SVFR and MFR have low-computational burden compared to FR. The CG factor for updating the search direction in FORM formulation can be provided a stable normalized sensitivity vector that it can be enhanced the computational burden for this problem. Two factors as conjugate scaler and participation factors provide a line search direction which is enhanced the computational efforts of FORM. The results of this problem exposed that the directional conjugate NSV computed by CG vector and finite step length is a suitable analytical reliability method as a robust formulation provided acceptable efficiency.

Table 6.9 The variables for conical structure applied in Example 6.11

COV	Mean	PDF	Description	Variables
0.05	70,000	Lognormal	Young's modulus (MPa)	E
0.02	0.524	Normal	Slop angle (rad)	α
0.05	0.0025	Normal	Thickness (m)	t
0.025	0.9	Normal	Internal radius (m)	r_1
0.08	70,000	Gumbel	Axial load (N)	P
0.08	80,000	Gumbel	Bending moment (N-m)	M

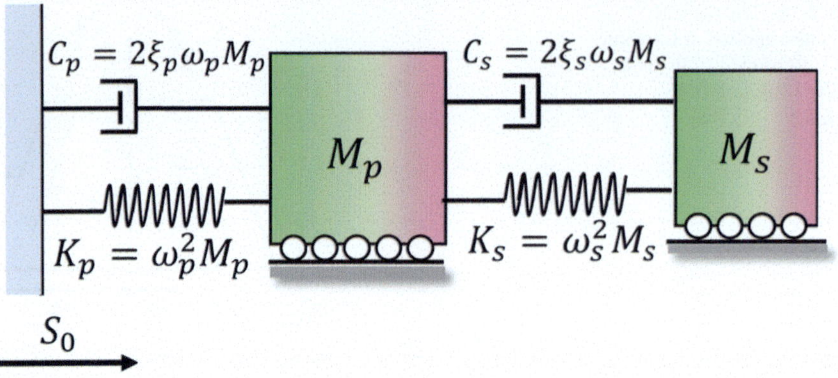

Fig. 6.17 A view of dynamic system applied in Example 6.12. Reprinted from [16], with permission from Elsevier

Example 6.12 As shown in Fig. 6.17, a nonlinear LSF is employed as below [10, 12],

$$g = Fs - Ks \times 3(E[x_s^2])^{1/2} \tag{6.53}$$

And,

$$E[x_s^2] = \frac{\pi S_0}{4\xi_s\omega_s}\left[\frac{\xi_a\xi_s}{\xi_p\xi_s(4\xi_a^2 + \theta^2) + \gamma\xi_a^2} \times \frac{\left(\xi_p\omega_p^3 + \xi_s\omega_s^3\right)\omega_p}{4\xi_a\omega_a^4}\right] \tag{6.54}$$

in which $\gamma = \frac{M_s}{M_p}$, $\omega_a = \frac{\omega_p + \omega_s}{2}$, $\xi_a = \frac{\xi_p + \xi_s}{2}$ and $\theta = \frac{\omega_p - \omega_s}{\omega_a}$ as variables followed by Lognormal distributions gives form Table 6.10.

Extracted results for FORM-based reliability analysis from Keshtegar [10] showed $\beta = 2.016348$ and $\boldsymbol{X}^* = [1.00191, 0.01009, 1.10209, 0.01115, 0.02799, 0.01211, 103.7171, 13.7360]$. Using the LFR method after 23 iterations, we have $\boldsymbol{X}^* = [1.00184, 0.01009, 1.10186, 0.01115, 0.0280, 0.01212, 103.7115, 13.7343]$ and $\beta = 2.014604$ which is are very closely followed to extracted results of Ref. [10, 12].

Figure 6.18 displays iterations for computing the reliability index based on FR, LFR, SVFR, and MFR algorithms. It is obvious that FR, LFR and MFR robustly converged ($\beta = 2.014604$). The MFR is an efficient approach among others.

Table 6.10 Variables applied in dynamic system of Example 6.12

Random variable	M_s	M_p	K_s	K_p	ξ_s	ξ_p	F_s	S_0
Mean	0.01	1	0.01	1	0.02	0.05	15	100
Standard deviation	0.001	0.1	0.001	0.2	0.01	0.02	1.5	10

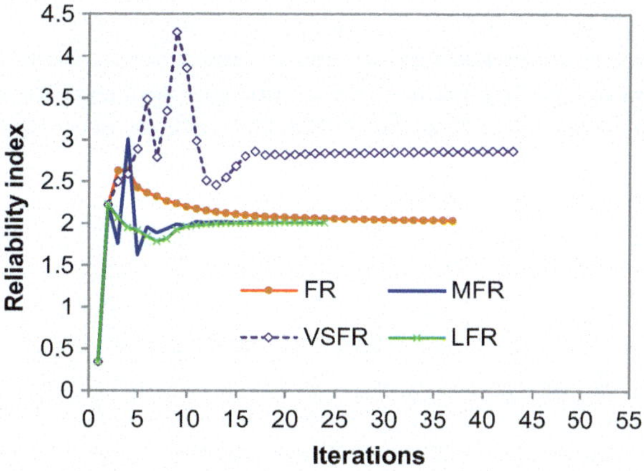

Fig. 6.18 Results of reliability index for different modified FR in Example 6.12

The reliability index obtained by SVFR i.e. $\beta = 2.869749$ is differed with other CFORMs and literature results. The conjugate search direction for SVFR developed by CG factor θ_k^{VFR} while other methods are a type of FR CG factor. Consequently, a premature convergence is provided based on scalar factor θ_k^{VFR} applied in SVFR. It can be concluded that the θ_k^{FR} is an acceptable CG factor for computing conjugate NSV. It is illustrated that MFR is more efficient than others with the convergent rate about twice faster than FR.

The effects of adaptive FSS and CG vector as factors in computing NSDV are discussed that it is shown that CFORM is robust formulated. The effect of step size applied in line search direction for computing NSDV is compared by HL-RF [13], FR [5], and LFR with constant intimal FSS = 20. The results of HL-RF, FR and LFR are arranged in Table 6.11 which is applied to demonstrate the effect of NSDV. The HL-RF method is not converged as presented in Examples 6.8, 6.9, 6.11, and 6.12. The FRand LFR methods are formulated with robust iterations by comparing with HL-RF formulated by SDSD. This means that the CG vector for computing NSDV reduces the risk of instability of analytical reliability methods.

Table 6.11 Results for HL-RF, CHL-RF, and LFR with a constant step size of 20

Examples	HL-RF ($\lambda = \infty$) β(Iter)	CHL-RF (FR) ($\lambda = 20$) β(Iter)	LFR ($\lambda = 20$) β(Iter)
6.8	Periodic-2 (1.5370, 1.5355)	2.283895 (57)	2.297514 (18)
6.9	Periodic-4 (3.1757, 2.0351, 1.9180, 1.2957)	3.485290 (47)	3.495399 (18)
6.10	2.064514 (5)	2.065667 (7)	2.064536 (4)
6.11	Chaos	4.700608 (117)	4.627777 (34)
6.12	Periodic-2 (4.2170, 4.9807)	2.044173 (36)	2.001147 (34)

The NSDV at new and previous iterations are not the same by using conjugate CG vector in FR and LFR methods. Consequently, the FR and LFR methods are robust analytical approaches. By comparing iterations for convergence in LFR and FR methods, the efficiency of FORM is strongly dependent on the conjugate normalized sensitivity vector. the efficiency of the LFR is improved by compassion with the FR method. The FSS in LFR and FR methods is not effective parameters in computing the line search while the CG vector is an effective factor to achieve an MPP by enhancing the efficiency of FORM.

Figures 6.19 and 6.20 show the convergence of the reliability index using HL-RF, FR, and LFR for Examples 6.9 and 6.12, respectively. The FR (CHL-RF) as well as LFR schemes are converged, robustly, but, LFR shows low-computational burden. The HL-RF method provided periodic-4 i.e. $\beta = (3.1757, 2.0351, 1.9180, 1.2957)$ solutions for Example 6.9 and periodic-2 i.e. $\beta = (4.2170, 4.9807)$ solutions for Example 6.12. Furthermore, the LFR is more robust. The results of these figures demonstrated that FR and LFR with $\lambda = 20$ are the robust methods. Thus, the line direction vector is more important for computing the NSDV.

The effects of FSS are also investigated as well as CG vector using adaptive FSS. Table 6.12 presents the reliability results for FR, and LFR that these results are obtained based on CFORM formulated by adaptive FSS. By comparing the results in Table 6.11 and 6.12, the adaptive FSS improved the efficiency of FR in Examples 6.9 and 6.11 and LFR methods almost examples. Consequently, the efficiency of CFORM can be improved by applying the FSS but it is not significant. As seen in Examples 6.11 and 6.12, LFR with adaptive FSS is twice faster than LFR with constant $\lambda = 20$.

Tables 6.11 and 6.12 showed that FR and LFR are robust for both FSS as adaptive formulation given by SDC and constant step size. The adaptive FSS is increased ability for high-convergence rate. Therefore, the conjugate SDV and FSS are two significant factors affected on abilities of analytical reliability methods.

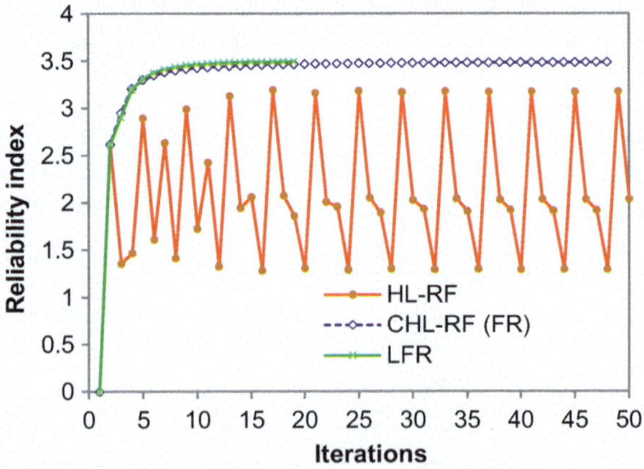

Fig. 6.19 Iterations applied for computing reliability index given by CFORM with $\lambda = 20$ for Example 6.9

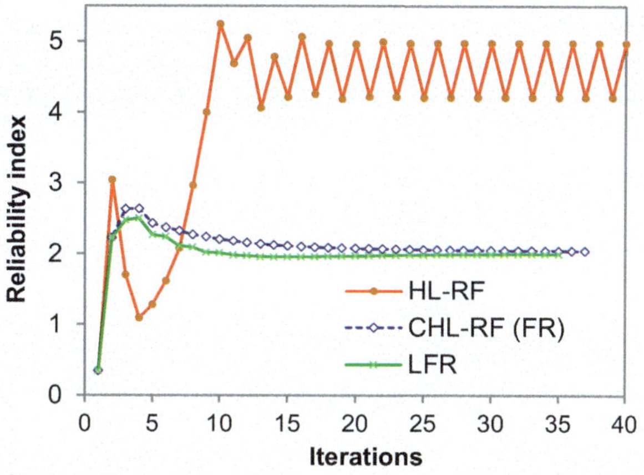

Fig. 6.20 Iterations applied for computing reliability index given by CFORM with $\lambda = 20$ for Example 6.12

Table 6.12 Comparing self-adaptive step length applied in CHL-RF and LFR

Examples	FR (CHL-RF) β (Iter)	LFR β (Iter)
6.8	2.283895 (57)	2.298247 (13)
6.9	3.472649 (31)	3.497464 (17)
6.10	2.065667 (7)	2.064536 (4)
6.11	4.723984 (101)	4.627765 (17)
6.12	2.04508 (36)	2.015282 (23)

Three improved CFORM methods for SRA are formulated in this section named as LFR, MFR and SVFR coupled with adaptive FSS. Their abilities are compared with FR [5] (with $\lambda_0 = 50$ and $c = 0.95$) and FSL-based SDSD method (with $\lambda_0 = 50$ and $c = 0.95$) [8]. The reliability index (β), Iterations, and the Central Processing Unit (CPU)-run times (T) as β/T(s)/Iter are utilized for comparing performances of these studied methods in this part. The results of studied methods are listed in Table 6.13. As presented results in Table 6.13, modified FORM using conjugate SDV are robustly converged but the different efficiencies are obtained for CFORM. As presented results for Examples 6.8, 6.9, and 6.12, FR, LFR, SVFR, and MFR are more computationally efficient than FSL. In Example 6.11, the LFR is a more efficient method, while for Examples 6.8 and 6.12, the MFR is more efficient. The adaptive FSS can improve the efficiency of CFORM as the FR method that is formulated by a constant initial step $\lambda_0 = 50$.

Three modified search direction versions using conjugate FR search direction called as LFR, MFR, and SVFR are compared. An adaptive FSS obtained by the Armijo rule and SDC is extended in MFR, SVFR and LFR schemes. The results for reliability problems displayed that the LFR, SVFR and MFR are successfully applied for SRA and these CFORMs as well as FR are robust when a step size is selected without line search rules as SDC. The adaptive FSS enhances the efficiency

Table 6.13 Results of FSL, FR, LRF, SVFR and MFR for Examples 6.8, 6.9, 6.10, 6.11, and 6.12

Examples	FSL [8]	FR [5]	LFR	SVFR	MFR
6.8	2.2983/1.93/ 646	2.2955/0.14/ 28	2.2983/0.10/13	2.2979/0.09/11	2.2980/0.11/ 14
6.9	3.4975/6.99/ 1293	3.4746/0.51/ 28	3.4975/0.38/17	3.4975/0.39/16	3.5113/0.43/ 21
6.10	2.0645/0.11/ 5	2.0657/0.14/ 7	2.0657/0.08/4	2.0657/0.14/7	2.0657/0.14/7
6.11	4.6278/0.81/ 54	4.7029/1.71/ 122	4.6277/0.44/17	4.6353/0.82/30	4.6355/0.64/ 30
6.12	2.0164/20.14/ 331	2.0455/2.61/ 36	2.0152/1.72/23	2.8688/2.86/43	2.0178 /1.24/ 16

of FORM in modified versions of FR. On the other hand, by using SDC in the FSS of the FSL method, this FORM-based SDSD has also a robust formulation but its efficiency is depended on nonlinearity of LSFs.

6.5 Chaotic Conjugate CC Method

The instabilities of FORM for MPP search are improved by adjusting a step size in analytical methods. In these cases, the step size is adjusted by a convergence rule, for example in IHL-RF-based steepest descent sensitivity vector applied as merit function [11], Armijo rules and Wolfe conditions. However in STM [3] and DSTM [17]. The step size is given a smaller positive constant value. An adapting step size with SDC is applied for adjusting and computing the FSS in FSL [8] and FR (CHL-RF) [5], respectively. The robustness of FORM is improved when a step size with the adaptive formulation is applied in modified FORM algorithms compared to the HL-RF method. As shown in Chaps. 2 and 4, STM and DSTM are as simple as HL-RF but are robust enough to search the MPP. However, two methods for convex or linear problems may provide inefficient time-consuming iterations. In this section, an iterative FORM algorithm is introduced formulated by chaotic conjugate dynamical map and combined with chaotic FSS which is adjusted by SDC that this CFORM-based MPP search is named chaotic conjugate CC (CCC) method.

6.5.1 Formulation and Algorithm of the CCC Method

In the CCC method, the iterative FORM conjugate vector is applied for MPP as below,

$$\boldsymbol{U}_{k+1} = \frac{\nabla^T g(\boldsymbol{U}_k)U_k - g(\boldsymbol{U}_k)}{\nabla^T g(\boldsymbol{U}_k)\boldsymbol{\alpha}_k^{c\lambda}}\boldsymbol{\alpha}_k^{c\lambda} \tag{6.55}$$

where, $\boldsymbol{\alpha}_k^{c\lambda}$ is conjugate NSV determined as $\boldsymbol{\alpha}_k^{c\lambda} = \frac{U_{k+1}^{c\lambda}}{\|U_{k+1}^{c\lambda}\|}$ and $\boldsymbol{U}_{k+1}^{c\lambda} = \boldsymbol{U}_k + \lambda_k \boldsymbol{d}_k$. Where, $\boldsymbol{d}_k$ named CG vector is approximated by FR as $\boldsymbol{d}_k = -\nabla g(\boldsymbol{U}_k) - \frac{\|\nabla g(\boldsymbol{U}_k)\|^2}{\|\nabla g(U_{k-1})\|^2}\boldsymbol{d}_{k-1}$. The major different in this method with FR and its modified formulation is the step size λ_k which is defined as a FSS i.e. $\lambda \geq 1$. It is supposed that λ satisfies the SDC i.e. $\|\boldsymbol{U}_{k+1} - \boldsymbol{U}_k\| < \|\boldsymbol{U}_k - \boldsymbol{U}_{k-1}\|$. On the other hand, Armijo-type line search as well as enriched FR method is approximated by largest step size which is computed by initial Armijo rule as below,

$$\lambda_{\max} \leq M\frac{\|\boldsymbol{d}_k\|^2}{\|\nabla g(\boldsymbol{U}_k)\|^2} \tag{6.56}$$

in which, $M >> 1$. Therefore, initial finite-step size is suggested as $\lambda_0 = \frac{25}{\|\nabla g(\boldsymbol{U}_{|U=\mu})\|}$.

The SDC implies as $\lambda_k = c_m\lambda_{k-1}$ for control instability of FORM in CCC algorithm by condition of $\left\|\boldsymbol{U}_{k+1}^C - \boldsymbol{U}_k^C\right\| \geq \left\|\boldsymbol{U}_k^C - \boldsymbol{U}_{k-1}^C\right\|$. Therefore, the step size is adjusted as follows,

$$\lambda_k = \begin{cases} \lambda_{k-1} & \left\|\boldsymbol{U}_{k+1}^C - \boldsymbol{U}_k^C\right\| < \left\|\boldsymbol{U}_k^C - \boldsymbol{U}_{k-1}^C\right\| \\ c_m\lambda_{k-1} & o.w \end{cases} \tag{6.57}$$

where, c_m is a chaotic adjusting factor in interval $0 < c_m < 1$. Thus, we have $\lambda_k \leq \lambda_{k-1}$. This means that when applied a nonlinear function, for $k \to \infty$ we have $\lambda_k \approx 0$. The Logistic map can apply for computing c_m with pseudo-random chaotic behavior by the following discrete map,

$$c_m = ac_m(1 - c_m) \tag{6.58}$$

where, $c_m \in (0, 1)$ is named as chaotic adjusted factor given by Logistic map with parameter of $0 < a \leq 4$. For a selected in interval 3.57 ~ 4, Logistic map simulates chaotic data and also $c_0 \notin \{0.0, 0.25, 0.5, 0.75, 1.0\}$. By chaotic adjusted factor, it can be given a positive value as $0 < c_m < 1$, thus, $\lambda_k \leq \lambda_{k-1}$.

The instability in FORM provides at the point $\boldsymbol{U}_k$ when, the spectral radius of the Jacobian matrix $[J]\left([\boldsymbol{J}] = \frac{\partial f_i(u)}{\partial u_j}|_{U_k}, f(u) = \boldsymbol{U}_{k+1}\right)$ by iterative FORM formula is obtained larger than 1 as $\rho(J) > 1$. The iterative analytical method can be shown the stable solutions when $\rho(J) < 1$ and for $\rho(\boldsymbol{J}) = 1$, it is obtained a fixed point related to new and previous points are same points. In an iterative FORM scheme, for $\rho(\boldsymbol{J}) > 1$, the analytical method provides unstable results thus it showed be computed a point that it provided $\rho(\boldsymbol{J}) < 1$ to control the instability of FORM. Therefore, chaos feedback control can be applied to achieving a $\rho(\boldsymbol{J}) \leq 1$ as below,

$$\boldsymbol{U}_{k+1} = \boldsymbol{U}_k + c_m\left[f^C(u_k) - \boldsymbol{U}_k\right] \tag{6.59}$$

in which, $0 < c_m < 1$ relates chaotic control factor that it is generated by Logistic map. $f^C(u_k)$ is the solution of the k_{th} iteration for CFORM at point U_k that it is determined as below,

$$f^C(u_k) = \frac{\nabla^T g(\boldsymbol{U}_k)\boldsymbol{U}_k - g(\boldsymbol{U}_k)}{\nabla^T g(\boldsymbol{U}_k)\boldsymbol{\alpha}_k^{c\lambda}}\boldsymbol{\alpha}_k^{c\lambda} \tag{6.60}$$

Using chaotic CFORM by using c_m, the $\rho(J)$ is improved as $\left[\overline{\boldsymbol{J}}\right] = I + c_m([\boldsymbol{J}] - \boldsymbol{I}))$. The stability condition as $\rho\left(\overline{\boldsymbol{J}}\right) \leq 1$ is defined for adjusting the step size in the iterative CFORM formula as below,

$$\boldsymbol{U}_{k+1} = \begin{cases} \boldsymbol{U}_k + c_m\left[f^C(u_k) - \boldsymbol{U}_k\right] & \rho\left(\overline{\boldsymbol{J}}\right) > 1 \\ f^C(u_k) & o.w \end{cases} \tag{6.61}$$

The chaotic map is applied in this formulation of FORM. As seen from the CCC formulation, it is given the SDC for updating finite-step size with the initial value computed by the Armijo rule in the first stage. In the second stage, the stability condition defined by $\rho(\boldsymbol{J})$ is also controlled by the instability of this method by chaotic adjusting factor c_m. The stability condition defined as $\rho\left(\overline{\boldsymbol{J}}\right) > 1$ for CFORM is given by FR approach. The CCC method can be coded based on the following steps in a computer program:

Step 1: Define ε (stopping criterion), $g(\boldsymbol{U}) = 0$, give variables with distributions, $\boldsymbol{\mu}$ and $\boldsymbol{\sigma}$, set $c_0 \notin \{0.0, 0.25, 0.5, 0.75, 1.0\}$, $a \in [3.6, 4]$, $k = 0$ and initial value as $\boldsymbol{X}_0 = \boldsymbol{\mu}$ and $d_0 = 0$;

Step 2: Transform variables into $\boldsymbol{U}$-space as $u_k = \frac{x_k - \mu_x^e}{\sigma_x^e}$, where $\sigma_x^e = \frac{1}{f_X(x_k)}\varphi\left[\Phi^{-1}\{F_X(x_k)\}\right]$ and $\mu_x^e = x_k - \sigma_x^e\Phi^{-1}[F_X(x_k)]$;

Step 3: Iterative process of modified version of FR methods by the following relations;

(i) Determine $\nabla g(\boldsymbol{U})$ and $g(\boldsymbol{U})$ at pint $\boldsymbol{U}_k$;
(ii) If $k = 0$ then compute $\lambda_0 = \frac{25}{\left\|\nabla g\left(\boldsymbol{U}|_{U=\mu}\right)\right\|}$;
(iii) Compute $\theta_k^{FR} = \frac{\|\nabla g(\boldsymbol{U}_k)\|^2}{\|\nabla g(\boldsymbol{U}_{k-1})\|^2}$ and determine $\boldsymbol{d}_k = -\nabla g(\boldsymbol{U}_k) - \theta_k^{FR}\boldsymbol{d}_{k-1}$;
(iv) Determine $\boldsymbol{U}_{k+1}^{c\lambda} = \boldsymbol{U}_k + \lambda\boldsymbol{d}_k$ and then $\boldsymbol{\alpha}_{k+1}^{c\lambda} = \frac{\boldsymbol{U}_{k+1}^{c\lambda}}{\left\|\boldsymbol{U}_{k+1}^{c\lambda}\right\|}$;
(v) Determine $f^C(u_k) = \frac{\nabla^T g(\boldsymbol{U}_k)\boldsymbol{U}_k - g(\boldsymbol{U}_k)}{\nabla^T g(\boldsymbol{U}_k)\alpha_k^{c\lambda}}\alpha_k^{c\lambda}$;

(vi) Adjust step length as $\lambda_k = \begin{cases} \lambda_{k-1} & \left\|f^C(u_k) - \boldsymbol{U}_k\right\| < \|\boldsymbol{U}_k - \boldsymbol{U}_{k-1}\| \\ c_m\lambda_{k-1} & \text{otherwise} \end{cases}$
where c_m generated positive value by Logistic map as $c_m = ac_m(1 - c_m)$;
(vii) Compute $\rho(\boldsymbol{J})$ as $f^C(u_k) = [\boldsymbol{J}]\boldsymbol{U}_k$;
(viii) Control new point as $\boldsymbol{U}_{k+1} = \begin{cases} \boldsymbol{U}_k + c_m[f^C(u_k) - U_k] & \rho(\boldsymbol{J}) > 1 \\ f^C(u_k) & o.w \end{cases}$;

Let the next iterate using $\boldsymbol{X}_{k+1} = \mu_X^e + \sigma_X^e \boldsymbol{U}_{k+1}$ and $\beta_{k+1} = \|\boldsymbol{U}_{k+1}\|$ and set $k = k + 1$;

Step 4: If $\|U_{k+1} - U_k\| < \varepsilon$, then presents results as $P_f \approx \Phi(-\beta)$ and $\boldsymbol{\alpha} = \boldsymbol{\alpha}_{k+1}^{c\lambda}$, else go to Step 2.

The flowchart of the CCC method for reliability analysis which is a manner of better coding this method in a computer program is presented in Fig. 6.21. It is shown that two controlling processes are given for updating the new point using SDC and proposed stability conditions that in the first case, the Logistic map is utilized to pitch adjust the finite-step size by sufficient descent condition i.e. $\left|\left|f^C(u_k) - \boldsymbol{U}_k\right|\right| < ||\boldsymbol{U}_{k-1} - \boldsymbol{U}_k||$ while the new point obtained by CFORM formulation is updated using a chaotic random number based on stability condition i.e.$\rho(\boldsymbol{J}) > 1$ in the second stage. The stability condition is controlled Jacobian matrix $\rho(\boldsymbol{J})$ for stabilization.

6.5.2 Illustrative Performances of the CCC Method

The performances of CCC-based FORM algorithms are investigated using several reliability problems. Efficiency and robustness are two main preformed factors for evaluating the CCC iterative method. The CCC performances are compared with FORM-based SDSD of HL-RF and STM (with $\lambda = 0.1$) and $\varepsilon = 10^{-5}$ as stopping criterion is selected in all algorithms.

Example 6.13 A non-normal and nonlinear LSF,

$$g(\boldsymbol{X}) = x_1^4 + x_2^2 - 50 \tag{6.62}$$

where, Lognormal x_1 has $\mu = 5$ and $\sigma = 1$, Gumbel x_2 has $\mu = 10$ and $\sigma = 10$. The obtained result using FORM from Ref [18] as $\beta = 3.254$. The average and STD for the reliability index obtained after 30- a time run using MCS with 1.2×10^6 samples are $\beta = 3.5009$ and STD $= 0.0133$). By using CCC based on 11 iterations, the reliability index is $\beta = 3.259315$ which is followed the results of literature.

The iterations for reliability index STM, HL-RF, and CCC are presented in Table 6.14. The HL-RF not robustly converged, as seen, we have periodic-4 $\beta = [2.5118, 1.8550, 2.3857, 1.5966]$. Unlike HL-RF, other methods are robust. However, the CCC is significantly more efficient formulation and it is about seven-time efficient than STM.

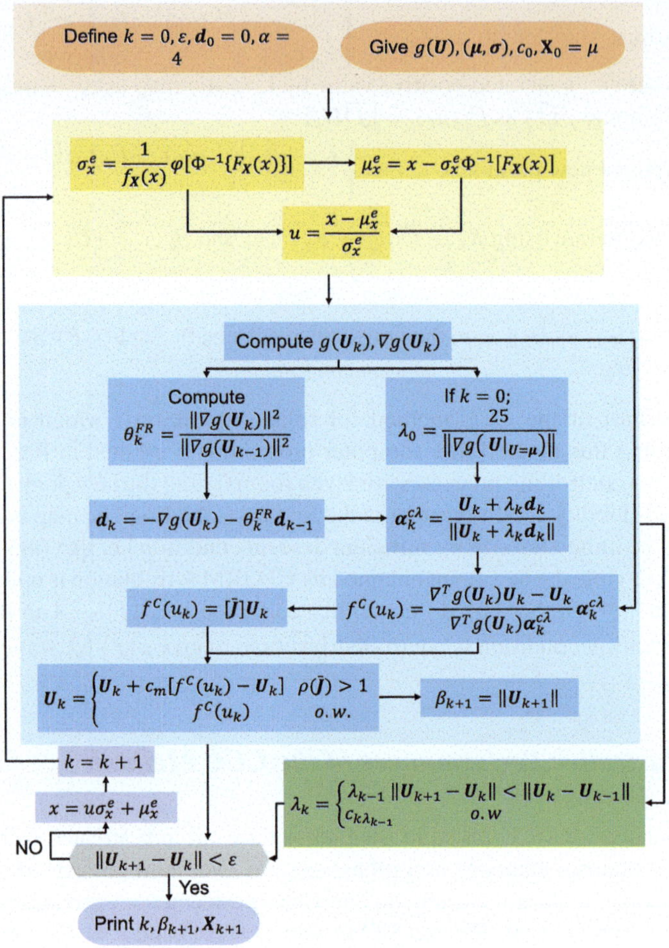

Fig. 6.21 Flowchart of CCC for iterative algorithm of FORM

Example 6.14 A artificial high-frequency LSF as [13],

$$g(\boldsymbol{X}) = x_1 + 2x_2 + 2x_3 + x_4 - 5x_5 - 5x_6 + 0.001\sum_{i=1}^{6}\sin(100x_i) \tag{6.63}$$

The Lognormal $x_1 - x_6$ variables are defined as $x_1 - x_4$ with mean of 120 and STD of 12; x_5 with mean of 50 and STD of 15; and x_6 with mean of 40 and STD of 12.

Table 6.14 Results for CC, HL-RF and STM applied in Example 6.13

Iterations	HL-RF	STM	CCC
1	0.202789	0.202789	0.202789
2	1.281802	0.119649	1.115648
…	…	…	…
10	1.668143	0.935938	3.259315
11*	2.476948	1.046539	**3.259315**
12	1.671091	1.154875	Converged
13	2.499926	1.260821	
…	…	…	
113	2.511743	3.259177	
114	1.85489	3.25919	
115	2.385844	3.259213	
116	1.596655	3.259215	
117	2.511745	**3.259215**	
…	…	Converged	
297	**2.511747**		
298	**1.855042**		
299	**2.385726**		
300	**1.596633**		
	Periodic-4		

We use extracted results from literatures as:

(i) Liu and Der Kiureghian [13], $\beta = 2.3482$ and $\boldsymbol{X}^* = [117.3, 115.3, 115.3, 117.3, 83.62, 55.54]$;
(ii) Obtained by Yang [3], $\beta = 2.3482$ and $\boldsymbol{X}^* = [117.29, 115.16, 115.16, 117.29, 83.38, 55.65]$;
(iii) By Ref. [1] with $\beta = 2.348257$ and $\boldsymbol{X}^* = [117.35, 115.28, 115.28, 117.35, 83.52, 55.64]$, and computed using MCS with 1×10^6 samples $\beta = -\Phi^{-1}(P_f) = 2.25191$.

The iterations of the reliability index for FORM-based STM, HL-RF, and CCC are demonstrated in Fig. 6.22. As seen, the CFORM obtained by SDC proposed in CCC provides stable solution with low computational burden after 23 iterations. Nevertheless, the STM and HL-RF exhibit chaotic solutions. The reliability results based on CCC are obtained as $\boldsymbol{U}^* = [-0.1863, -0.3646, -0.3646, -0.1863, 1.8726, 1.2936]$, $\boldsymbol{X}^* = [117.206, 115.140, 115.149, 117.206, 82.984, 56.011]$, and $\beta = 2.348443$. As is obvious, these results are confirmed by the extracted results from literatures.

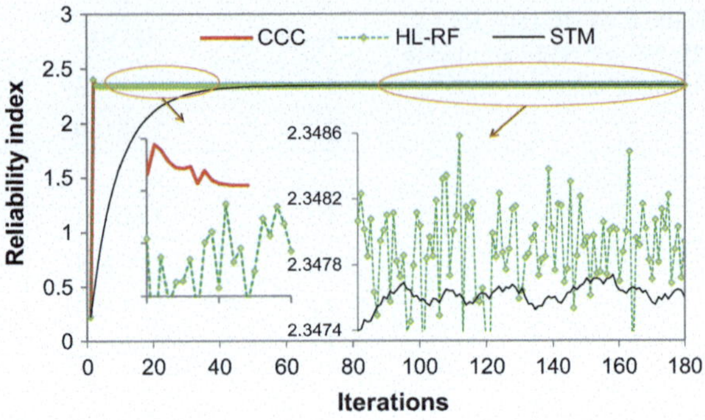

Fig. 6.22 Iteration for reliability index for comparing the CCC method in Example 6.14

Example 6.15 A Conical Structure as presented in Fig. 6.16 is considered with below LSF,

$$g = 1 - \frac{\sqrt{3(1 - v^2)}}{\pi E t^2 \cos^2 \alpha} \times \left(\frac{P}{0.66} + \frac{M}{0.41 r_1} \right) \tag{6.64}$$

The buckling LSF is considered in this Example where, $v = 0.3$ and LSF presented in Eq. (6.64) is evaluated by variables listed in Table 6.15.

By MCS with a ten-million samples, the reliability index is evaluated as 4.4163. The reliability index using CCC after 22 iterations is 4.630749 with $\boldsymbol{X}^* = [65442.63, 0.00208, 0.5275, 0.8887, 106{,}149.59, 73{,}112.66]$.

Figure 6.23 shows the iterations for the reliability index given from STM, HL-RF and CCC. The results illustrated that CCC-based CFORM formulated based on conjugated NSV is a stable framework. However, the HL-RF and STM are unstable iterative approaches.

Table 6.15 Variables for structure applied in Example 6.15

Variables	Description	PDF	Mean	COV
E	Young's modulus (MPa)	Lognormal	70,000	0.05
t	Thickness (m)	Normal	0.0025	0.05
α	Slop angle (rad)	Normal	0.524	0.02
r_1	Internal radius (m)	Normal	0.9	0.025
P	Axial load (N)	Gumbel	70,000	0.08
M	Bending moment (N-m)	Gumbel	80,000	0.08

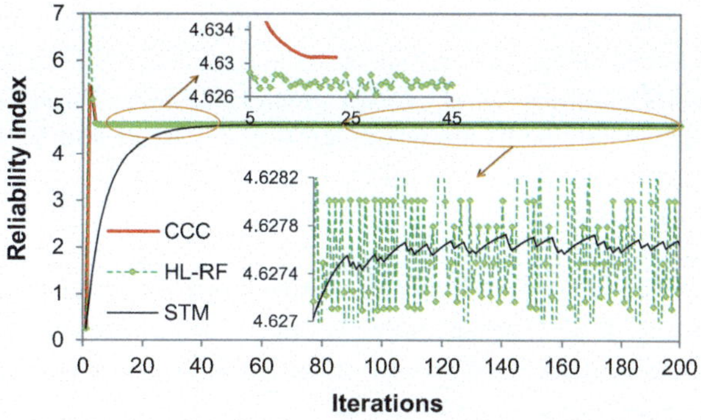

Fig. 6.23 Iteration for reliability index for comparing the CCC method Example 6.15

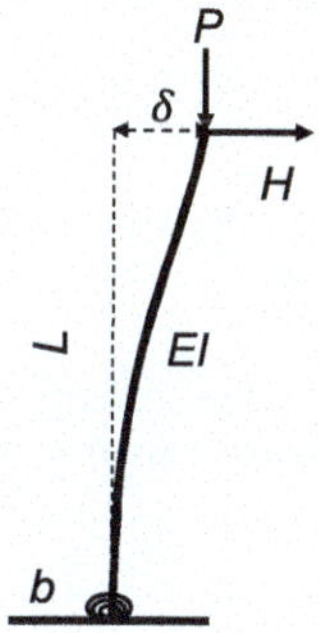

Fig. 6.24 Column applied in Example 6.16. Reprinted from [11], with permission from Springer Nature

Example 6.16 A slender column as shown in Fig. 6.24 with displacement LSF by the following function,

$$g = 40 - \delta \tag{6.65}$$

where δ is the horizontal displacement at top of column that it is measured as follows,

$$\delta = t[(\tan(\psi) - \psi) + \frac{b^2\left(1 + \sqrt{1 - 4\left(HEI\tan^2(\psi)\right)/b^2}\right)^2}{4HEI\tan(\psi)} \tag{6.66}$$

in which,

$$\psi = L\sqrt{\frac{P}{EI}}, t = \frac{H}{EI\left(\frac{P}{EI}\right)^{1.5}} \tag{6.67}$$

The column is connected with a rotational spring with parameter b and nonlinear relation between moment M and rotation α as $M = b\sqrt{\alpha}$.

Six normal variables of this example are listed in Table 6.16. The extracted MPP from Ref [19] is $U^* = (0.550, 1.181, -1.238, 1.407, -0.562, 0.0003)$. By MCS with 1×10^6 samples, the reliability index is 2.35434. Based on CCC after 14 iterations, the converged results are presented $\beta = 2.350831$ and $U^* = (0.5608, 1.2274, -1.2315, 1.3696, -0.5594, 0.00021)$. It is concluded that the CCC is accurately converged with results similar to MPP extracted from Ref. [19] and reliability index given from MCS.

The iteration for the reliability index using STM, HL-RF, and CCC are plotted in Fig. 6.25. The unstable chaotic solutions are captured by STM and HL-RF. The CCC-based CFORM with adaptive FSS provides a suitable line search for MPP search with robust properties.

Example 6.17 A frame as plotted in Fig. 6.26 with below LSF,

$$g = 0.05 - \Delta \tag{6.68}$$

The variables for computing horizontal displacement are presented in Table 6.17.

The results obtained by the MCS with 1.2×10^6 samples are $P_f = 1.08 \times 10^{-3}$ related to $\beta = 3.0673$. Using CCC, the obtained results after 17 iterations is a reliability index of 3.168196. Figure 6.27 shows the iterations for the reliability index obtained by FORM-based SDSD of HL-RF, STM, and CFORM-based CCC methods.

Table 6.16 Variables for slender column applied in Example 6.16

Variable	Description	Mean	COV
P	Axial load (N)	1,000,000	0.2
E	Young's modulus (MPa)	20,000	0.1
I	moment of inertia (mm^4)	1.0417×10^{10}	0.05
H	Horizontal load (N)	10,000	0.2
L	Column of length (mm)	10,000	0.05
b	Spring parameter (N mm)	0.4167	0.1

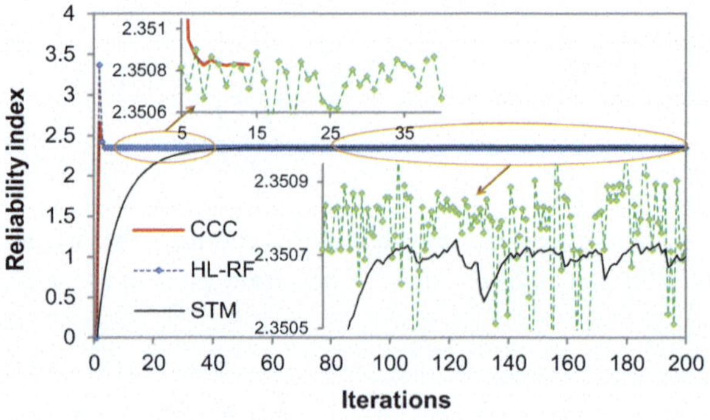

Fig. 6.25 Iteration for reliability index for comparing the CCC method for Example 6.16

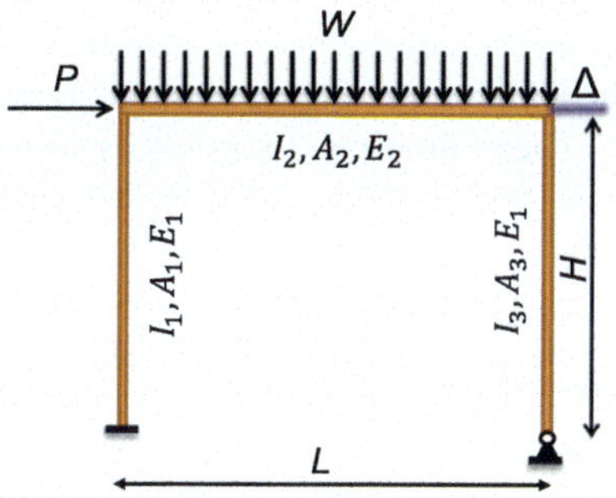

Fig. 6.26 The steel frame applied in Example 6.17. Reprinted from [10], with permission from Elsevier

Table 6.17 Variables for steel frame applied in Example 6.17

Variable	Mean	STD	PDF	Variable	Mean	STD	PDF
A_1 (m^2)	91×10^{-4}	9.1×10^{-4}	Lognormal	E_1 (kg/m^2)	2.38×10^{10}	1.9×10^9	Normal
A_2 (m^2)	53.8×10^{-4}	5.38×10^{-4}	Lognormal	E_2 (kg/m^2)	2.17×10^{10}	1.74×10^9	Normal
A_3 (m^2)	106×10^{-4}	10.6×10^{-4}	Lognormal	H (m)	4.5	0.36	Normal
I_1 (m^4)	80.9×10^{-6}	8.09×10^{-6}	Lognormal	L (m)	6	0.48	Normal
I_2 (m^4)	83.6×10^{-6}	8.36×10^{-6}	Lognormal	P (kg)	2000	350	Gumbel
I_3 (m^4)	112.6×10^{-6}	11.26×10^{-6}	Lognormal	W (kg/m)	4500	630	Gumbel

The STM as well as HL-RF are unstable relation for this problem, nonetheless CCC computed by conjugate NDSV with chaotic FSS is robust.

The results from Figs. 6.22, 6.23, 6.25, and 6.27 illustrated that the HL-RF yields an unstable formulation of FORM. The STM with step size $\lambda = 0.1$ is exhibited unstable iterations as chaotic solutions. On the other hand, the STM with $\lambda = 0.1$ is

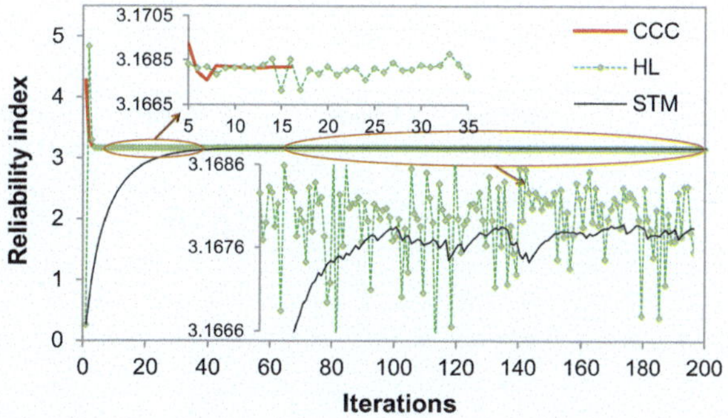

Fig. 6.27 Iteration for reliability index for comparing the CCC method for Example 6.17

an inefficient method as seen in Example 6.13. However, unlike HL-RF and STM, the CCC can converge to stable results.

The evaluations for gradient vector (Iter) and β for CCC, RHL-RF [4], STM (with $\lambda = 0.05$ and $\lambda = 0.5$) [3], FSL (with $\lambda = 15$ and $c = 1.4$) [8], and FR (with λ_{max} and $c_1 = 0.75$) [5] are listed in Table 6.18.

By results in Table 6.18, STM with $\lambda = 0.05$ shows the chaotic results for Examples 6.15 and 6.17. The STM with $\lambda = 0.5$ periodically converged as [2.510031, 2.572364] in Example 6.13. Therefore, STM and HL-RF are two FORM-based analytical approaches these procedures are formulated by SDSD.

The FSL, CCC and FR formulated based on SDC are robust but FR is efficient. Unlike FR and FSL methods, RHL-RF does not provide stable results for some case of reliability problems (see Example 6.14). The FSL scheme is an inefficient approach compared to CFORM-based CCC and FR.

The iterative CCC algorithm as CFORM is as stable as FR, and more efficiently. Therefore, the CCC is more robust than the FORM-based SDSD method of STM and RHL-RF and more efficient than CFORM-based FR while it is significantly better than the FSL algorithm.

The iterative SDSD applied in FORM using HL-RF without control, and STM with control provided the unstable iterations as chaotic or periodic solutions. The dynamical map for computing MPP can be stably enhanced by a chaotic control-based conjugate SDV named CCC in this section. It was demonstrated that HL-RF as well as STM and RHL-RF produced numerical instabilities. The CCC formulated by chaotic FSS is converged to results without instability in its iteration, efficiently. The CCC method is faster FORM approach by comparing with FSL and FR presented as robust formulations.

Table 6.18 Results for RHL-RF, STM, FSL, FR and CCC

Examples	STM		FSL	RHL-RF	FR	CCC
	$\lambda = 0.05$	$\lambda = 0.5$				
6.13	3.259127 (224)	Periodic-2	3.259287 (37)	3.259315 (105)	3.244598 (54)	3.259315 (11)
6.14	2.348114 (207)	Chaos	Chaos	2.348149 (96)	2.348127 (46)	2.348443 (24)
6.15	Chaos	Chaos	4.627784 (51)	4.627849 (63)	4.627742 (45)	4.630749 (22)
6.16	2.350689 (782)	Chaos	2.350663 (56)	2.350819 (91)	2.355773 (38)	2.350831 (14)
6.17	Chaos	Chaos	3.167918 (1391)	3.169266 (118)	3.168134 (44)	3.168197 (17)

6.6 Chaotic Conjugate STM

As presented results in Sect. 6.5, the STM is an inefficient method and provides unstable results for some cases of the performance functions. In this section, an iterative CFORM algorithm is discussed which is formulated using conjugate FR SDV applied in STM strategy. A CC factor simulated by a Logistic map is applied in CFORM-based STM. A transformation-rotational matrix is suggested in formulation of CFORM-named chaotic conjugate STM (CCSTM) [10]. To improve both the efficiency and the robustness of STM, the CCSTM is formulated in this section.

6.6.1 CCSTM Formulation and Algorithm

A novel iterative algorithm is introduced by the idea of conjugate SDV and STM in FORM. Therefore, the normalized SDV at $\boldsymbol{U}_k$ is suggested as follows,

$$\alpha_{k+1}^{c} = \frac{\boldsymbol{U}_{k+1}^{C}}{\left\|\boldsymbol{U}_{k+1}^{C}\right\|} \tag{6.69}$$

where

$$\boldsymbol{U}_{k+1}^{C} = \boldsymbol{U}_k + \boldsymbol{d}_k \tag{6.70}$$

where, $\boldsymbol{d}_k$ is CG vector given by FR as $\boldsymbol{d}_k = -\nabla g(\boldsymbol{U}_k) - \theta_k^{FR}\boldsymbol{d}_{k-1}$ with CG factor of $\theta_k^{FR} = \frac{\|\nabla g(\boldsymbol{U}_k)\|^2}{\|\nabla g(\boldsymbol{U}_{k-1})\|^2}$. The new point is calculated by CG line search direction as,

$$f^{c}(u_k) = \frac{\nabla^T g(\boldsymbol{U}_k)\boldsymbol{U}_k - g(\boldsymbol{U}_k)}{\nabla^T g(\boldsymbol{U}_k)\alpha_{k+1}^{c}}\alpha_{k+1}^{c} \tag{6.71}$$

As presented in this chapter, the FORM-based conjugate NSV has a global convergence, when $d_k \approx 0$, thus $\boldsymbol{U}_{k+1}^{C} \approx \boldsymbol{U}_k$ as a fixed position obtained for $\boldsymbol{U}_{k+1}$. Also, it showed that $\|\boldsymbol{d}_k\| < \|\boldsymbol{d}_{k-1}\|$ for FR formulation, thus we $\lim_{k\to\infty} \|\boldsymbol{d}_k\| \approx 0$ for a large computational burden i.e. $k \to \infty$ that it is obtain based on $\|\boldsymbol{d}_k\| < \theta_k \|\boldsymbol{d}_{k-1}\|$. For $\theta_k = 0$, the iterative CFORM is unrobust relation without stabilization because we have,

$$\|\nabla g(\boldsymbol{U}_k)\| = \left\|\boldsymbol{U}_{k+1}^{HL} - \boldsymbol{U}_k\right\| > \|\nabla g(\boldsymbol{U}_{k-1})\| \tag{6.72}$$

Thus, $\theta_k = \frac{\|\nabla g(\boldsymbol{U}_k)\|^2}{\|\nabla g(\boldsymbol{U}_{k-1})\|^2} > 0$. The CFORM strategy is combined with a STM formulation for controlling its instability by the following iterative formulation,

$$\boldsymbol{U}_{k+1} = \boldsymbol{U}_k + \lambda C\left[f^{c}(u_k) - \boldsymbol{U}_{\boldsymbol{k}}\right] \tag{6.73}$$

where, $f^c(u_k)$ is discrete conjugate map determined by Eq. (6.71). A chaotic control factor is introduced using Logistic map for determining λ as $\lambda = a\lambda(1-\lambda)$, where, λ is step size with a chaotic number $\lambda \in (0, 1)$ and a is the control parameter as $0 < a \leq 4$. The sequence of the Logistic map has chaotic behavior when $a = 4$ and $\lambda_0 \notin \{0.0, 0.25, 0.5, 0.75, 1.0\}$. The chaotic control factor is defined by Logistic map as follows:

$$\lambda_{k+1} = 4\lambda_k(1-\lambda_k)\lambda_0 \in (0, 1), \notin \{0.25, 0.5, 0.75\} \tag{6.74}$$

A chaotic CC factor (λ^C) is introduced for CCSTM as below relation,

$$\lambda_{k+1}^C = \min\left\{4\lambda_k^C\left(1-\lambda_k^C\right),\ 0.98^k\right\} \tag{6.75}$$

The chaotic CC factor in Eq. (6.75) is driven in Fig. 6.28 for a 10-time random CC factor. The CC factors are chaotically performances in iterations less than 10, a same stable value is obtained at iterations > 30. λ^C varied from 0 to 0.8 when $k <$ *30* but is a small value at final iterations (e.g. $\lambda_{100}^C = 0.135$, $\lambda_{500}^C = 4.19 \times 10^{-5}$). When $k \to \infty$, we give a $\lambda_k \to 0$ consequently, $\lambda C\left[f^c(u_k) - \boldsymbol{U}_k\right] \approx 0$ and we have $\boldsymbol{U}_{k+1} \approx \boldsymbol{U}_k$.

The involutory matrix i.e. $\boldsymbol{C}$ is as important as the CC factor. The matrix $\boldsymbol{C}$ is transferred and rotated new point. A chaotic transformation $\boldsymbol{C}$ matrix is proposed for CCSTM which is used for transferring of $f^c(u_k)$ as follows,

$$[\boldsymbol{H}_{ij}]_k = \delta_{ij} + \eta_k\xi_k\left[\frac{\delta_{ij}}{\beta_k} - \frac{u_{i,k} \times u_{j,k}}{\beta_k^3}\right] \tag{6.76}$$

where, $0 < \eta_k < 1$ which is generated using Logistic map, $u_{i,k}$ is i_{th} element of point $\boldsymbol{U}_k$ at k_{th} iteration.

$$\xi_k = -\frac{\nabla^T g(\boldsymbol{U}_k)\boldsymbol{U}_k}{\nabla^T g(\boldsymbol{U}_k)\nabla g(\boldsymbol{U}_k)} \tag{6.77}$$

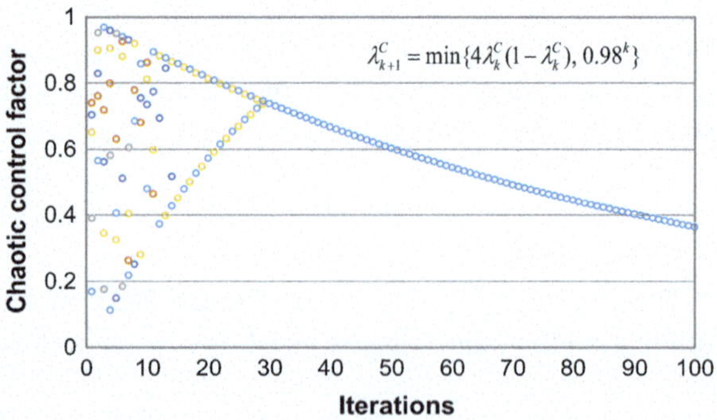

Fig. 6.28 The control factor applied in CCSTM

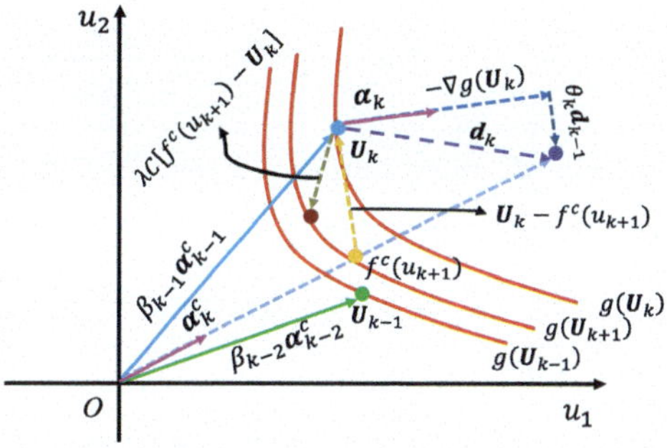

Fig. 6.29 Iterative procedure of FORM using CCSTM

$$\delta_{ij} = \begin{cases} 1 i = j \\ 0 i \neq j \end{cases} \tag{6.78}$$

It can be seen from Eqs. (6.76), (6.77), (6.78), The a symmetric is defined for $\boldsymbol{H}$ as $[\boldsymbol{H}]_k = [\boldsymbol{H}]_k^T$. The main diagonal of $\boldsymbol{H}$ is generally as more than 1 but others are small as $[\boldsymbol{H}_{ii}]_k > [\boldsymbol{H}_{ij}]_k$ Accordingly, this transformation matrix is positive definite. Thus, maximum spectral radius of H is larger than 1 i.e. $\rho(\boldsymbol{H}) > 1$, and we have $\rho\left(\boldsymbol{H}^{-1}\right) \leq 1$. Thus, it is defined, $[\boldsymbol{C}]_k = [\boldsymbol{H}]_k^{-1}$. If $\rho(\boldsymbol{J}) = 1$ for $[\boldsymbol{J}]\left([\boldsymbol{J}] = \frac{\partial f_i(u)}{\partial u_j}|_{U_k}\right)$, then we have a fixed point.

It can be found for $k \to \infty$ that $\lambda^C_{k+1} \approx 0$ thus $\left[\boldsymbol{J}\right]_k \approx \boldsymbol{I}$ and it is computed $\rho\left(\boldsymbol{J}\right) \approx 1$ thus, it is given a fixed point using CCSTM.

The reliability index and NSDVs for an iteration of CFORM are schematically plotted in Fig. 6.29. As seen, d_k s not along the direction of $\nabla g(\boldsymbol{U}_k)$, thus $\boldsymbol{\alpha}_{k+1} \neq \boldsymbol{\alpha}^C_{k+1}$. Consequently, NSDV at new and previous points are not same, thus solutions are without oscillating or predict results.

The iterative FORM algorithm using CCSTM is introduced in this section. In CCSTM, a chaotic CC factor and a transformation-rotational matrix are formulated for achieving the stabilization of analytical FORM. The CCSTM is described for SRA by the following relations,

$$\boldsymbol{U}_{k+1} = \boldsymbol{U}_k + \lambda^C_k [\boldsymbol{H}]_k^{-1} \left[f^c(u_k) - \boldsymbol{U}_k\right] \tag{6.79}$$

where,

$$f^c(u_k) = \frac{\nabla^T g(\boldsymbol{U}_k)\boldsymbol{U}_k - g(\boldsymbol{U}_k)}{\nabla^T g(\boldsymbol{U}_k)\boldsymbol{\alpha}^c_{k+1}} \boldsymbol{\alpha}^c_{k+1} \tag{6.80}$$

where, $\boldsymbol{\alpha}^c_{k+1} = \frac{\boldsymbol{U}^C_{k+1}}{\|\boldsymbol{U}^C_{k+1}\|}$, and $\boldsymbol{U}^C_{k+1} = \boldsymbol{U}_k + \boldsymbol{d}_k$ with conjugate search direction of,

$$\boldsymbol{d}_k = \begin{cases} -\nabla g(\boldsymbol{U}_k) k = 0 \\ -\nabla g(\boldsymbol{U}_k) - \theta^{FR}_k \boldsymbol{d}_{k-1} k \geq 1 \end{cases}$$

$$\theta_k^{FR} = \frac{||\nabla g(\boldsymbol{U}_k)||^2}{\|\nabla g(\boldsymbol{U}_{k-1})\|^2} \tag{6.81}$$

Using $\boldsymbol{C}$, there introduced two CCSTM algorithm as below:

Algorithm 1: CCSTM-1.

The chaos CC factor are determined by λ_{k+1}^C and $[\boldsymbol{C}]_k = [\boldsymbol{H}]_k^{-1}$;

Algorithm 2: CCSTM-2.

The chaos CC factor are determined by λ_{k+1}^C and $[\boldsymbol{C}] = [\boldsymbol{I}]$.

The framework of CCSTM is shown in Fig. 6.30. The CCSTM method can be coded based on the following steps in a computer program.

Step 1: Define ε (stopping criterion), $g(\boldsymbol{U}) = 0$, give random variables as information of μ and σ, set $\lambda_0 \notin \{0.0, 0.25, 0.5, 0.75, 1.0\}$, $a \in [3.6, 4]$, $k = 0$ and apply $\boldsymbol{X}_0 = \boldsymbol{\mu}$ and $\boldsymbol{d}_0 = 0$;

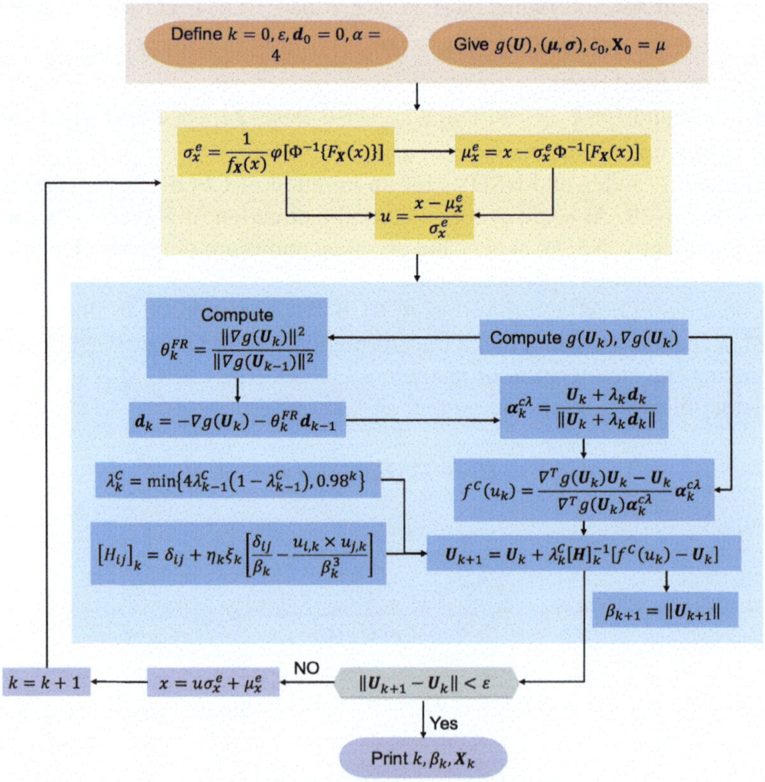

Fig. 6.30 Flowchart of CCSTM for codding in a computer program

Step 2: Transform variables into U-space as $u_k = \frac{x_k - \mu_x^e}{\sigma_x^e}$, where $\sigma_x^e = \frac{1}{f_X(x_k)}\varphi[\Phi^{-1}\{F_X(x_k)\}]$ and $\mu_x^e = x_k - \sigma_x^e \Phi^{-1}[F_X(x_k)]$;

Step 3: Iterative process of modified version of FR methods by the following relations;

(i) Determine $\nabla g(\boldsymbol{U})$ and $g(\boldsymbol{U})$ at pint $\boldsymbol{U}_k$;

(ii) Compute $\theta_k^{FR} = \frac{\|\nabla g(\boldsymbol{U}_k)\|^2}{\|\nabla g(\boldsymbol{U}_{k-1})\|^2}$ and determine $\boldsymbol{d}_k = -\nabla g(\boldsymbol{U}_k) - \theta_k^{FR}\boldsymbol{d}_{k-1}$;

(iii) Determine $\boldsymbol{U}_{k+1}^C = \boldsymbol{U}_k + \lambda \mathbf{d}_k$ and then $\boldsymbol{\alpha}_{k+1}^c = \frac{\boldsymbol{U}_{k+1}^C}{\|\boldsymbol{U}_{k+1}^C\|}$;

(iv) Determine $f^c(u_k) = \frac{\nabla^T g(\boldsymbol{U}_k)\boldsymbol{U}_k - g(\boldsymbol{U}_k)}{\nabla^T g(\boldsymbol{U}_k)\boldsymbol{\alpha}_k^c}\boldsymbol{\alpha}_k^c$;

(v) Determine control factor as $\lambda_{k+1}^C = \min\{4\lambda_k^C(1-\lambda_k^C), 0.98^k\}$ and $[\boldsymbol{H}_{ij}]_k = \delta_{ij} + \eta_k \xi_k \left[\frac{\delta_{ij}}{\beta_k} - \frac{u_{i,k} \times u_{j,k}}{\beta_k^3}\right]$;

(vi) Compute transformation matrix where $\xi_k = -\frac{\nabla^T g(\boldsymbol{U}_k)\boldsymbol{U}_k}{\nabla^T g(\boldsymbol{U}_k)\nabla g(\boldsymbol{U}_k)}$ and $\delta_{ij} = \begin{cases} 1 i = j \\ 0 i \neq j \end{cases}$;

(vii) Determine new point as $\boldsymbol{U}_{k+1} = \boldsymbol{U}_k + \lambda_k^C [\boldsymbol{H}]_k^{-1}\left[f^c(u_k) - \boldsymbol{U}_k\right]$; Let $\boldsymbol{X}_{k+1} = \mu_X^e + \sigma_X^e \boldsymbol{U}_{k+1}$ and $\beta_{k+1} = \|\boldsymbol{U}_{k+1}\|$ and set $k = k + 1$;

Step 4: If $\|\boldsymbol{U}_{k+1} - \boldsymbol{U}\|_k < \varepsilon$, then presents results as $P_f \approx \Phi(-\beta)$ and $\boldsymbol{\alpha} = \boldsymbol{\alpha}_{k+1}^c$, else Go to Step 2.

CCSTM has two differences with FORM-based SDSD as i) conjugate SDV and ii) chaotic control factor with transportation-rotation matrix. The CC factor is randomly determined by a Logistic map. As shown in the previous results, the conjugate SDV is a robust strategy of FORM which is applied in CCSTM. As seen from the CCC algorithm, this method is developed by SDC with finite-step length. On the other hand, the transportation-rotation matrix $\boldsymbol{H}$ in CCSTM is a novel empirical relation for the rotation of a new point for determining the new position as shown in Fig. 6.29.

6.6.2 *Illustrative Performances of CCSTM*

The performances of CCSTM are compared with HL-RF, STM) with $\lambda = 0.1$ (, FSL)with $\lambda = 15$ and $c = 1.4$(, RHL-RF. The CCSTM are computed by $[\boldsymbol{C}]_k = [\boldsymbol{H}]_k^{-1}$, $\lambda_0 = 0.375$ in CCSTM-1, and $[\boldsymbol{C}]_k = [\boldsymbol{I}]$, $\lambda_0 = 0.375$ in CCSTM-2. The performances of FORM algorithms are examined using efficiency as the number of evaluations for LSF and robustness as converged to a fixed point.

$\varepsilon = 10^{-6}$ is given in these analytical algorithms and the efficiency by using iterations and robustness by using converged reliability index are evaluated for each method.

Example 6.18 LSF as below,

$$g(\boldsymbol{X}) = x_1^3 + x_1^2 x_2 + x_2^3 - 18 \tag{6.82}$$

where, normal x_1 and x_2 are simulated by $\mu_1 = 10$, $\mu_2 = 9.9$ and $\sigma_1 = \sigma_2 = 5$. Using MCS with 10^6 samples, reliability index is $\beta = 2.5139$. After 29 iterations using CCSTM-1, we have stable results as $\beta = 2.298222$, $\boldsymbol{X}^* = [1.68463, 1.96919]$, and $U^* = [-1.66307, -1.58616]$. The it is computed as MPP $[\boldsymbol{H}]_{MPP} = \begin{bmatrix} 1.0045 & -0.0047 \\ -0.0047 & 1.0049 \end{bmatrix}$ where $\rho(\boldsymbol{H}) = 1.0093$. As seen, main diagonal elements are larger than 1 while other elements are computed negative smaller value.

As presented in Chap. 2, using map $f(\boldsymbol{U}_k) = f\left(\boldsymbol{X}_k, \mu_X^e, \sigma_X^e\right)$ for computing MPP, the statistical properties of basic random variables are performed on convergent abilities of FORM thus, for HL-R, the reliability index with various μ_1 and σ_1 are showed in Fig. 6.31. It is given that unstable reliability index as chaotic solutions are extracted in μ_1 between 17 and 21 and $9 < \sigma_1 < 10$. Consequently, the HL-RF formulation should be improved to achieve robustness.

Figure 6.32 shows the reliability index with respect to various μ_1 and σ_1 for Example 6.18. This Figure is plotted based on extracted results from CCSTM-1. It obvious that the stabilizations are achieved for various μ_1 and σ_1. The CG vector applied in STM with a novel formulation of FORM-based CCSTM enhances the instability of FORM-based HL-RF.

Example 6.18 is 10-time randomly analyzed with various initial values for iterative of the CCSTM-1 method. The spectral radius and iterations to the MPP search are plotted in Fig. 6.33. As given a $\rho(\boldsymbol{J}) < 1$ thus this presented FORM formulation has stable results for various initial points.

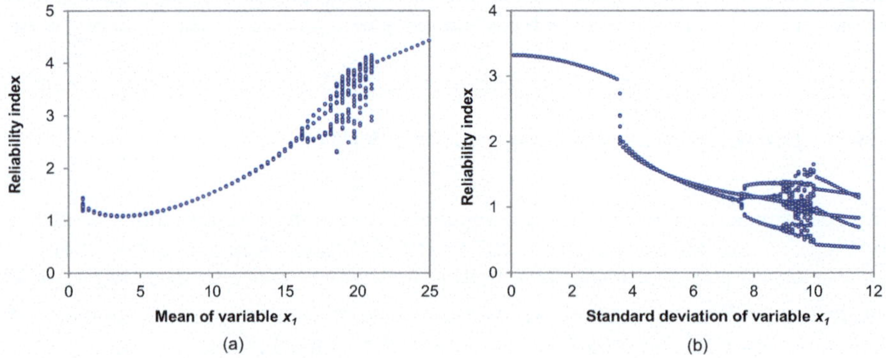

Fig. 6.31 Unstable solutions with various μ_1 and σ_1.using HL-RF for Example 6.18; **a** μ_1, **b** σ_1. Reprinted from [10], with permission from Elsevier

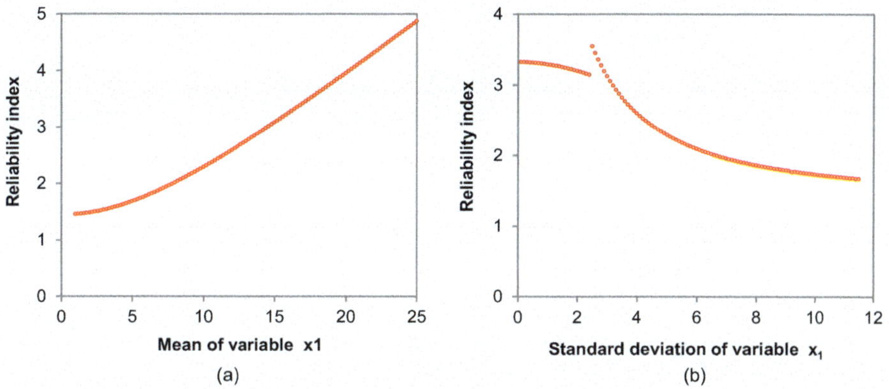

Fig. 6.32 The reliability index with various μ_1 and σ_1.using CCSTM-1 for Example 6.18; **a** μ_1, **b** σ_1. Reprinted from [10], with permission from Elsevier

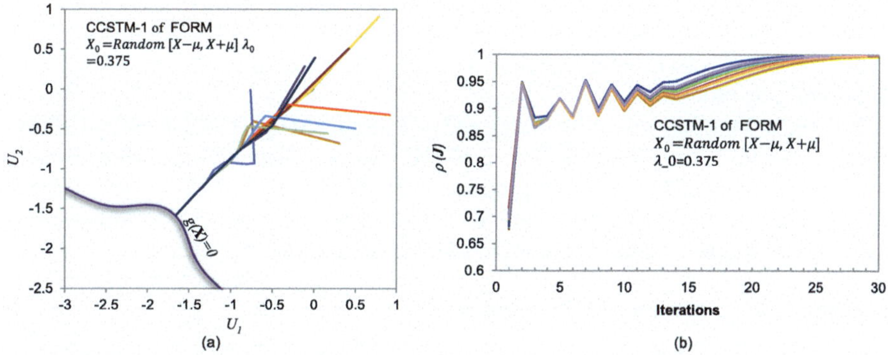

Fig. 6.33 Iterations of CCSTM-1 for various random initial points in Example 6.18; **a** spectral radius to the MPP search, **b** iterations to the MPP search

Example 6.19 A non-normal LSF as,

$$g(\boldsymbol{X}) = x_1^4 + x_2^2 - 50 \tag{6.83}$$

where. Lognormal x_1 with mean of 5 and STD of 1, Gumbel x_2 with mean of 10 and STD of 10.

Using MCS computed by 1.2×10^6 samples and FORM given form Ref [11], it extracted $\beta_{MCS} = 3.5009$ and $\beta_{FORM} = 3.259315$. After 29 iterations using CCSTM-1 with $\boldsymbol{X}_0 = \mu$, results of reliability analysis are as $\beta = 3.259175$, $\boldsymbol{X}^* = [2.64768, 0.93018]$, $\boldsymbol{U}^* = [-3.11118, -0.97078]$ with eigenvalues of matrix $\boldsymbol{C} = \boldsymbol{H}^{-1}$ at MPP equal to [0.99978, 1.00000] which are less than 1 that it seen maximum $\rho(\overline{\boldsymbol{J}}) \approx 1$.

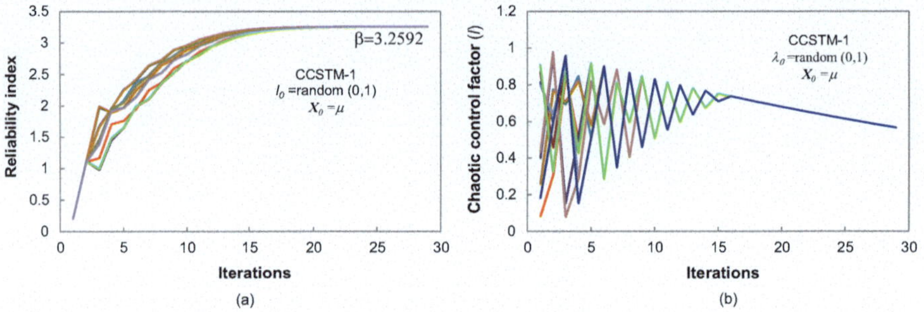

Fig. 6.34 Reliability index and CC factor obtained by CCSTM-1 for Example 6.19; **a** reliability index, **b** chaotic CC factor for 10-time analysis using CCSTM-1

Figure 6.34 demonstrates the iterations for reliability index and chaotic CC factor for 10-time analysis using CCSTM-1 where initial CC factor $\lambda_0 \in (0, 1)$ is randomly selected. Between 27 and 29 iterations require for the stabilization of the analytical method. Thus, this method is not sensitive to chaotic CC factor. From Fig. 6.34, CCSTM-1 provides a reliability index of $\beta = 3.2592$ for all random initial CC factor with iterations more than 15.

The iterations of the reliability index and $\rho(\boldsymbol{J})$ obtained by STM, HL-RF, CCSTM-1 and CCSTM-2, are plotted in Fig. 6.35. HL-RF is a discrete map with unstable iteration due to its $\rho(\boldsymbol{J}) > 1$ and it produces periodic-4 $\beta = [2.5118, 1.8550, 2.3857, 1.5966]$. However, the CCSTM-1 and CCSTM-2 showed $\rho(\boldsymbol{J}) < 1$ in all iterations that these methods are effectively more efficient than STM.

Example 6.20 A nonlinear LSF as below,

$$\begin{aligned} g(\mathrm{X}) = 1.1 - 0.00115x_1x_2 + 0.00157x_2^2 + 0.00117x_1^2 + 0.0135x_2x_3 \\ -0.0705x_2 - 0.00534x_1 - 0.0149x_1x_3 - 0.0611x_2x_4 + 0.0717x_1x_4 \\ -0.226x_3 + 0.0333x_3^2 - 0.558x_3x_4 + 0.998x_4 - 1.339x_4^2 \end{aligned} \tag{6.84}$$

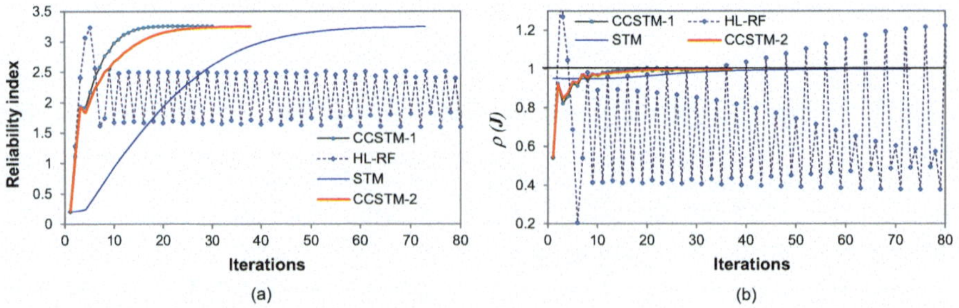

Fig. 6.35 Comparing performances of CCSTM for Example 6.19; **a** iterations of the reliability index, **b** iterations of the $\rho(\boldsymbol{J})$

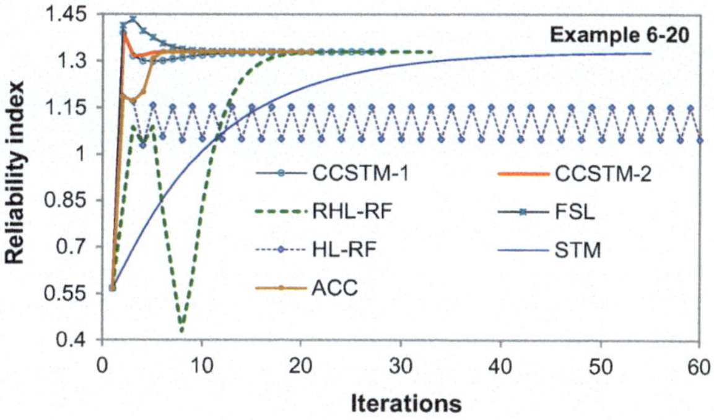

Fig. 6.36 Comparing performances of CCSTM for Example 6.20

where, type-II largest x_1 with mean of 10 and STD of 5. Normal x_2 with means of 25 and STD of 5, Normal x_3 with mean of 0.8, and STD of 0.2. Lognormal x_4 with mean of 0.0625 and STD of 0.0625.

Extracted results of reliability index is 1.36 and $\boldsymbol{X}^* = [15.09, 25.027, 0.8653, 0.03582]$. Using CCSTM-1, it is obtained $\boldsymbol{X}^* = [14.9004, 25.0689, 0.8598, 0.0463]$,$\boldsymbol{U}^* = [1.29516, 0.01379, 0.29888, 0.05604]$ and $\beta = 1.330479$. As seen, these results are closely followed the results given from literature. At MPP, we also have $\boldsymbol{H} = \begin{bmatrix} 1.0357 & -0.0068 & -0.1500 & -0.0267 \\ -0.0068 & 1.6869 & -0.0016 & -0.0003 \\ -0.1500 & -0.0016 & 1.6525 & -0.0061 \\ -0.0267 & -0.0003 & -0.0061 & 1.6859 \end{bmatrix}$ with $\rho(\boldsymbol{H}) = 1.68701$ and eigenvalues of $\boldsymbol{H}^{-1}$ are [0.32800, 0.32800, 0.32800, 0.99997] that all of them are less than 1.

The reliability index of different FORM algorithms is presented in Fig. 6.36 for Example 6.20. It can be given that CCSTM algorithms provide stable results while the instable results are obtained by HL-RF. The efficiency of STM is last among other while the CCSTM-2 has the lowest iteration among others.

Example 6.21 A vehicle side-impact performance function is used as below LSF,

$$g = 0.489x_3x_7 + 0.843x_5x_6 - 0.0432x_9x_{10} + 0.0556x_9x_{11} + 0.000786x_{11}^2 - 0.75 \tag{6.85}$$

where, $i = 1 \sim 7 x_i \sim N(1, 0.005)$, $i = 8 \sim 9 x_i \sim N(0.3, 0.006)$, $i = 10 \sim 11 x_i \sim N(0, 10.0)$. The MPP with $\beta = 3.4910495$ using CCSTM-1 in $\boldsymbol{U}$-space and $\boldsymbol{X}$-space are presented as below,

$\boldsymbol{X}^* = [0.97225, 0.95286, 0.95286, 0.97225, 0.30116, 30.07688, -8.37131]$.

$\boldsymbol{U}^* = [-0.55495, -0.94286, -0.94286, -0.55495, 0.19358, 3.00769, -0.83713]$.

The reliability index of $\beta = 3.6968$ is given by using MCS based on 1.5×10^6 samples.

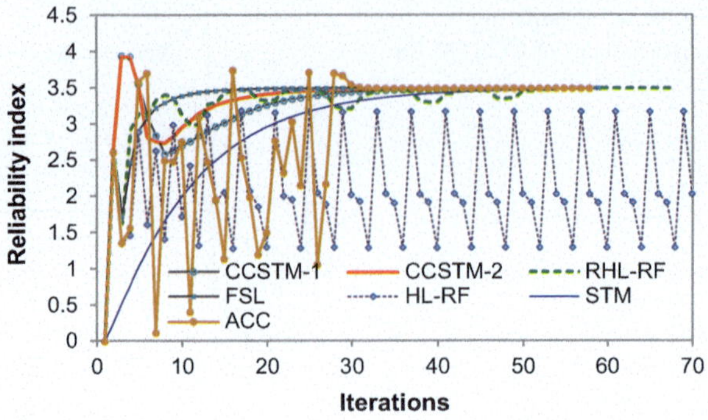

Fig. 6.37 Comparing performances of CCSTM for Example 6.21

Figure 6.37 shows the reliability index for different methods in Example 6.21. It obvious that the stability of the iteration is controlled based on STM, FSL and RHL-RF. The RHL-RF and STM has the similar efficiency but they are more computational inefficient than the FSL method. The superior computational burden is given by CCSTM which are formulated by chaotic control factor using Logistic map and conjugate sensitivity vector.

Example 6.22 A frame with LSF as below,

$$g = 0.05 - \Delta \tag{6.86}$$

The schematic view of the frame is plotted in Fig. 6.38 the above LSF are computed based on variables listed in Table 6.19.

Using MCS with 1.2×10^6 samples, $\beta = 3.168196$. Based on SRA using CCSTM-1 obtained after 43 iterations, $\beta = 3.1679244$.

The reliability results for different FORM methods are tabulated in Table 6.20 for Example 6.22. The STM produces unstable results. The results from this table showed that the HL-RF, and STM methods yield chaotic solutions, while the CCSTM, RHL-RF and FSL are robustly converged. This problem is confirmed the line search direction effect in FORM formulas including as normalized sensitivity vector and related step size.

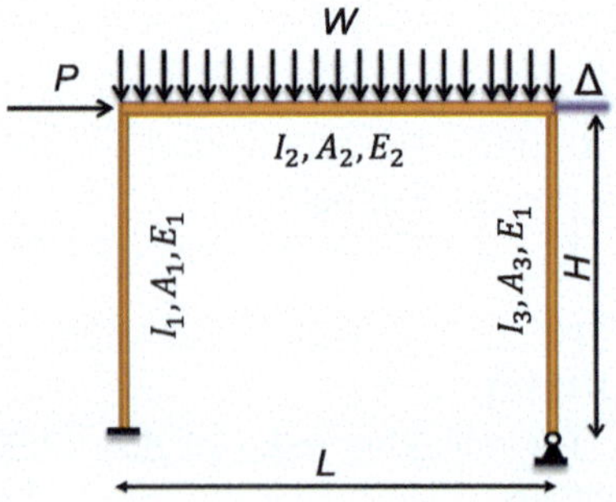

Fig. 6.38 Steel frame structure for Example 6.22. Reprinted from [10], with permission from Elsevier

Table 6.19 Variables of frame structure applied in Example 6.22

Variable	Mean	STD	PDF	Variable	Mean	STD	PDF
A_1 (m^2)	91×10^{-4}	9.1×10^{-4}	Lognormal	L (m)	6	0.48	Normal
A_2 (m^2)	53.8×10^{-4}	5.38×10^{-4}	Lognormal	E_1 (kg/m^2)	2.38×10^{10}	1.9×10^9	Normal
A_3 (m^2)	106×10^{-4}	10.6×10^{-4}	Lognormal	E_2 (kg/m^2)	2.17×10^{10}	1.74×10^9	Normal
I_1 (m^4)	80.9×10^{-6}	8.09×10^{-6}	Lognormal	H (m)	4.5	0.36	Normal
I_2 (m^4)	83.6×10^{-6}	8.36×10^{-6}	Lognormal	P (kg)	2000	350	Gumbel
I_3 (m^4)	112.6×10^{-6}	11.26×10^{-6}	Lognormal	W (kg/m)	4500	630	Gumbel

Table 6.20 Extracted results from reliability methods applied in Example 6.22

Iteration	CCSTM-1	CCSTM-2	FSL	HL-RF	STM	RHL-RF
1	0.264762	0.264762	0.264762	0.264762	0.264762	0.264762
2	5.361999	5.361999	5.938913	4.838155	0.640181	2.536320
3	3.497421	3.642948	5.223139	3.305996	1.015995	2.558270
...	...	...	...	...	...	...
36	3.167811	3.168311	3.292420	3.168350	3.121851	3.164063
37	3.167831	3.168311	3.281511	3.167791	3.126474	3.164319
38	3.167855	3.168311	3.271725	3.168098	3.130639	3.164347
39	3.167940	Converged	3.262579	3.167837	3.134372	3.164393
41	3.167861		3.246756	3.167841	3.140705	3.165009
42	3.167861		3.239998	3.168047	3.143458	3.165626
43	3.167862		3.233753	3.168070	3.145862	3.165709
...	Converged		...	...	...	...
89			3.169725	3.168080	3.167508	3.167763
90			3.169676	3.168130	3.167568	3.167760
91			3.169620	3.168134	3.167625	3.167761
...			...	...	...	Converged
154			3.169307	3.168136	3.167733	
155			3.169307	3.168273	3.167776	
156			3.169307	3.167467	3.167772	
...			Converged	...	...	
1997				3.167857	3.167378	
1998				3.166834	3.167457	
1999				3.167832	3.167521	
2000				3.167495	3.167585	

The evaluating gradient as iterations and β are respectively utilized to consider efficiency and robustness. The results of FORM schemes are presented in Table 6.21. Results from Table 6.21 indicated that CCSTM methods provide stable results. Nevertheless, the HL-RF is an unstable formulation as periodically results for Examples 6.18, 6.19, 6.20, and 6.21 and chaotically iterations for Example 6.22. Among FORM methods by comparing the efficiency, the RHL-RF, STM, and FSL are inefficient formulations compared to CCSTM and STM is worst. The CCSTM produces the stable solution, more efficiently. The CCSTM-2 is performed better than CCSTM-1. As can be seen, CCSTM-2 is provided stable results twice faster than STM. The CCSTM iterative FORM showed superior performances in terms of high-convergent rate-based with low computational burden and robustness.

The iterations for stabilization using two CFROM-based CCSTM-1 and CCSTM-2 are discussed for different Logistic map parameters as a where results are listed in Table 6.22. Extracted results from Table 6.22 showed that CCSRM methods are inefficiently converged when it is given $a = 2$. For parameters larger than 3.5 i.e. $a > 3.5$, the CCSRM methods provide stable solutions, more efficiently. As seen for $a = 3.75$ compared with $a = 2$ and 3, CCSRM methods are converged, efficiently.

Table 6.21 Reliability results i.e. β (iterations/call function) for studied FORM algorithms

Examples	CCSTM-1	CCSTM-2	HL-RF	RHL-RF	STM	FSL
6.18	2.29824 (26/130)	2.29822 (29/145)	Piriodic-2	2.29825 (47/235)	2.29794 (111/555)	2.29825 (185/925)
6.19	**3.25918 (29/145)**	3.25495 (37/185)	Priodic-4	3.25931 (46/230)	3.25839 (96/480)	3.25931 (79/395)
6.20	1.33048 (27/243)	**1.33054 (15/135)**	Priodic-2	1.33053 (33/297)	1.33035 (82/738)	1.33054 (17/153)
6.21	3.49105 (46/690)	**3.49407 (36/540)**	Priodic-4	3.49747 (68/1020)	3.49111 (62/930)	3.49748 (58/870)
6.22	3.16786 (43/1075)	**3.16831 (38/950)**	Chaos	3.16776 (91/1365)	Chaos	3.16931 (156/2340)

Bold numbers are minimum iterations obtained by a method among others

Table 6.22 Effect of control parameter of Logistic map for iterations of CCSTM-1 and CCSTM-2 methods

Examples	CCSTM-1					CCSTM-2				
	$a = 2$	3	3.5	3.75	3.95	$a = 2$	3	3.5	3.75	3.95
6.18	33	27	**25**	27	26	58	38	34	31	**27**
6.19	33	29	**27**	28	29	33	29	**28**	31	34
6.20	35	42	33	32	**27**	**15**	**15**	**15**	**15**	**15**
6.21	79	73	63	59	**47**	38	33	**29**	30	32
6.22	78	65	54	**38**	43	83	75	71	49	38

Bold numbers are minimum iterations obtained by a method among other

The FORM as the steepest descent method (HL-RF) may be not robustly converged to stable solutions. Therefore, one of the main challenges in FORM formulation is to find an analytical reliability method which has a robust formulation while its stabilization for MPP search is iterated by low-computational burden. In this section, an iterative formula for CFORM is developed using STM named chaotic conjugate STM i.e. CCSTM. Two factors as chaotic CC factor and adaptive transformation-rotational matrix are introduced in CCSTM applied in SRA-based CFORM. Generally, the CCSTM methods are controlled instability of iterative formulation which are applied for MPP search as shown a spectral radius of the Jacobian matrix less than 1. The CCSTM has a complicated iterative formulation compared to HL-RF, FSL and STM due to computing $\boldsymbol{H}$. However, the CCSTM is robust while they improve the efficiency of FORM compared to FSL, STM, and RHL-RF.

6.7 Discussion for FR-Based CG Methods

The abilities of FORM-based conjugate search direction are strongly depended on the conjugate line search which is used to compute the normalized conjugate vector. In the conjugate line search, two factors of the CG factor and step length are the main parameters for determining the normalized conjugate vector. As seen from the methods of the CG method, the FR scalar factor and variant type of FR CG methods are the main contribution in this section due to satisfying the sufficient descent condition. The step length is adaptively adjusted based on the SDC and the Armijo-type condition which is applied to the initial step length. In this section, the line search effects as step length and CG vector are discussed through several examples. By this compression, the methods-based formulation of FORM is also discussed by using two formulations for the reliability index as below,

$$\beta_{k+1} = \frac{\boldsymbol{d}_k^T \boldsymbol{U}_k - g(\boldsymbol{U}_k)}{\boldsymbol{d}_k^T \boldsymbol{\alpha}_k^c} \tag{6.87}$$

$$\beta_{k+1} = \frac{\nabla^T g(\boldsymbol{U}_k)\boldsymbol{U}_k - g(\boldsymbol{U}_k)}{\nabla^T g(\boldsymbol{U}_k)\boldsymbol{\alpha}_k^c} \tag{6.88}$$

Equation (6.88) is applied for computing the reliability index using the CG vector while the steepest descent vector is applied for computing the reliability index in Eq. (6.87). Based on Eqs. (6.87) and (6.88), the new point is given as $\boldsymbol{U}_{k+1} = \beta_{k+1}\boldsymbol{\alpha}_k^c$. Where, $\boldsymbol{\alpha}_k^c = \frac{\boldsymbol{U}_k + \lambda_k \boldsymbol{d}_k}{\|\boldsymbol{U}_k + \lambda_k \boldsymbol{d}_k\|}$. denotes normalized conjugated sensitivity vector.

It is obvious that $\boldsymbol{\alpha}_k^c$ is just computed by CG vector that this is main different between FORM formula-based CG and steepest descent. Step length λ_k is important as well as the CG vector $\boldsymbol{d}_k$, which its initial value is computed using Armijo condition as below,

$$\lambda_0 = \frac{M}{\left\| \nabla g\left(\boldsymbol{U}|_{U=\mu} \right) \right\|} \tag{6.89}$$

In which, $10 \leq M \leq 100$ and $\nabla g\left(\boldsymbol{U}|_{U=\mu} \right)$ is gradient vector at the mean point. In this dissection, $M = 30$. Then, the step length is adjusted by using the sufficient descent condition as below,

$$\lambda_k = \begin{cases} \lambda_{k-1} & \|\boldsymbol{U}_{k+1} - \boldsymbol{U}_k\| < \|\boldsymbol{U}_k - \boldsymbol{U}_{k-1}\| \\ \frac{\lambda_{k-1}}{c} & \|\boldsymbol{U}_{k+1} - \boldsymbol{U}_k\| \geq \|\boldsymbol{U}_k - \boldsymbol{U}_{k-1}\| \end{cases} \tag{6.90}$$

where, $c = 1.4$. Using above Eqs. (6.87), (6.88), (6.89), and (6.90), three scenarios for comparative CFORM are evaluated as below,

Scenario 1: $\beta_{k+1} = \frac{\nabla^T g(\boldsymbol{U}_k)\boldsymbol{U}_k - g(\boldsymbol{U}_k)}{\nabla^T g(\boldsymbol{U}_k)\boldsymbol{\alpha}_k^c}$ and $\lambda_k = \begin{cases} \lambda_{k-1} & \|\boldsymbol{U}_{k+1} - \boldsymbol{U}_k\| < \|\boldsymbol{U}_k - \boldsymbol{U}_{k-1}\| \\ \frac{\lambda_{k-1}}{1.4} & \|\boldsymbol{U}_{k+1} - \boldsymbol{U}_k\| \geq \|\boldsymbol{U}_k - \boldsymbol{U}_{k-1}\| \end{cases}$ with $\lambda_0 = \frac{30}{\left\| \nabla g\left(\boldsymbol{U}|_{U=\mu} \right) \right\|}$ named SCEI.

Scenario 2: $\beta_{k+1} = \frac{\nabla^T g(\boldsymbol{U}_k)\boldsymbol{U}_k - g(\boldsymbol{U}_k)}{\nabla^T g(\boldsymbol{U}_k)\boldsymbol{\alpha}_k^c}$ and $\lambda_k = 1$ named SCEII.

Scenario 3: $\beta_{k+1} = \frac{\boldsymbol{d}_k^T \boldsymbol{U}_k - g(\boldsymbol{U}_k)}{\boldsymbol{d}_k^T \boldsymbol{\alpha}_k^c}$ and $\lambda_k = \begin{cases} \lambda_{k-1} & \|\boldsymbol{U}_{k+1} - \boldsymbol{U}_k\| < \|\boldsymbol{U}_k - \boldsymbol{U}_{k-1}\| \\ \frac{\lambda_{k-1}}{1.4} & \|\boldsymbol{U}_{k+1} - \boldsymbol{U}_k\| \geq \|\boldsymbol{U}_k - \boldsymbol{U}_{k-1}\| \end{cases}$ with $\lambda_0 = \frac{30}{\left\| \nabla g\left(\boldsymbol{U}|_{U=\mu} \right) \right\|}$ named SCEIII.

The CG vectors and main iterative procedure for determining new points are given in Table 6.23 for different CFORM formulas.

The formulation presented in Table 6.23 for CFORM showed that, the CCC and the conjugate stability transformation method (CSTM) are established based on the FR method but the new point is adjusted using a chaotic control factor and transformation matrix. Two variant-type of CFORM-based FR with participation factor of gradient vector are presented in MFR and SVFR methods. The limited FR (LFR) is established based on the CG vector by a CG factor less than the FR for satisfying the sufficient descent condition. Finally, a different CG factor is considered with the FR method. By giving the main formulation of CFORM in CHL-RF (FR), it can be given three main categories of CFORM as:

(i) general FR-based CFORM including FR, CCC and CSTM;
(ii) variant FR-based CFORM including MFR, SVFR;
(iii) modified FR-based CFORM including LFR.

As presented results in this chapter, The robust formulations are extracted based on all CFORM algorithms but the efficiency is the essential challenge to develop the CFORM algorithms. The efficiency of CFORM is compared by six reliability problems with $\varepsilon = 10^{-4}$ that their LSFs with statistical properties are given in Examples 6.23, 6.24, 6.25, 6.26, 6.27, and 6.28.

Table 6.23 Methods of CFORM for computing CG vector ***d***

Method	CG vector	Conjugate parameters
FR (CHL-RF)	$\boldsymbol{d}_k = -\nabla g(\boldsymbol{U}_k) - \theta_k \boldsymbol{d}_{k-1}$	$\theta_k^{FR} = \frac{\|\nabla g(\boldsymbol{U}_k)\|^2}{\|\nabla g(\boldsymbol{U}_{k-1})\|^2}$
LFR	$\boldsymbol{d}_k = -\nabla g(\boldsymbol{U}_k) - \theta_k \boldsymbol{d}_{k-1}$	$\theta_k^{LFR} = \min\{\delta, \theta_k^{FR}\}$, $0.5 \le \delta \le 1.0$ where, $\delta = 0.75$
MFR	$\boldsymbol{d}_k = -\eta_k \nabla g(\boldsymbol{U}_k) - \theta_k^{FR} \boldsymbol{d}_{k-1}$	$\eta_k = \frac{d_{k-1}^T[\nabla g(\boldsymbol{U}_k) - \nabla g(\boldsymbol{U}_{k-1})]}{\|\nabla g(\boldsymbol{U}_{k-1})\|^2}$
SVFR	$\boldsymbol{d}_k = -\eta_k \nabla g(\boldsymbol{U}_k) - \theta_k^{VSFR} \boldsymbol{d}_{k-1}$	$\theta_k^{VFR} = \frac{\|\nabla g(\boldsymbol{U}_k)\| \cdot \|\nabla^T g(\boldsymbol{U}_{k-1}) \nabla g(\boldsymbol{U}_k)\|}{\|\nabla g(\boldsymbol{U}_{k-1})\|^3}$
CCC	$\boldsymbol{d}_k = -\nabla g(\boldsymbol{U}_k) - \theta_k^{FR} \boldsymbol{d}_{k-1}$ $f^C(u_k) = \frac{\nabla^T g(\boldsymbol{U}_k)\boldsymbol{U}_k - g(\boldsymbol{U}_k)}{\nabla^T g(\boldsymbol{U}_k)\boldsymbol{\alpha}_k^c} \boldsymbol{\alpha}_k^c$	If $\rho(J) > 1$, $\boldsymbol{U}_{k+1}^{CCC} =$ $\boldsymbol{U}_k^{CCC} + c_m\left[f^C(u_k) - \boldsymbol{U}_k^{CCC}\right]$ $c_m = 4c_m(1 - c_m)$ and $c_m = 0.375$
CSTM	$\boldsymbol{d}_k = -\nabla g(\boldsymbol{U}_k) - \theta_k^{FR} \boldsymbol{d}_{k-1}$ $[H_{ij}]_k = \delta_{ij} + \eta_k \xi_k \left[\frac{\delta_{ij}}{\beta_k} - \frac{u_{i,k}^{CG} \times u_{j,k}^{CG}}{\beta_k^3}\right]$	$\boldsymbol{U}_{k+1}^{CSTM} =$ $\boldsymbol{U}_k^{CSTM} + \alpha \boldsymbol{C}\left[f^C(u_k) - \boldsymbol{U}_k^{CSTM}\right]$ $\alpha = \min\left(4c_m(1 - c_m), 0.98^k\right)$ $\xi_k =$ $-\frac{\nabla^T g(\boldsymbol{U}_k)\boldsymbol{U}_k}{\nabla^T g(\boldsymbol{U}_k)\nabla g(\boldsymbol{U}_k)}$, $[C]_k = [H]_k^{-1}\delta_{ij} =$ $\begin{cases} 1 i = j \\ 0 i \ne j \end{cases}$, $\eta = 4\eta(1 - \eta)$, $c_m = \eta = 0.375$

Example 6.23 $g = x_1^3 + 2x_1^2 x_2 + x_2^3 - 18$ where x_1 and x_2 are normal with $\mu_1 = 10$, $\mu_2 = 9.9$ and $\sigma_1 = \sigma_2 = 5$.

Example 6.24 $g(\boldsymbol{X}) = x_1^4 + x_2^2 - 50$ where, x_1: LN (5,1) and x_2: Gumbel (10,10).

Example 6.25 $g = 1.1 - 0.00115x_1x_2 + 0.00157x_2^2 + 0.00117x_1^2 + 0.0135x_2x_3 - 0.0705x_2 - 0.00534x_1 - 0.0149x_1x_3 - 0.0611x_2x_4 + 0.0717x_1x_4 - 0.226x_3 + 0.0333x_3^2 - 0.558x_3x_4 + 0.998x_4 - 1.339x_4^2$, where, x_1: extreme value II (10,5), x_2: N(25,5), x_3: N(0.8,0.2), x_4: LN(0.0625,0.0625).

Example 6.26 A conical structure as indicated in Fig. 6.16 by LSF as $g = 1 - \frac{\sqrt{3(1-v^2)}}{\pi E t^2 \cos^2 \alpha} \times \left(\frac{P}{0.66} + \frac{M}{0.41 r_1}\right)$ where, $v = 0.3$ and data of Table 6.9 is the random variables of conical example.

Example 6.27 A dynamic system as presented in Fig. 6.17 with LSF of $g = Fs - Ks \times 3\left(\frac{\pi S_0}{4\xi_s \omega_s}\left[\frac{\xi_a \xi_s}{\xi_p \xi_s (4\xi_a^2 + \theta^2) + \gamma \xi_a^2} \times \frac{(\xi_p \omega_p^3 + \xi_s \omega_s^3)\omega_p}{4\xi_a \omega_a^4}\right]\right)^{1/2}$ where, $\gamma = \frac{M_s}{M_p}$, $\omega_a = \frac{\omega_p + \omega_s}{2}$, $\xi_a = \frac{\xi_p + \xi_s}{2}$ and $\theta = \frac{\omega_p - \omega_s}{\omega_a}$ that their characteristics of Lognormal random variables is tabulated in Table 6.24.

Table 6.24 Variables applied in Example 6.27

Variables	Mean	STD	Variables	Mean	STD
M_P	1	0.1	ξ_P	0.05	0.02
M_S	0.01	0.001	ξ_s	0.02	0.01
K_p	1	0.2	S_0	100	10
K_S	0.01	0.002	F_s	15	0.15

Example 6.28 A planar ten-bar structure presented in Fig. 6.39 by LSF of $g = 0.004 - \left|\Delta_3^V\right|$ that the statistical properties of variables are listed in Table 6.25.

These three categories of CFORM named as (1) general FR, (2) variant FR and (3) modified FR are applied for Example 6.23 in SECI. The convergences in normal standard space are plotted in Figs. 6.40, 6.41, and 6.42 for Example 6.23 which are computed based on three categories of CFORM. It can be conducted that all methods are converged robustly. The CFORM in the initial iterations is not similar to the FORM-based steepest descent. Based on a nonlinear line search, it provides a stable MPP point by almost methods.

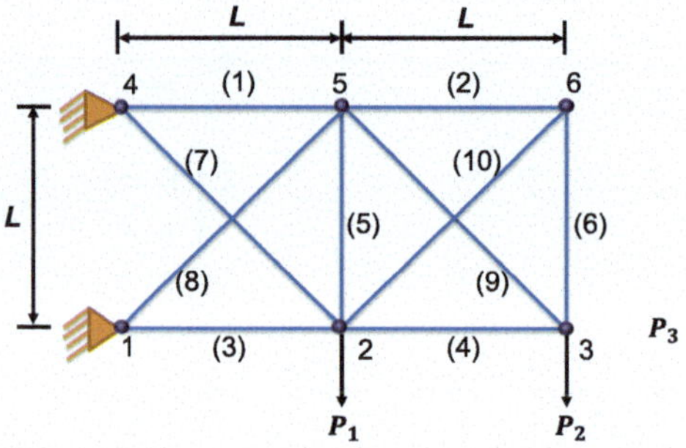

Fig. 6.39 A view of ten-bar structure applied in Example 6.28. Reprinted from [10], with permission from Elsevier

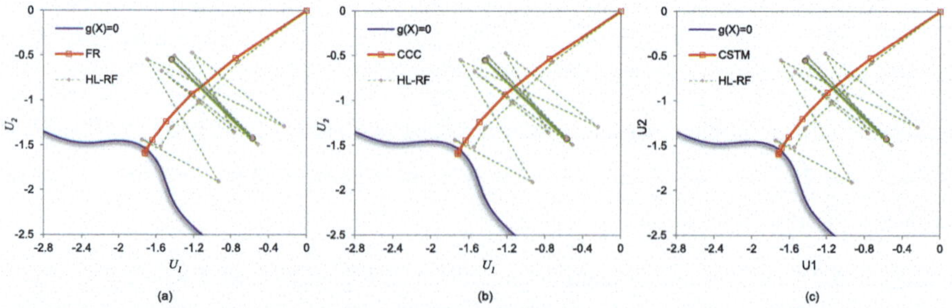

Fig. 6.40 Iterations of General FR-based FR, CCC and CSTM methods for Example 6.23; **a** FR, **b** CCC, **c** CSTM

Table 6.25 Variables applied for computing LSF of Example 6.28

Variable	A_i (m^2)	E (Gpa)	L (m)	P_1 (kN)	P_2 (kN)	P_3 (kN)
Mean	0.001	100	1	80	10	10
COV	0.15	0.05	0.05	0.1	0.1	0.1
PDF	Normal	Normal	Log-normal	Gumbel	Gumbel	Gumbel

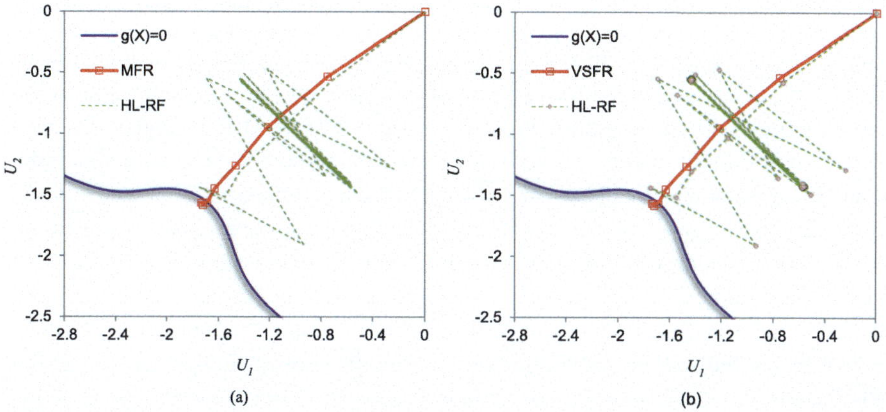

Fig. 6.41 Iterations of variant FR-based MFR, SVFR methods for Example 6.23; **a** MFR, **b** VSFR

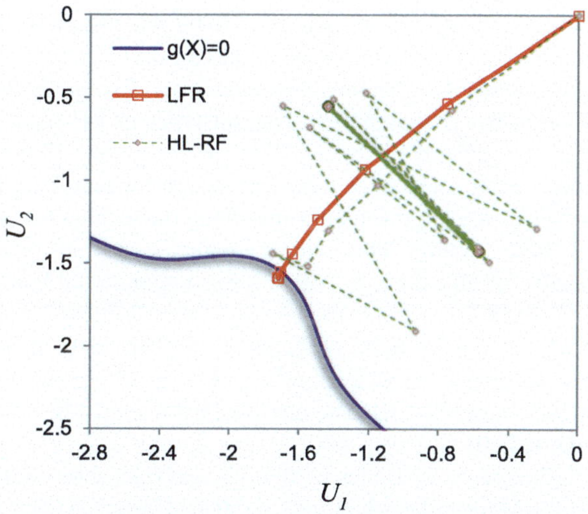

Fig. 6.42 Iterations of modified FR-based LFR method for Example 6.23

The results of the reliability index and iteration for convergence as β (iter) are computed for CFORM using three categories of FR formulation in three scenarios as:

(1) SCEI: $\beta_{k+1} = \frac{\nabla^T g(U_k)U_k - g(U_k)}{\nabla^T g(U_k)\alpha_k^c}$ and λ_k with $\lambda_0 = \frac{30}{\|\nabla g(U|_{U=\mu})\|}$,

(2) SCE II: $\beta_{k+1} = \frac{\nabla^T g(U_k)U_k - g(U_k)}{\nabla^T g(U_k)\alpha_k^c}$ and $\lambda_k = 1$ and

(3) SCE III: $\beta_{k+1} = \frac{d_k^T U_k - g(U_k)}{d_k^T \alpha_k^c}$ and λ_k with $\lambda_0 = \frac{30}{\|\nabla g(U|_{U=\mu})\|}$.

The results of different CFORM-based general FR formulations are presented in Table 6.26 for Examples 6.23, 6.24, 6.25, 6.26, 6.27, and 6.28. As seen, all methods are converged, robustly. The step size is an important factor as well as the CG vector for providing efficient results. The CCC method is more efficient approach compared to the CSTM and FR method. As seen by comparing the results of SCE I and SCEII, the change in the formulation of CG for computing reliability index, the computational burden for stabilization using the FR method is strongly increased. Thus, using FR method, the step size not affective parameters but the reliability index should be computed as $\beta_{k+1} = \frac{\nabla^T g(U_k)U_k - g(U_k)}{\nabla^T g(U_k)\alpha_k^c}$ for best efficiency.

On the other hand, the FR method in all comparative algorithm, provides the efficient results for highly concave performance functions as Examples 6.23 and 6.24, but it is an inefficient formulation for the convex problems. By using controlling the instability in CCC, the efficiency of the FR method is improved for convex problems as seen in Example 6.25. Thus, in these results, the best method is CCC in SCE I and II.

The results of reliability analysis using variant-types of FR formulation including MFR, and SVFR are given in Table 6.26. All scenarios of these method are converged but the local or premature results is given in MFR and SVFR method when applied $\beta_{k+1} = \frac{d_k^T U_k - g(U_k)}{d_k^T \alpha_k^c}$ in SEC III. Therefore, the variant formulations of FR are just applied for the reliability index based on SCE I or II. The MFR and SVFR provide efficient results for convex problems see Examples 6.27 and 6.28. For SCE I, the MFR is better than SVFR. Variant-types of FR-based CFORM are more efficient than the general FR-based CFORM for almost Examples. Adjusting the gradient vector can provide a suitable line search for searching MPP but the step size should be selected a large value for removing the local convergence or slow convergence in the MFR method. This category named of variant-types of FR provides acceptable results for all examples with a low-computational burden.

Table 6.28 presents the results of the CFORM-based modified FR for Examples 6.23, 6.24, 6.25, 6.26, 6.27, and 6.28. This method is converged for all scenarios. The abilities of LFR are similar to FR but it is generally efficient in SCE I compared to SCEII and SCEIII. The modified FR using LFR in SCE I is better than the SCE II and SCE III.

The results in Tables 6.26, 6.27, and 6.28 showed the acceptable formulation for CFORM formulas which are established based on FR method in general FR, modified FR and variant FR methods that, they are a robust iterative formulation with selected step size without line search rules as Armijo or sufficient descent condition.

The efficiency of variant type FR method of MFR and SVFR method is more than other methods when a finite step size is selected with adjusting procedure by sufficient descent condition. By comparing the compactional burden, it can be given

Table 6.26 results of different general FR reliability methods for SCE I to III

Examples	FR			CCC			CSTM		
	SCE I	SCE II	SCE III	SCE I	SCE II	SCE III	SCE I	SCE II	SCE III
6.23	2.3406 (15)	2.3406 (10)	2.3324 (97)	2.3406 (17)	2.3406 (10)	2.3324 (97)	2.3406 (17)	2.3405 (10)	2.3057 (91)
6.24	3.2593 (11)	3.2593 (11)	3.2508 (92)	3.2593 (14)	3.2593 (11)	3.2508 (92)	3.2593 (13)	3.2593 (13)	3.2301 (80)
6.25	1.3306 (90)	1.3309 (312)	1.3295 (95)	1.3308 (28)	1.3368 (74)	1.329547 (95)	1.3306 (57)	1.3333 (113)	1.3258 (61)
6.26	4.6295 (115)	4.6281 (368)	4.6589 (304)	4.6358 (64)	4.6301 (71)	5.1463 (72)	4.6307 (80)	4.6289 (111)	4.7889 (141)
6.27	2.0165 (82)	2.0165 (73)	2.0254 (99)	2.0165 (16)	2.0165 (22)	2.1651 (26)	2.0165 (49)	2.0165 (61)	2.0441 (80)
6.28	1.7366 (51)	1.7372 (436)	1.7423 (65)	1.7367 (17)	2.53246 (72)	1.7787 (25)	1.7388 (80)	1.7466 (188)	1.7787 (25)

Table 6.27 Results of different variant-type FR methods for SCE I to III

Examples	MFR			SVFR		
	SCE I	SCE II	SCE III	SCE I	SCE II	SCE III
6.23	2.3406 (9)	2.3406 (9)	2.3401 (55)	2.3406 (9)	2.3406 (9)	2.3401 (55)
6.24	3.2803 (13)	3.2775 (11)	3.2594 (58)	3.3157 (13)	3.2838 (11)	3.2594 (58)
6.25	1.369 (12)	1.415 (150)	1.3706 (9)	1.3591 (7)	1.414 (115)	1.3587 (9)
6.26	4.6731 (8)	4.7506 (1969)	4.6745 (97)	4.6734 (8)	4.7668 (2565)	4.6773 (37)
6.27	2.1021 (22)	2.0897 (9)	2.2133 (43)	2.7682 (55)	2.11 (10)	2.2716 (42)
6.28	1.7963 (43)	1.7896 (197)	1.7992 (39)	1.7947 (33)	3.7652 (47)	Diverged

Table 6.28 results of different modified FR reliability methods for SCE I to III

Examples	LFR		
	SCE I	SCE II	SCE III
6.23	2.3406 (16)	2.3406 (13)	2.3401 (37)
6.24	3.2593 (13)	3.2593 (12)	3.2589 (30)
6.25	1.3306 (28)	1.3305 (71)	1.3305 (28)
6.26	4.6277 (60)	4.6279 (119)	4.6277 (74)
6.27	2.0164 (87)	2.0164 (103)	2.0164 (96)
6.28	1.7513 (103)	1.6703 (421)	1.7405 (82)

for the first category the variant FR method, then modified FR as well as the general FR methods.

The CCC in general FR, and MFR in variant FR are the best methods but it is suggested that the finite step size with adjusting procedure is applied for better computational burden. In general, the MFR is best related by comparing the computational burden for CFROM formulation.

6.8 Summary

In this chapter, directional conjugate gradient approaches are introduced. Several conjugate FORM formulations are collected which have been developed for instability control of the iterative formula of FORM. These methods are collected based on the conjugate sensitivity vector and the main formulation of conjugate FORM named FR. The basic formulation of conjugate FORM-based FR is presented and then several improvements in the conjugate search direction vectors are established for FR methods. These methods are named as (i) limited FR, (ii) Enhanced FR-based variant type in gradient vector, (iii) Chaotic conjugate chaos control method, (iv) chaotic conjugate stability transformation method. These formulations are extended based on the FR conjugate scalar factor but their formulation is different for computing line search. The abilities of FORM-based conjugate search direction are strongly dependent on the conjugate line search two factors of the conjugate scalar factor and step length are the main parameters for computing the conjugate line search. The step size is adaptively adjusted based on the sufficient descent condition and the Armijo-type condition in these formulations. Finally, two terms of conjugate search direction vector and step size affected on line search are discussed through several examples. The FR method provided robust results for highly concave performance functions but it is an inefficient formulation for the convex problems. However, the efficiency of the FR method is improved for convex problems using its improved versions for convex problems.

References

1. B. Keshtegar, M. Miri, Introducing Conjugate gradient optimization for modified HL-RF method. Eng. Comput. **31**(4), 775–790 (2014)
2. T. Santosh, R. Saraf, A. Ghosh, H. Kushwaha, Optimum step length selection rule in modified HL-RF method for structural reliability. Int. J. Press. Vessels Piping **83**(10), 742–748 (2006)
3. D. Yang, Chaos control for numerical instability of first order reliability method. Comm. Nonlinear Sci. Num. Simulat. **15**(10), 3131–3141 (2010)
4. B. Keshtegar, Z. Meng, A hybrid relaxed first-order reliability method for efficient structural reliability analysis. Struct. Saf. **66**, 84–93 (2017)
5. B. Keshtegar, M. Miri, Reliability analysis of corroded pipes using conjugate HL–RF algorithm based on average shear stress yield criterion. Eng. Fail. Anal. **46**, 104–117 (2014)
6. R. Melchers, M. Ahammed, A fast approximate method for parameter sensitivity estimation in Monte Carlo structural reliability. Computers Struct. **82**(1), 55–61 (2004)
7. B. Echard, N. Gayton, M. Lemaire, AK-MCS: an active learning reliability method
8. J.X. Gong, P. Yi, A robust iterative algorithm for structural reliability analysis. Structural Multidiscip. Optimiz. **43**, 519–527 (2011)
9. B. Keshtegar, I. Lee, Relaxed performance measure approach for reliability-based design optimization. Struct. Multidiscipl. Optim. **54**, 1439–1454 (2016)
10. B. Keshtegar, Chaotic conjugate stability transformation method for structural reliability analysis. Comp. Methods Appl. Mech. Eng. **310**, 866–885 (2016)
11. B. Keshtegar, Stability iterative method for structural reliability analysis using a chaotic conjugate map. Nonlinear Dyn. **84**, 2161–2174 (2016)
12. B. Keshtegar, A hybrid conjugate finite-step length method for robust and efficient reliability analysis. Appl. Math. Model. **45**, 226–237 (2017)
13. P.L. Liu, A. Der Kiureghian, Optimization algorithms for structural reliability. Struct. Saf. **9**(3), 161–177 (1991)
14. B. Keshtegar, Limited conjugate gradient method for structural reliability analysis. Eng. Comp. **33**, 621–629 (2017)
15. B. Keshtegar, Enriched FR conjugate search directions for robust and efficient structural reliability analysis. Eng. Comp. **34**, 117–128 (2018)
16. C. Luo, S.P. Zhu, B. Keshtegar, X. Niu, O. Taylan, An enhanced uniform simulation approach coupled with SVR for efficient structural reliability analysis. Rel. Eng. Syst. Safety **237**, 109377 (2023)
17. Z. Meng, G. Li, D. Yang, L. Zhan, A new directional stability transformation method of chaos control for first order reliability analysis. Struct. Multidiscipl. Optim. **55**, 601–612 (2017)

18. Y.G. Zhao, Z.H. Lu, Fourth-moment standardization for structural reliability assessment. J. Struct. Eng. **133**(7), 916–924 (2007)
19. J.E. Hurtado, D.A. Alvarez, The encounter of interval and probabilistic approaches to structural reliability at the design point. Comp. Methods in Appl. Mech. Eng. **22**, 74–94 (2012)

Chapter 7
Improved Conjugate FORM

Abstract In this chapter, the extended different types of conjugate FORM which are formulated with different conjugate scalar factors. Several conjugate FORM formulations are developed as: (1) hybrid conjugate finite-step length method; (2) hybrid self-adaptive directional conjugate method and (3) three-term conjugate gradient method. The parameters applied in computing the line search including (i) step length and (ii) conjugate gradient vectors are discussed through several examples. It is conducted that the methods for conjugate FORMs established based on sufficient descent condition provided the stable results without applying the step size.

Keywords Conjugate FORM · Conjugate scalar factor · Hybrid self-adaptive · Three-term conjugate gradient

7.1 Introduction

The first-order reliability method (i.e. FORM) is formulated based on the MPP (i.e. most probable point) search that commonly used for approximating the reliability index. In the FORM, the conjugate search direction vector (i.e. CSDV) can be used for determining the line search by introducing a directional normalized sensitivity vector based on the FR (i.e. Fletcher-Reeves) method. Generally, the most probable point (i.e. MPP) is computed in the conjugate FR method based on the below relation [1, 2],

$$\boldsymbol{U}_{k+1} = \beta_{k+1}\boldsymbol{\alpha}_{k+1}^{\lambda} \tag{7.1}$$

where, α_{k+1}^{λ} is normalized CSDV which is computed as,

$$\boldsymbol{\alpha}_{k+1}^{\lambda} = \frac{\boldsymbol{U}_{k+1}^{C\lambda}}{\left\|\boldsymbol{U}_{k+1}^{C\lambda}\right\|} \tag{7.2}$$

S.-P. Zhu and B. Keshtegar, *Structural Reliability Analysis*, Structural Integrity 31,
https://doi.org/10.1007/978-3-031-75197-4_7

$$\beta_{k+1} = \frac{g(\boldsymbol{U}_k) - \nabla^T g(\boldsymbol{U}_k)\boldsymbol{U}_k}{\nabla^T g(\boldsymbol{U}_k)\boldsymbol{\alpha}_{k+1}^{\lambda}} \tag{7.3}$$

In which, $\boldsymbol{U}_{k+1}^{C\lambda} = \boldsymbol{U}_k + \lambda \boldsymbol{d}_k$ is applied for computing the sensitivity vector using conjugate gradient (i.e. CG) vector ($\boldsymbol{d}_k$) and finite-step size (i.e. FSS) $\lambda > 0$. $\boldsymbol{d}_k$ is given by the following relation,

$$\boldsymbol{d}_k = -\nabla g(\boldsymbol{U}_k) - \frac{||\nabla g(\boldsymbol{U}_k)||^2}{||\nabla g(\boldsymbol{U}_{k-1})||^2}\boldsymbol{d}_{k-1} \tag{7.4}$$

The conjugate factor of $\theta_k^{FR} = \frac{||\nabla g(\boldsymbol{U}_k)||^2}{||\nabla g(\boldsymbol{U}_{k-1})||^2}$ is given by Fletcher-Reeves (FR) method in the general formulation of conjugate FORM (i.e. CFORM). As mentioned in Chap. 6, the search direction has an effective rule on the convergence performances of FORM, septically efficiency. The FSS and CSDV are the two main factors for controlling the normalized sensitivity vector. The FR method is enhanced by enriched FR which is presented in Chap. 6 [1]. The CCC (i.e. chaotic conjugate chaos control method) and CCSTM (i.e. chaotic conjugate stability transformation method) [3] were also developed by using FR formulation with a directional conjugate line search. The global convergence properties are given based on the FR method without the step size but the CFORM in the FR method should be enhanced by improving in the CSDV for achieving efficient results.

The existing reliability method-based CFORM applied for MPP search was investigated in Chap. 6. However, it can be extended different types of the conjugate line search which are an improvement of FR method. Several methods of the CFORM method are established based on a novel CG are presented in this chapter. There is collected three formulations of the analytical method by CFORM with different formulations by FR method. These formulations are named as (1) Hybrid conjugate finite-step length method; in this method, two conjugate scalar factors are presented and they are combined, dynamically; (2) Hybrid self-adaptive directional conjugate method, the FR formulation is enhanced by using a dynamical conjugate scalar factor (i.e. CSF) and a combined CSF is presented by using the FR and proposed adjusting FR scalar factor; (3) three-term CG method; the scalar factor is computed based on three terms as gradient, previous CG and previous gradient vectors. The search direction is determined based on the scalar factor given from FR for the CG vector while a new scalar factor is presented for a combination of the previous gradient vector. These three methods are investigated for convergence performances for both accuracy and efficiency in this chapter.

7.2 Hybrid Conjugate Finite-Step Length (CFSL-H) Method

Generally, Armijo rule [2] and sufficient descent condition (i.e. SDC) have been utilized to select a suitable step size for controlling the instability of analytical CFORM in SRA (i.e. structural reliability analysis). Consequently, it is extended an adaptive formula by applying in CFORM including CCSTM [2], limited conjugate search direction [3], and the conjugate HL-RF methods. Consequently, the efficiency and robustness of FORM given by CSDV are the vital performances to develop an iterative analytical formulation in SRA.

In this section, a hybrid CSDV with FSS is introduced to enhance efficiency and robustness of FORM-based steepest descent search direction (i.e. SDSD) in SRA. The CSFs are determined by Rivaie, Mustafa, Ismail, and Leong (RMIL) [4] and Conjugate Descent (CD) methods. Finally, a hybrid formulation is presented by the combination of CD and RMIL methods that hybrid conjugate method is named the conjugate finite-step length (CFSL-H) [5]. For controlling the instability of CFORM in the hybrid formulation of CD and RMIL, the FSS is applied.

7.2.1 Formulation and Framework of CFSL-H

The main effort in the CFORM is to compute the point along the CG vector. This point in the iterative formula of the CFORM is given by,

$$\boldsymbol{U}_{k+1}^{c\lambda} = \boldsymbol{U}_k + \lambda_k \boldsymbol{d}_k \tag{7.5}$$

where, $\boldsymbol{d}_k$ denotes the CG vector, which is computed by the following relation [6],

$$\boldsymbol{d}_k = \begin{cases} -\nabla g(\boldsymbol{U}_k) & k = 0 \\ -\nabla g(\boldsymbol{U}_k) + \theta_k \boldsymbol{d}_{k-1} & k \geq 1 \end{cases} \tag{7.6}$$

where, θ_k is CSF. The CSF as shown in the enhanced FR method in Sect. 5.3 is a sensitive factor for reducing the computational burden of the CFORM. Generally, the various CG algorithms can be established by demining the scalar θ_k. The major differences between the CFORM are (1) finite step length and (2) the conjugate sensitivity vectors. A parameter that is strongly affected on the CSDV, is CSF. The CSF can be controlled the direction of the conjugate vector and its length. By using the FSS, the length of the conjugate vector is not an important factor but the direction of the conjugate factor is a sensitive factor in the CFORM formulation. In this work, two CSFs of CD and RMIL are given for investigation of direction in CFORM formulas as follows,

$$\theta_k^{CD} = -\frac{\|\nabla g(\boldsymbol{U}_k)\|^2}{\boldsymbol{d}_{k-1}^T \nabla g(\boldsymbol{U}_{k-1})} \tag{7.7}$$

$$\theta_k^{RMIL} = \frac{\nabla^T g(\boldsymbol{U}_k)\left[\nabla g(\boldsymbol{U}_k) - \nabla g(\boldsymbol{U}_{k-1})\right]}{\|\boldsymbol{d}_{k-1}\|^2} \tag{7.8}$$

Iterative CFORM relations named as CFSL (i.e. conjugate finite-step length) approach is rewritten as below,

$$\begin{aligned} \boldsymbol{U}_{k+1} &= \frac{\boldsymbol{d}_k^T \boldsymbol{U}_k - g(\boldsymbol{U}_k)}{\boldsymbol{d}_k^T \boldsymbol{\alpha}_k^{c\lambda}} \boldsymbol{\alpha}_k^{c\lambda} \\ \beta_{k+1} &= \frac{\boldsymbol{d}_k^T \boldsymbol{U}_k - g(\boldsymbol{U}_k)}{\boldsymbol{d}_k^T \boldsymbol{\alpha}_k^{c\lambda}} \end{aligned} \tag{7.9}$$

where, $\boldsymbol{\alpha}_k^{c\lambda}$ is normalized CSDV, which is given by,

$$\boldsymbol{\alpha}_k^{c\lambda} = \frac{\boldsymbol{U}_{k+1}^{c\lambda}}{\left\|\boldsymbol{U}_{k+1}^{c\lambda}\right\|} \tag{7.10}$$

Two nonlinear LSFs (i.e. limit state functions) are utilized to illustrate the performances of FORM using conjugated search directions of CFSL and FSL (i.e. finite step length) methods by using the below LSFs,

Example 7.1 $G_1 = x_1^4 + 2x_2^4 - 20$ where, x_1 and x_2 are normally variables with $\mu_1 = \mu_2 = 10$ and $\sigma_1 = \sigma_2 = 5$;

Example 7.2 $G_2 = x_1^3 + x_1^2 x_2 + x_2^3 - 18$, where, x_1 and x_2 are normally variables with $\mu_1 = 10$, $\mu_2 = 9.9$ and $\sigma_1 = \sigma_2 = 5$. The reliability indexes by using MCS (i.e. Monte Carlo simulation) for Examples 7.1 and 7.2 as presented in the above function are approximated by 2×10^5 samples as $\beta_1 = 2.8861$ and $\beta_2 = 2.5139$. Number of iterations for obtaining the stable results by using the finite step length (i.e. FSL) method and two CFORM methods using CFSL extended by CD and RMIL scalar factors are listed in Table 7.1. Note that the FSL and CFSL schemes are stably converged to the same reliability index of 2.3654 for (i) Example 7.1 and 2.2982 for (ii) Example 7.2, but CFSL-based CD and RMIL formulations provided efficient results. Therefore, the CSDV applied for computing the conjugate sensitivity vector in FORM as well as FSS is a significant factor to improve the efficiency. The different iterations are extracted in Examples 7.1 and 7.2 by iterative formulation of CD and RMIL methods. The robust results are given by these methods while the computational burden of these methods is different by changing the initial step size. Consequently, for improving the efficiency, it can be combined two methods of CD and RMIL.

As presented in the results of Table 7.1, the efficiency of CFORM is depended CSFs and the step length. In the smaller step length, the computational burden is

Table 7.1 Iterations of different FORM for various λ in Examples 7.1 and 7.2

Examples	Methods	λ														β
		50	30	20	15	10	5	1	0.5	0.2	0.1	0.05	0.01	0.005	0.001	
7.1	CFSL-CD	55	51	46	49	46	48	38	39	36	33	34	27	24	23	2.36545
	CFSL-RMIL	54	53	53	52	47	43	37	34	34	33	29	29	23	21	
	FSL	119	114	110	108	107	101	90	86	74	72	68	56	50	39	
7.2	CFSL-CD	68	613	40	609	100	29	26	22	21	19	19	16	16	16	2.29824
	CFSL-RMIL	47	63	43	44	39	41	27	26	30	24	25	21	18	18	
	FSL	112	155	122	189	104	92	356	118	88	102	114	125	29	25	

directly changed by two methods of RMIL and CD. The direction of the conjugate sensitivity vector computed using RMIL and CD methods provide a distributed number of iterations for these two examples. Therefore, the RMIL and CD are combined to determine a new CSF in CFORM-based SRA as follows,

$$\theta_k^H = \eta\theta_k^{CD} + (1-\eta)\theta_k^{RMIL} \tag{7.11}$$

where, θ_k^H is a hybrid CSF with a positive parameter $0 \leq \eta \leq 1$ which is determined as below,

$$\eta = -\frac{\|\nabla g(\boldsymbol{U}_{k-1}^{CG})\|^2}{4\nabla^T g(\boldsymbol{U}_k^{CG})\boldsymbol{d}_{k-1}} \tag{7.12}$$

The factor η is empirically extended for the best combination of two methods of CD and RMIL in this work. For other hybrid CFORM methods, this factor may be changed, dynamically. If $\eta = 0$ then $\theta_k^H = \theta_k^{RMIL}$ and, for $\eta = 1$, we have $\theta_k^H = \theta_k^{CD}$, this means that CD and RMIL conjugate methods are adaptively combined to approximate the CSDV in CFSL-H.

The results given from Table 7.1 demonstrated that FSS as well as NSCSV are affected on the computational burden of FORM. The conversance rates of CFSL (i.e. conjugate finite-step length) methods are shown high efficiency for smaller step length i.e. $\lambda_0 \approx 0.001$ for Examples 7.1 and 7.2. Thus for an acceptable computational burden, it is conducted that the initial FSS should be given as $\lambda_0 < \lambda_{\max}$ in which $\lambda_{\max}$ is the maximum FSS which can be controlled by λ_0 obtained using Armijo condition as good rule to find an acceptable $\lambda_{\max}$ which can be given by the as follows [7],

$$\lambda_{\max} \leq M\frac{\|\boldsymbol{d}_k\|^2}{\|\nabla g(\boldsymbol{U}_k)\|^2} \tag{7.13}$$

where $M >> 1$. It is suggested that $\lambda_{\max}$ as follows,

$$\lambda_{\max} = \frac{M}{\|\nabla g(\boldsymbol{U}|_{U=\mu})\|} \tag{7.14}$$

in which $10 \leq M \leq 100$ and $\nabla g(\boldsymbol{U}|_{U=\mu})$ is gradient of LSF at mean point. It is assumed SDC is applied for adjusting the as follows,

$$\lambda_k = \begin{cases} \lambda_{k-1} & \|\boldsymbol{U}_{k+1} - \boldsymbol{U}_k\| < \|\boldsymbol{U}_k - \boldsymbol{U}_{k-1}\| \\ \frac{\lambda_{k-1}}{c} & \|\boldsymbol{U}_{k+1} - \boldsymbol{U}_k\| \geq \|\boldsymbol{U}_k - \boldsymbol{U}_{k-1}\| \end{cases} \tag{7.15}$$

in which $1.2 < c < 1.5$ and $\boldsymbol{U}_{k+1}$ is point computed by CFSL at new iteration. $\lambda = 0$ when $k \to \infty$ due to applying condition of $\|\boldsymbol{U}_{k+1} - \boldsymbol{U}_k\| < \|\boldsymbol{U}_k - \boldsymbol{U}_{k-1}\|$ for highly nonlinear functions. Consequently, it is computed a fixed point as $\boldsymbol{U}_{k+1}^{c\lambda} = \boldsymbol{U}_k$, this means that U_k is located on the new iterations and this method is converged or it is provided stable results. As seen from SDC of $\|U_{k+1} - U_k\| < \|U_k - U_{k-1}\|$, we have,

$$\begin{aligned} &\|U_2 - U_1\| = t_1\|U_1 - U_0\|, \|U_3 - U_2\| = t_1 t_2\|U_1 - U_0\|, \\ &\Rightarrow U\|_{k+1} - U_k\| = \|U_1 - U_0\| \prod_{i=1}^{k-1} t_i \end{aligned} \tag{7.16}$$

in which $0 \le t < 1$ thus $\lim\limits_{k\to\infty} \prod_{i=1}^{k-1} t_i \approx 0$. Consequently, $U_{k+1} - U_k \approx 0$. This means that the CFSL is converged as a fixed point. The steps applied in the CFSL scheme with the hybrid formula of CFORM named CFSL-H are summarized by relations of CFORM with CD and RMIL as follows,

Step 1: Define $\varepsilon << 0$ (stopping criterion), $g(X) = 0$, Given μ, σ and distributions of variables, set $c \in (1.2,\ 1.5)$, $10 \le M \le 100$, $k = 0$, and initial point $X_0 = \mu$;

Step 2: For $k = 0$, compute $\lambda_{\max} = \frac{M}{\|\nabla g(U|_{U=\mu})\|}$;

Step 3: Compute θ_k^{CD}, θ_k^{IMLM}, and $\eta = -\frac{\|\nabla g(U_{k-1})\|^2}{4\nabla^T g(U_k) d_{k-1}}$, Compute $\theta_k^H = \eta\theta_k^{CD} + (1-\eta)\theta_k^{RMIL}$;

Step 4: Determine d_k;

Step 5: Compute $U_{k+1}^{c\lambda} = U_k^{CG} + \lambda_k d_k$ and $\alpha_k^{c\lambda} = \frac{U_{k+1}^{c\lambda}}{\|U_{k+1}^{c\lambda}\|}$,
Determine $U_{k+1} = \frac{d_k^T U_k - g(U_k)}{d_k^T \alpha_k^{c\lambda}} \alpha_k^{c\lambda}$, $\beta_{k+1} = \frac{d_k^T U_k - g(U_k^{CG})}{d_k^T \alpha_k^{c\lambda}}$;

Step 6: Update $\lambda_k = \begin{cases} \lambda_{k-1} & \|U_{k+1} - U_k\| < \|U_k - U_{k-1}\| \\ \frac{\lambda_{k-1}}{c} & \|U_{k+1} - U_k\| \ge \|U_k - U_{k-1}\| \end{cases}$;

Step 7: If $\|U_k - U_{k-1}\| < \varepsilon$ then iterative process is stopped, print $U^* = U_{k+1}$, $X^* = X_{k+1}$ and $\beta = \beta_{k+1}$;else $k = k + 1$ and go to Step 2.

The CFSL as well as other CFORM is utilized to improve the robustness of FORM formulation using sufficient descent conditions. As seen from Chap. 3, the modified method of FORM which is established based on the steepest descent vector applied a step size for controlling the instability of FORM. In analytical iterative CFLS-H formulation, the CSDV is determined by using an adjusting dynamical CSF combined with FSS which is computed by the Armijo rule as the maximum value and FSS is adjusted by SDC for control instability. Thus, the search direction vector computed by CG vector and step size adjusted by SDC are major differences of CFLS-H with improved FORM extracted by SDSD vector.

7.2.2 Examples for Illustrating Performances CFSL-H

In this section, several mathematical/structural examples are given to demonstrate the robustness and efficiency of CFOR-based CFSL-H iterative formulation. The CFSL-H method is compared with several FORM algorithms named as HL-RF (i.e. Hasofer-Lind and Rackwitz-Fiessler), STM (i.e. stability transformation method) which is iterated by parameters of $\lambda = 0.1$ and $\boldsymbol{C} = \boldsymbol{I}$, RHL-RF (i.e. relaxed HL-RF), and FSL which is determined by parameters of $\lambda = 15$ and $c = 1.4$. The parameters of CFSL-based FSS are given as $M = 35$ for initial step length and $c = 1.4$ for adjusting step length in all examples. Tolerance level for stopping iteration is given as $\|\boldsymbol{U}_{k+1} - \boldsymbol{U}_k\| \leq 10^{-9}$.

Example 7.3 A highly nonlinear LSF as: $g_1 = x_1^4 + 2x_2^4 - 20$ where, normal x_1 and x_2 have $\mu_1 = \mu_2 = 10$ and $\sigma_1 = \sigma_2 = 5$.

The reliability index, run-times as consuming times, Iteration as evaluating ∇g and call functions as evaluating the LSF and are listed in Table 7.2 for different studied reliability methods. It is shown that HL-RF produces the periodic-2 solution as [0.9267, 0.9863] for reliability index, while CFSL-H as well as the FSL method is converged to $\beta = 2.36545$, robustly. The results of CFSL-H are closely followed with the extracted results from Ref. [6] as $\beta = 2.365373$. The CFSL-H is improved the computational burden about three times and five times faster than RHL-RF and STM methods, respectively. The robustness with an acceptable computational burden of FORM is enhanced using hybrid CG formulation.

Example 7.4 A nonlinear LSF as: $g_2 = x_1^3 + x_1^2 x_2 + x_2^3 - 18$, where, normal x_1 and x_2 have $\mu_1 = 10$, $\mu_2 = 9.9$ and $\sigma_1 = \sigma_2 = 5$.

The iterations of the reliability index for HL-RF, STM, RHL-RF, FSL, and CFSL-H are listed in Fig. 7.1 for Example 7.4. The results of this figure show that the CFSL-H approach is converged, stably (i.e. reliability index of $\beta = 2.29824$). However, the HL-RF provides the unstable periodic-2 solutions as [1.5370, 1.5355]. The CFSL-H as well as the STM, RHL-RF, and FSL methods have a robust formulation for computing the MPP of this example. However, it is more efficient than other methods for this example. The efficiency and robustness of the analytical iterative reliability method for finding the MPP is improved using CFORM given by hybrid formulation.

The iterations of CFORM-based CFSL-H formulation for different M factors which is applied in maximum FSS as Eq. (7.14) are listed in Table 7.3 for Examples

Table 7.2 Results of different FORMs for Example 7.3

Methods	HL-RF	STM	FSL	RHL-RF	CFSL-H
Reliability index	[0.9267, 0.9863]	2.36516	2.36545	2.36537	2.36545
Iterations	Periodic-2	130	108	67	15
Call functions	–	2450	2356	1192	288
Run-times (sec)	–	0.56	0.53	0.31	0.09

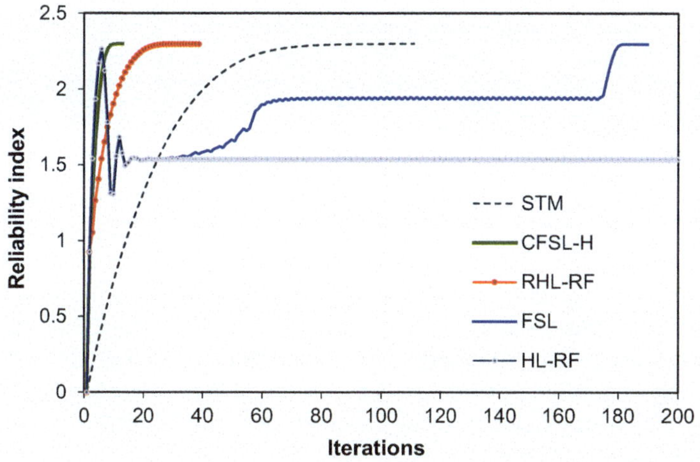

Fig. 7.1 Iterative result of reliability index applied for comparing CFLS-H in Example 7.4

7.3 and 7.4. The results presented that factor M is not a significant factor for improving the efficiency of CFSL which is applied for searching the MPP. However, it provides stable results with iterations between 12 and 20 with changing M in the range from 20 to 40.

Example 7.5 A dynamic system as: $g_3 = Fs - Ks \times 3(E[x_s^2])^{1/2}$,where Fs denotes the force capacity and Ks for the stiffness of secondary spring, and

$$E[x_s^2] = \frac{\pi S_0}{4\xi_s\omega_s}\left[\frac{\xi_a\xi_s}{\xi_p\xi_s(4\xi_a^2 + \theta^2) + \gamma\xi_a^2} \times \frac{\left(\xi_p\omega_p^3 + \xi_s\omega_s^3\right)\omega_p}{4\xi_a\omega_a^4}\right] \tag{7.17}$$

in which, $\gamma = \frac{M_s}{M_p}$, $\omega_a = \frac{\omega_p+\omega_s}{2}$, $\xi_a = \frac{\xi_p+\xi_s}{2}$, and $\theta = \frac{\omega_p-\omega_s}{\omega_a}$ that these relations are determined by lognormal variables presented in Table 7.4. STD is the standard deviation of each variable.

The reliability index computed by HL-RF, FSL, STM, RHL-RF, and CFSL-H are plotted in Fig. 7.2 for this example. It can be conducted, periodic-2 $\beta = [4.9807, 4.2170]$ result is captured by HL-RF method. However, the stable results as $\beta = 2.01644$ is extracted by the STM, FSL, RHL-RF, as well as CFSL-H schemes, but CFSL-H is faster than other FORM-based SDSD methods. As presented results

Table 7.3 Iterations for CFSL method with different M in Examples 7.3 and 7.4

M	10	30	50	100
Example 7.3	21	15	17	21
Example 7.4	13	12	18	19

Table 7.4 Variables for dynamic system applied in Example 7.5

Variable	Mean	STD	Variable	Mean	STD
M_P	1	0.1	ξ_P	0.05	0.02
M_S	0.01	0.001	ξ_s	0.02	0.01
K_p	1	0.2	S_0	100	10
K_S	0.01	0.002	F_s	15	0.15

for Examples 7.3, 7.4, and 7.5, the CSDV by hybrid adjusting scalar factor is improved the computational burden as minimum iterations of FORM. The CFSL-H approach also provided results without chaotic or periodic solutions. Thus, the CFORM using finite step length adjusted by sufficient descent conditions can be controlled the instability of FORM with efficient results. The application of CFORM is acceptable for nonlinear concave problems. The efficiency and robustness are achieved by using an adjusting dynamical CSF. The nonlinear effects of LSF at point are directly used for computing the cooperative factor. This adjustment is applied for computing a dynamical search direction with a discrete nonlinear map which is provided a stable MPP is reliability analysis.

Figure 7.3 shows iterations for reliability index which are extracted form iterative results of CSDVs-based CFSL-H, CFSL-CD and CFSL-RMIL. It is extracted from Fig. 7.3 that, CFSL-H is better than CFSL-RMIL and CFSL-CD. The hybrid adjusting CSF applied in CSDV can be reduced computational burden of FORM which is formulated by SDSD including RHL-RF, FSL, and STM in Example 7.5.

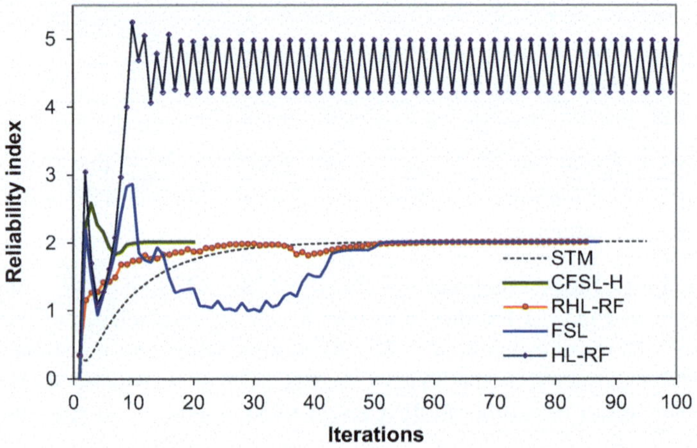

Fig. 7.2 Iterative results of reliability index applied for comparing CFLS-H in Example 7.5

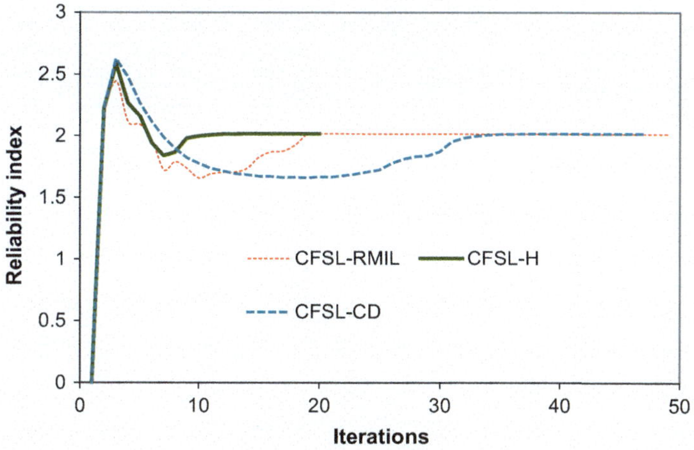

Fig. 7.3 Iterative results for different CFSL algorithms in Example 7.5

Example 7.6 An I-beam as: $g_4 = S - \frac{12PA}{2L}\left[\frac{(L-A)d}{b_f d^3-(b_f-t_w)(d-2t_f)^3}\right]$.

An I-beam with simply-support as plotted in Fig. 7.4 is used in this example, the LSF is computed based on the maximum stress obtained by the bending moment. The random variables which applied to compute LSF are arranged in Table 7.5.

The extracted reliability index using the MCS from Ref. [8] is 2.81962. The reliability index, run-time, iterations, call functions, and relative errors with MCS ($Rel\text{-}error = (\beta_{MC} - \beta)/\beta \times 100$) of studied methods are recorded in Table 7.6. The FORM algorithms are converged to $\beta = 2.811767$ with similar relative errors (i.e. 0.28%) while the CFSL-H is more efficient among other reliability methods. The STM has the least efficient property among the all methods but the HL-RF is a more efficient scheme than STM, FSL, and RHL-RF. The CFORM as well as the HL-RF provided an efficient result for this example.

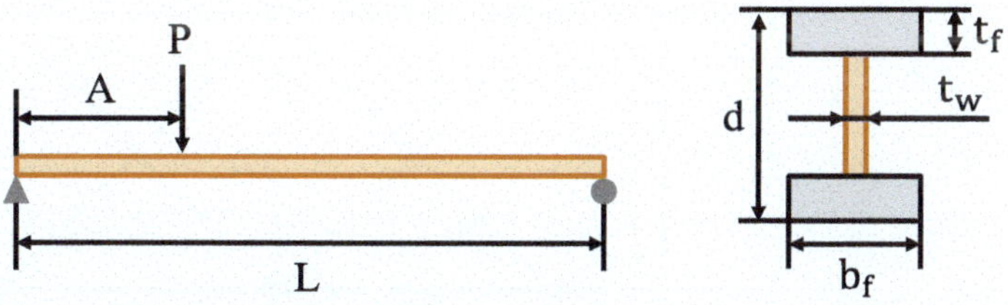

Fig. 7.4 An I-beam with simply-support and a constrained load for Example 7.6. Reprinted from [5], with permission from Elsevier

Table 7.5 Random variables for I-beam with simply-support example

Variable	Distribution	Mean	Standard deviation	Dimension
P	Normal	6500	300	N
L	Normal	120	5	mm
A	Log-normal	72	5	mm
S	Gumbel	170,000	4760	MPa
d	Normal	2.3	0.04	mm
b_f	Weibull	2.97	0.075	mm
t_w	Normal	0.16	0.02	mm
t_f	Normal	0.26	0.02	mm

Table 7.6 The results for different reliability methods of FORM for I- simply-support beam

Methods	HL-RF	RHL-RF	STM	FSL	CFSL-H
Iterations	12	34	114	17	7
Call functions	1187	3539	11,034	1729	607
Reliability index	2.811767	2.810887	2.811767	2.811767	2.811767
Run-time (sec)	0.53	1.46	4.33	0.79	0.36
Rel-error %	0.28	0.31	0.28	0.28	0.28

Example 7.7 A planar ten-bar structure as: $G_5 = 0.004 - \left|\Delta_3^V\right|$

A planar ten-bar truss as structural view presented in Fig. 7.5, is considered as an implicit LSF for evaluating the convergence performances of the CFSL-H method.

The LSF of this truss structure as point 3-vertical displacement is computed by normal/ non-normal variables listed in Table 7.7.

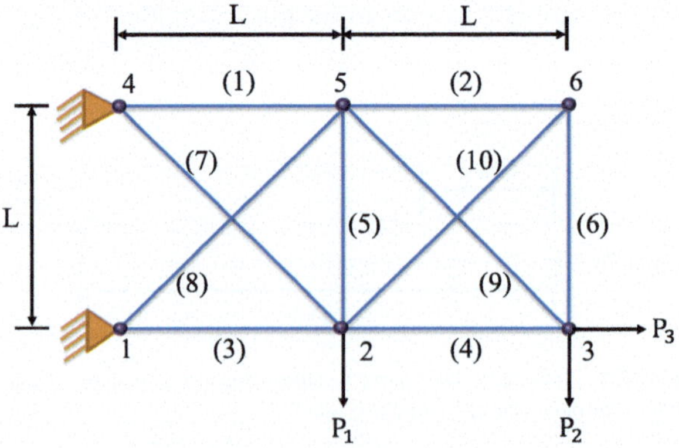

Fig. 7.5 2-D ten-bar structure. Reprinted from [5], with permission from Elsevier

Table 7.7 Statistical properties of random variables for Example 7.7

Variable	A_i (m^2)	E (Gpa)	L (m)	P_1 (kN)	P_2 (kN)	P_3 (kN)
Mean	0.001	100	1	80	10	10
COV[a]	0.15	0.05	0.05	0.1	0.1	0.1
Description	Sectional area of bars	Young's modulus	Horizontal and vertical length	Point loads		
PDF[b]	Normal	Normal	Lognormal	Gumbel	Gumbel	Gumbel

[a] COV (coefficient of variation)
[b] PDF (probability density function)

Using a finite element method, the node 3- vertical displacement is computed at the mean of variable that this displacement is captured as 0.00291501 m, which is closely followed by extracted results Ref. [9] as 0.002915 m.

The approximated reliability index using MCS with 5×10^5 simulated points is $\beta = 1.5224$. The chaotic solutions as unstable results are obtained using iterative HL-RF, RHL-RF, and STM methods-based SDSD. The SDSD method with dynamical step size applied in RHL-RF and a constant small step size applied in STM can be not provided a robust result for FORM while the FSL and CFLS-H formulated with SDC are respectively converged to $\beta = 3.43463$ (after 65 iterations) and $\beta = 1.77024$ (after 49 iterations). As seen, the convergence control criterion is an essential factor for the warranty of stabilization of the FORM analytical reliability method. The CFSL-H with CSDV is robust by comparing with HL-RF, STM, and RHL-RF and is efficient by comparing the results of FSL.

7.3 Hybrid Self-adaptive Directional Conjugate Method

The FORM-based SDSD and CFORM are simple iterative approaches that they can be widely applied for SRA by considering a suitable balance for giving efficiency and accuracy. In FORM/CFORM, P_f is estimated as $P_f \approx \Phi(-\beta)$ where reliability index $= \beta$ which is iterated for searching the MPP i.e. $\boldsymbol{U}^*$ which is on limit state surface i.e. $g(\boldsymbol{U}^*) = 0$. It is shown that the directional normalized vector is provided an efficient solution for FORM in SDSD given by STM [10] and the conjugate search direction given by CCSTM [3]. In Sect. 7.2 showed that the FSL [11], the CFSL [5] can be enhanced the efficiency of FORM by using the SDC adjusted for the step length. It is showed that the modified FORM algorithms are more robust than HL-RF method; however, the efficiency improved analytical methods is major challenge for structural reliability analyses. Recently, the conjugate search direction applied for computing normalized sensitivity vector has been extended by Keshtegar [1–3, 5, 12] that CFORM is utilized to enhance the abilities of analytical iterative formulas by using Armijo [2, 3] and SDCs [5, 12]. The CFORM compared to the steepest descent developed FORM including RHL-RF [13] and FSL [11] provided better efficiency. The literature researches indicated that robustness as well as efficiency are important factors for developing an analytical reliability method. Almost research efforts have been focused on the convergence stability of FORM for applying in practical/engineering reliability problems.

In this section, a robust FORM algorithm is introduced without line search rules such as merit function [14], Armijo [2, 5], Wolfe [15], and SDC [12]. In this conjugate method, an iterative formulation is discussed using hybrid self-adaptive CSF named as HSAC (i.e. hybrid self-adaptive conjugate) method [16] which is applied to estimate MPP and reliability index.

7.3.1 *Algorithm of Hybrid Conjugate Line Search*

The CSF applied for CSDV is presented to improve the efficiency of FORM. In this section, two adaptive CSFs are introduced that these factors are applied to enhance the conjugate FR scalar factor with a dynamical relation. The CSF of self-adaptive FR is computed by the following factor,

$$\theta_k^{SA} = \theta_k^{FR} - \frac{\sqrt{\theta_k^{FR}}\left|\nabla^T g(\boldsymbol{U}_k)\nabla g(\boldsymbol{U}_{k-1})\right|}{\|\boldsymbol{d}_{k-1}\|^2} \tag{7.18}$$

Using self-adaptive θ_k^{SA} in Eq. (7.18), the FORM formula using CSDV is named self-adaptive CSDV (SAC). By determining the CSF of FR as presented in Chap. 6 and SAC, it can be denoted that the previous conjugate and gradient vector are applied in scalar factor Eq. (7.18). On the other hand, we supposed that $\theta_k^{SA} < \theta_k^{FR}$. Consequently, the SAC scalar factor as well as LFR (i.e. limited FR method) scalar factor should be adjusted less than FR scalar factor. Thus, the efficiency of CFORM can be controlled as well as the LFR method. The SAC and FR methods are dynamically combined in FORM formula to attain a positive scalar factor less than the FR factor by below hybrid CSF,

$$\theta_k^{HS} = \min\{\theta_k^{FR}, \left|\theta_k^{SA}\right| \tag{7.19}$$

It can be obtained $0 \leq \theta_k^{HS} \leq \theta_k^{FR}$. This means that the hybrid SAC factor is a positive factor less than FR. The iterative formula of CFORM computed by hybrid SAC scalar factor is rewritten as follows,

$$\boldsymbol{U}_{k+1} = \left(\frac{\boldsymbol{U}_k^T\nabla g(\boldsymbol{U}_k) - g(\boldsymbol{U}_k)}{\nabla^T g(\boldsymbol{U}_k)\boldsymbol{\alpha}_k^c}\right)\boldsymbol{\alpha}_k^c \tag{7.20}$$

where, $\boldsymbol{\alpha}_k^c = \frac{\boldsymbol{U}_{k+1}^{c\lambda}}{\|\boldsymbol{U}_{k+1}^{c\lambda}\|}$, $\boldsymbol{U}_{k+1}^{c\lambda} = \boldsymbol{U}_k + \lambda_k \boldsymbol{d}_k$, in which, $\boldsymbol{d}_k$ is CG vector and the parameter λ_k is finite-step length at k_{th} iteration that it given as $\lambda = 1$. In this method, we do not give any conditions for adjusting the step size, thus, the CFORM formula given by HSAC with step length $\lambda = 1$ should hold the SDC for the global convergent results. By this condition for line search direction given from CSDV alone, we have $\boldsymbol{U}_k - \boldsymbol{U}_{k-1} \approx 0$ at final iteration that this means, it is obtained a fixed point.

Lemma 1 *The* $\boldsymbol{d}_k$ *computed by the HSAC factor* θ_k^{HS}*, holds SDC as* $\boldsymbol{d}_k^T \nabla g(\boldsymbol{U}_k) \leq -\|\nabla g(\boldsymbol{U}_k)\|^2$*;*

Proof For $k = 0$, SDC is true subsequently, $\boldsymbol{d}_0 = -\nabla g(\boldsymbol{U}_0)$. However, $k > 0$, we have

$$\begin{aligned}\nabla^T g(\boldsymbol{U}_k)\boldsymbol{d}_k &= -\|\nabla g(\boldsymbol{U}_k)\|^2 + \theta_k^{HS}\nabla^T g(\boldsymbol{U}_k)\boldsymbol{d}_{k-1} \\ &\leq -\|\nabla g(\boldsymbol{U}_k)\|^2 + \theta_k^{FR}\nabla^T g(\boldsymbol{U}_k)\boldsymbol{d}_{k-1} \\ &< -\|\nabla g(\boldsymbol{U}_k)\|^2 \end{aligned} \tag{7.21}$$

The above relation showed HSAC CSF holds SDC. Thus, proof is completed.

On the other hand, using the SDC $\boldsymbol{d}_{k+1} \leq \boldsymbol{d}_k$, we have

$$\|\boldsymbol{d}_2\| = t_1\|\boldsymbol{d}_1\|,\ \|\boldsymbol{d}_3\| = t_1 t_2\|\boldsymbol{d}_1\|,\ \ldots,\ \|\boldsymbol{d}_k\| = \|\boldsymbol{d}_1\|\prod_{i=1}^{k-1} t_i \tag{7.22}$$

where, $0 \leq t \leq 1$ and $\lim_{k\to\infty} \prod_{i=1}^{k-1} t_i \approx 0$. Thus, $\lim_{k\to\infty} \boldsymbol{d}_k = \lim_{k\to\infty} \boldsymbol{d}_1 \prod_{i=1}^{k-1} t_i \approx 0$. Therefore, we have $\boldsymbol{U}_k \approx \boldsymbol{U}_{k-1}$ and this method converged to $\boldsymbol{U}_k$ as fixed point. The steps for coding of CFORM scheme in computer program are listed as follows,

Step 1: Define stopping criterion as $\varepsilon = 10^{-6}$, $g(\boldsymbol{U}) = 0$, distributions of variables with μ, σ, set $k = 0$, $\boldsymbol{d}_0 = 0$ and $\boldsymbol{X}_0 = \mu$;
Step 2: Transfer variable in $\boldsymbol{U}$-space as $\boldsymbol{U} = \frac{(\boldsymbol{X}-\mu_X^e)}{\sigma_X^e}$ where, μ_x^e. and σ_x^e are obtained as,

$$\sigma_x^e = \frac{1}{f_X(x^*)}\varphi\left[\Phi^{-1}\{F_X(x)\}\right] \tag{7.23}$$

$$\mu_x^e = x - \sigma_x^e \Phi^{-1}[F_X(x) \tag{7.24}$$

Step 3: Compute $\nabla g(\boldsymbol{U}_k)$ and $g(\boldsymbol{U}_k)$;
Step 4: Compute θ_k^H and $\boldsymbol{d}_k$ using;
Step 5: Compute $\boldsymbol{\alpha}_k^c = \frac{\boldsymbol{U}_{k+1}^{c\lambda}}{\|\boldsymbol{U}_{k+1}^{c\lambda}\|}$, $\boldsymbol{U}_{k+1}^{c\lambda} = \boldsymbol{U}_k + \lambda_k \boldsymbol{d}_k$;
Step 6: Obtain $\boldsymbol{U}_{k+1} = \left(\frac{\boldsymbol{U}_k^T \nabla g(\boldsymbol{U}_k) - g(\boldsymbol{U}_k)}{\nabla^T g(\boldsymbol{U}_k)\alpha_k^c}\right)\boldsymbol{\alpha}_k^c$;
Step 7: Determine $\boldsymbol{X}_{k+1} = \mu + \sigma \boldsymbol{U}_{k+1}$ and $\beta_{k+1} = \|\boldsymbol{U}_{k+1}\|$;
Step 8: If $\|\boldsymbol{U}_k - \boldsymbol{U}_{k-1}\| < \varepsilon$ then iterative method is stopped and give, $\boldsymbol{U}^* = \boldsymbol{U}_{k+1}$, $\boldsymbol{X}^* = \boldsymbol{X}_{k+1}$ and, $\beta = \beta_{k+1}$; else set $k = k + 1$ and Go to Step 2.

7.3.2 Examples for Comparative Studies of HSAC

Two performances including the efficiency and robustness of the new algorithm of CFORM presented in Sect. 7.3.1 are investigated. These convergent performances are compared with methods formulated by SDSV as HL-RF and FSL and methods formulated CSDV as FR, SAC, and HSAC using mathematical/structural examples. The performances of these methods are compared using iterations and stable recallability index (i.e. β). The FR, FSL, SAC, and HSAC are iterated by the same parameters for adjusting FSS $\lambda = 30$ and $c = 1.41$. Same stopping criterion as $\varepsilon = 10^{-6}$ are applied for all the algorithms.

Example 7.8 A non-linear LSF defined as,

$$g = x_1^3 + x_1^2 x_2 + x_2^3 - 18 \tag{7.25}$$

where, normal x_1 and x_2 have $\mu_1 = 10$, $\mu_2 = 9.9$ and $\sigma_1 = \sigma_2 = 5$. The results showed that the periodic-2 $\beta = (1.53547, 1.53695)$ solution is obtained based on HL-RF however, other methods-based SDC accurately converged to reliability index of 2.298252. The HSAC method converges to $\boldsymbol{U}^* = [-1.66290, -1.58642]$ that it is almost equal to results of Ref. [10]. It is detected that HSAC provides a robust formulation for FORM while it reduces computational burden compared to FR and FSL methods.

Example 7.9 A nonlinear LSF is now considered [3],

$$g(X) = x_1^4 + x_2^2 - 50 \tag{7.26}$$

where, Lognormal x_1 has mean of 5 and STD of 1 and Gumbel x_2 has mean of 10 and STD 10. The extracted results of FORM-based MPP search from Ref. [3], $\beta = 3.259315$ which is obtained by CCC.

Results using HSAC given after 11 iterations are $\beta = 3.259315$, $\boldsymbol{X}^* = [2.64754, 0.93141]$ and $\boldsymbol{U}^* = [-3.11145, -0.97059]$. The robust results with stable reliability index and efficient computational burden with iteration about 10 are provided based on the HSAC method.

Figure 7.6 demonstrates iterations for achieving reliability index using HL-RF, FSL FR, SAC, and HSAC methods. The iterative analytical method formulated by CSDV is yielded to stable results but HL-RF as the SDSD method provided unstable periodically i.e. $\beta = (2.5117, 1.8546, 2.3861, 1.5967)$ reliability indexes. The FSL method is inefficient in comparison with CFORM formulations. The SAC and HSAC have shown an efficient computational burden compared to the FR approach. By comparing the results of FR and FSL with HSAC, the CFORM-based HSAC significantly improved the efficiency of FSL for this example.

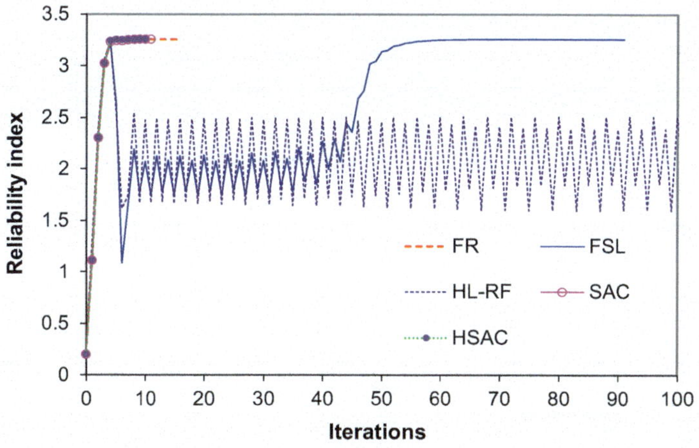

Fig. 7.6 Iterative results applied for comparing HSAC in Example 7.9

Example 7.10 A vehicle side-impact with response surface function as LSF is given by [1],

$$g = 0.489x_3x_7 + 0.843x_5x_6 - 0.0432x_9x_{10} + 0.0556x_9x_{11} + 0.000786x_{11}^2 - 0.75 \tag{7.27}$$

where $i = 1 \sim 7x_i \sim N(1, 0.005)$, $i = 8 \sim 9x_i \sim N(0.3, 0.006)$, $i = 10 \sim 11x_i \sim N(0, 10.0)$. The reliability index extracted from Ref. [1] is $\beta = 3.497464$.The iterations, MPP ($\boldsymbol{X}^*$) and reliability index obtained by CFORM using SAC, SAC, and FR are cited in Table 7.8. It displays that the CFORM methods provide stable results with the reliability index $\beta = 3.4975$ and agree more with the extracted reliability index from Ref. [1]. The SAC and HSAC are efficient CFORM by comparing with the FR method. As seen, HSAC converges about twice faster than the FR. By applying an acceptable CSF which is computed by adaptive hybrid relation less than FR CSF as presented in HSAC and LFR methods, it can be enhanced the efficiency of CFORM formulated by FR conjugate search direction such as CCC and CCSTM. Iter is the iteration of each method.

The iteration for achieving reliability index using HL-RF, FR, FSL, SAC, and HSAC are presented in Fig. 7.7. As seen, unstable results as periodic-4 β as (3.1757, 2.0351, 1.9180, 1.2957) are extracted by the HL-RF method. However, FSL as well as CFORM-based FR and SAC (HSAC) are a robust iterative analytical method; but it is shown that SAC (HSAC) is more efficient with a high-convergence rate (e.g. four times faster than FSL). Consequently, the CSDV applied in the HASC method is the essential key point for CFORM for achieving the stabilization of analytical MPP search algorithms.

Table 7.8 MPP results of CFORM algorithms for Example 7.10

Methods	x_3^*	x_5^*	x_6^*	x_7^*	x_9^*	x_{10}^*	x_{11}^*	β (Iter)
FR	0.9728	0.9539	0.9539	0.9728	0.3014	29.777	-8.365	3.4975 (30)
HSAC	0.9724	0.9534	0.9534	0.9724	0.30146	30.173	-8.390	3.4975 (18)
SAC	0.9725	0.9536	0.9536	0.9725	0.30144	30.050	-8.367	3.4973 (21)

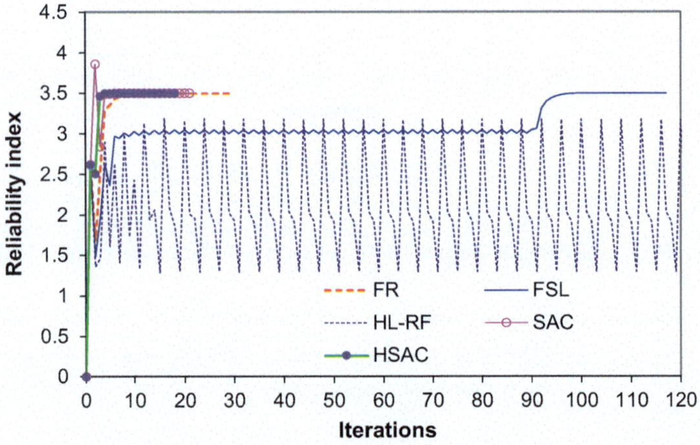

Fig. 7.7 Iterative result of reliability index applied for comparing HSAC in Example 7.10

Example 7.11 A dynamic system with log-normal random variables by LSF given by the force capacity of the secondary spring is defined as [3],

$$g = Fs - Ks \times 3\left(E\left[x_s^2\right]\right)^{1/2} \tag{7.28}$$

where,

$$E\left[x_s^2\right] = \frac{\pi S_0}{4\xi_s\omega_s}\left[\frac{\xi_a\xi_s}{\xi_p\xi_s\left(4\xi_a^2 + \theta^2\right) + \gamma\xi_a^2} \times \frac{\left(\xi_p\omega_p^3 + \xi_s\omega_s^3\right)\omega_p}{4\xi_a\omega_a^4}\right] \tag{7.29}$$

in which, Fs denotes the force capacity, $\gamma = \frac{M_s}{M_p}$, $\omega_a = \frac{\omega_p+\omega_s}{2}$ $\xi_a = \frac{\xi_p+\xi_s}{2}$ and $\theta = \frac{\omega_p-\omega_s}{\omega_a}$ that these factors applied in LSF are computed based on Lognormal PDF as listed in Table 7.9.

Extracted reliability index from Ref. [3], 2.01645 and it is given from HSAC after 26 iterations as $\beta = 2.016843$ with related MPP of $X^* =$

Table 7.9 Variables for computing LSF in Example 7.11

Variable	M_s	M_p	K_s	K_p	ξ_s	ξ_p	F_s	S_0
Mean	0.01	1	0.01	1	0.02	0.05	15	100
STD	0.001	0.1	0.001	0.2	0.01	0.02	1.5	10

[1.0015, 0.0101, 1.1002, 0.0111, 0.0281, 0.0121, 103.6687, 13.7103]. The iterations for reliability indexes with respect to different analytical methods are presented in Fig. 7.8. The periodic-2 solutions as $\beta = (4.980739, 4.217011)$ is determined by using the HL-RF method. The reliability methods of FSL and conjugate FR, SAC, and HSAC showed excellent convergence as stable results. However, the results from Fig. 7.8 denoted that conjugate schemes using scalar factors of SAC and HSAC are a robust formulation in compassion with HL-RF method. The computational burden of this method is also improved about twice faster than the FR. The limitation in the adaptive formulation of FR with dynamical CSF can provide stable results while it is strongly affected on the efficiency. These methods provide stable results without any conditions for computing the step size. The abilities of the adaptive CFORM formulation are vitally acceptable for nonlinear function but they should be examined for convex problems.

Example 7.12 A structure as plotted in Fig. 7.9 under buckling failure mode is considered by the following LSF [12],

$$g = 1 - \frac{\sqrt{3(1-v^2)}}{\pi E t^2 \cos^2\alpha} \times \left(\frac{P}{0.66} + \frac{M}{0.41 r_1}\right) \tag{7.30}$$

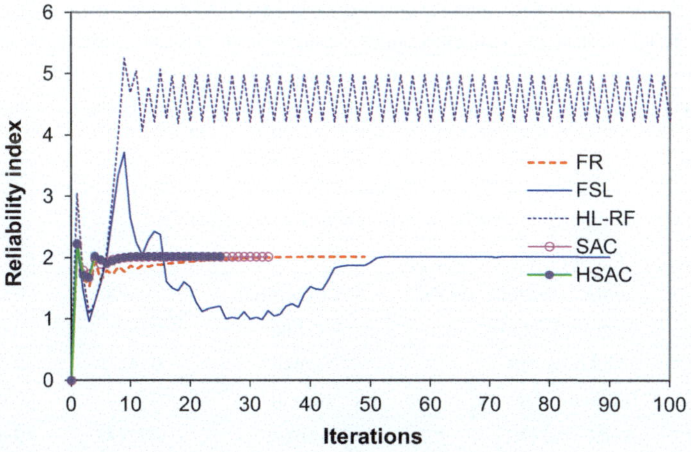

Fig. 7.8 Iterative results applied for comparing HSAC in Example 7.11

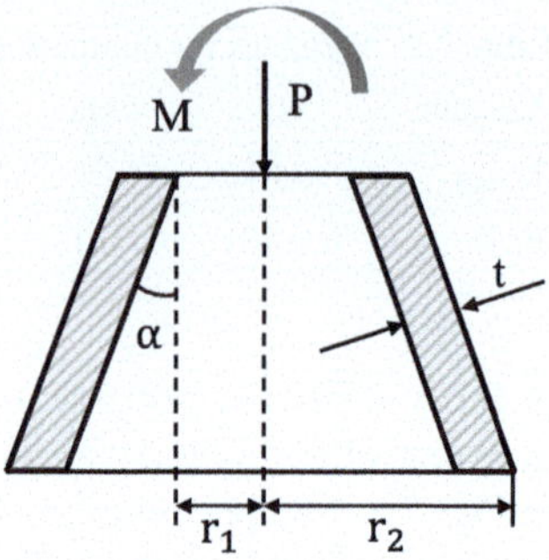

Fig. 7.9 A view of conical structure. Reprinted from [17], with permission from Elsevier

in which $\nu = 0.3$ [12]. Table 7.10 presents the variables as normal/non-normal distributions for Example 7.12.

The reliability index and MPP given from Ref. [12] are $\beta = 4.632489$ and $\boldsymbol{X}^* = [65442.63, 0.00208, 0.5275, 0.8887, 106149.59, 73112.66]$ which were obtained by CCC after 43 iterations. The reliability results given after 18 iterations using HSAC are presented as 4.670278 and $\boldsymbol{X}^* = [645975.57, 0.002054, 0.52812, 0.88743, 99013.14, 7421.56]$. Based on SRA using the CFORM, we have $\beta = 4.652248$ using FR obtained after 29 iterations and $\beta = 4.627818$ using SAC extracted after 23 iterations. As seen, the CSDV applied in FORM to search MPP is accurately provided stable results. Nevertheless, HSAC with limited SCF performs better than others as presented results in Fig. 7.10 for iterations of β in Example 7.12. Nevertheless, the HL-RF formulation produces chaos iterations.

By the results obtained in this section, the SDSD showed several issues as unstable results and inefficient convergence for reliability analysis of nonlinear problems. As the results of the HL-RF demonstrated its instability for both periodic and chaotic solutions. However, the FSL as a steepest descent FORM formulation is more robust than the HL-RF but it is computationally converged. These challenges of FORM formulation can be improved based on a novel CSDV applied in FORM formulation by a directional sensitivity vector with robust formulation and efficient computational method.

Table 7.10 The variables for conical structure applied in Example 7.12

Variable	Description	PDF	Mean	COV
E	Young's modulus (MPa)	Lognormal	70,000	0.05
α	Slop angle (rad)	Normal	0.524	0.02
t	Thickness (m)	Normal	0.0025	0.05
r_1	Internal radius (m)	Normal	0.9	0.025
P	Axial load (N)	Gumbel	70,000	0.08
M	Bending moment (N-m)	Gumbel	80,000	0.08

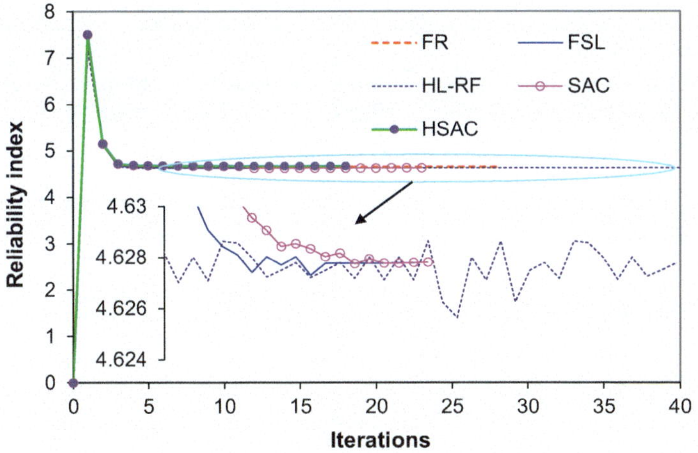

Fig. 7.10 Iterative result applied for comparing HSAC of reliability index in Example 7.12

As declared, the HSAC algorithm showed convergence with robust results while it is assumed finite step length in the main iterative formulation. Consequently, the normalized conjugate search direction is determined based on below point,

$$\boldsymbol{U}_{k+1}^{c} = \boldsymbol{U}_k + \boldsymbol{d}_k \tag{7.31}$$

in which, $\boldsymbol{d}_k$ is determined by CSF applied in the HSAC method. The step size is neglected in the above CSDV to validate the effects of CSDV without FSS given from SDC that it is named as HSAC$^+$. The results of HSAC are computed by FSS while iterative shame of HSAC$^+$ is without FSS. Also, CFORM is compared to FSL, HL-RF, FR, and SAC. Table 7.11 presents the results of reliability analysis for different studied methods for Examples 7.8, 7.9, 7.10, 7.11, and 7.12 in this section. It demonstrated that the FSL and other CFORM frameworks have the same robustness but the conjugate methods given by the adaptive CSF are more efficient than FSL while this method is not formulated by FSS. The SAC, HSAC, and HSAC$^+$ are as robust as FSL but they are most robust than HL-RF. The search direction, which is computed by the step size and conjugate (steepest descent) vectors in the FORM formula, is an important factor for achieving the robustness of FORM. It can be concluded that the effect of the CG vector is more significantly important than the selecting step size by comparing with FORM-based SDSD. The previous literature of research studies given from the works of Liu and Der Kiureghian [14], Yang [18], Santos et al. [15], and Keshtegar and Meng [13] showed a focus on computing the step size for giving the stabilization of FORM formulation. However, Keshtegar [2, 3, 12] pointed out that the SDSD applied in FORM formulation may yield unstable results. It is shown that the hybrid self-adaptive CSF can be applied for reliability analysis with efficient and robust results. The SAC, HSAC, and HSAC$^+$ provide an acceptable result for both robustness and efficiency. Therefore, reliability index can

Table 7.11 Effects of lines search (SDSD or CSDV) in FORM algorithms for Examples 7.8, 7.9, 7.10, 7.11, and 7.12

Examples	FR (Iter/CF) β	HSAC (Iter/CF) β	SAC (Iter/CF) β	HSAC$^+$ (Iter/CF) β	FLS (Iter/CF) β	HL-RF (Iter/CF) β
7.8	(10/134) 2.298252	(8/108)* 2.298252	(9/122) 2.298252	(8/108) 2.298252	(151/2082) 2.298252	Failed periodic-2 (1.5355, 1.5370)
7.9	(17/208) 3.259314	(10/124) 3.259315	(12/152) 3.259315	(9/116) 3.259314*	(91/1139) 3.259314	Failed periodic-4 (1.8538, 2.3874, 1.5969, 2.5117)
7.10	(30/450) 3.497481	(18/270) 3.497507	(21/315) 3.497483	(48/720) 3.541287	(117/1755) 3.497481	Failed periodic-4 (3.1757, 2.0351, 1.9180, 1.2957)
7.11	(49/4662) 2.016446	(26/2482) 2.016843	(33/3143) 2.016705	(23/2208) 2.0167664	(90/8494) 2.016511	Failed periodic-2 (4.2170, 4.9807)
7.12	(29/1007) 4.652248	(18/628) 4.670279	(23/806) 4.627818	(16/559) 4.6587242	(22/763) 4.627785	Failed chaos

be approximated by HSAC in the first stage and this method can be provided the stable results with an acceptable computational burden in the second stage. CF is the number of calling functions of each method.

An adaptive formulation of FR using dynamical limited CSF is presented for improving the efficiency and robustness of FORM. This method is not dependent on the step size thus it is iterated without sufficient descent properties. The adaptive conjugate search direction applied in the limited FR method is extended without line search rules including Armijo, Wolfe, or merit conditions. The CFORM given by adaptive FR conjugate method with adaptive limited CSF is named as self-adaptive conjugate (SAC) and hybrid SAC (i.e. HSAC) methods formulated in CFORM. As shown, the HSAC scalar factor is dynamically adjusted based on FR and SAC scalar factors which are given a minimum value from them. It is revealed that the HSAC method showed global convergence with SDC in FSS and without rule in step size. The performances (i.e. for efficiency and robustness as presented by stable convergence) of SAC and HSAC using FSS and HSAC without FSS (HCLS$^+$) are discussed. The results confirmed that the CFORM formulations provided a robust process while the HL-RF is instable. The CFORM is more efficient than the FSL algorithm. Consequently, the adaptive iterative approach by LFR applied in directional sensitivity vector can be improved the efficiency of FORM. The HSAC+ without FSS and SAC by applying the FSS are more efficient than FR and FSL methods.

7.4 Three-Terms Directional Conjugate Method

The major difference between FORM formulas with sensitivity vector computed by CSDV is to determine the CSFs which is applied for computing the conjugate sensitivity vector. The existing CFORM such as FR, limited FR [2], and enhanced FR [1] are formulated using the CSDV which is computed based on two terms as (i) new gradient vector and (ii) previous CG vector combined by a scalar factor. The CSF is applied to adjust an acceptable value of the CG vector that this line search it not parallel to the previous line search and it is also shown that the efficiency of reliability analysis in CFORM can be controlled by selecting a suitable scalar factor. However, it is hard to find an appropriate scalar factor which is applied in the formulation of CFORM. On the other hand, as presented in Chap. 6, the almost CFORM frameworks are developed based on the FR CSF to satisfy sufficient descent conditions by the FR method and they provide a robust result for MPP search. However, the efficiency of the CFORM by applying the FR CSF and two terms applied in the search direction vector is a major challenge in SRA. On the other hand, unlike the steepest descent methods applied for iterative FORM, the CFORM may be delivered an inefficient computational formulation for convex reliability problems. Consequently, the CFORM-based efficient formulation is major effort for approximating the failure probability that there should be extended an efficient sensitivity vector for the FORM iterative scheme.

In this section, a three-term CSF is introduced for CFORM formulation called as three-term conjugate finite-step length (TCFS) [19]. Two adjusted factors are applied in the TCFS method for computing the CG elements. These factors are applied to update the gradient and CG vectors applied in computing conjugate sensitivity vector. The conjugate sensitivity vector is computed by new CG vector determined by three terms as (i) new gradient vector without scalar factor, (ii) previous CG vector with a scalar factor, and (iii) previous gradient vector with a scalar factor. The scalar factor of the previous gradient and CG vectors are dynamically computed based on the convergence information of the FORM iterative formula.

7.4.1 Formulation and Algorithm of TCFS

The discrete map given from CSDV is directly computed based on new gradient and previous CG vectors that this nonlinear CSDV map can be provided a stable result for FORM and can be controlled convergent performances of FORM in SRA. For improving a robust formulation of FORM in TCFS, the CSDV should be satisfied the SDC i.e. $\nabla^T g(\boldsymbol{U}_k)\boldsymbol{d}_k \leq -c_1\|\nabla g(\boldsymbol{U}_k)\|^2$ with $0 < c_1 < 1$. Thus, there can be given as $\|\boldsymbol{d}_{k+1}\| < \|\boldsymbol{d}_k\|$. By using the SDC, a three-term CG vector is presented as below relation,

$$\boldsymbol{d}_k = -\nabla g(\boldsymbol{U}_k) + \theta_k \boldsymbol{d}_{k-1} + \eta_k \nabla g(\boldsymbol{U}_{k-1}) \tag{7.32}$$

where, θ_k and η_k denote the gradient and CG scalar factors, respectively. θ_k is computed using FR scalar factor as $\theta_k = -\frac{\|\nabla g(\boldsymbol{U}_k)\|^2}{\|\nabla g(\boldsymbol{U}_{k-1})\|^2}$, but the factor η_k is considered a positive value which is determined as follows,

$$\eta_k = \min\left\{0.9\frac{\|\nabla g(\boldsymbol{U}_k)\|^2}{\|\nabla g(\boldsymbol{U}_{k-1})\|^2}, \frac{\|\nabla g(\boldsymbol{U}_k)\|^2}{\|\boldsymbol{d}_{k-1}\|^2}\right\} \tag{7.33}$$

It can be given by a positive factor η_k that term $\eta_k \nabla g(\boldsymbol{U}_{k-1})$ in TCFS method provides a CSF which is less than CSF computed in FR, thus $\boldsymbol{d}_k < -\nabla g(\boldsymbol{U}_k) + \theta_k \boldsymbol{d}_{k-1}$.

Lemma *It is supposed that CG* $\boldsymbol{d}_k$ *holds* $\nabla^T g(\boldsymbol{U}_k)\boldsymbol{d}_k \le -c_1\|\nabla g(\boldsymbol{U}_k)\|^2$, *thus we have* $\|\boldsymbol{d}_{k+1}\| < \|\boldsymbol{d}_k\|$;

Proof For $k = 0$, then $\boldsymbol{d}_0 = -\nabla g(\boldsymbol{U}_0)$, thus SDC is satisfied but for $k > 0$, it can be given $\boldsymbol{d}_k < -\nabla g(\boldsymbol{U}_k) + \theta_k \boldsymbol{d}_{k-1}$ for TCFS therefore, it is computed as below relation,

$$\nabla^T g(\boldsymbol{U}_k)\boldsymbol{d}_k \le -c_1\|\nabla g(\boldsymbol{U}_k)\|^2 < -\|\nabla g(\boldsymbol{U}_k)\|^2 + \theta_k \nabla^T g(\boldsymbol{U}_k)\boldsymbol{d}_{k-1} \tag{7.34}$$

We have $\theta_k < 0$, therefore, $-\|\nabla g(\boldsymbol{U}_k)\|^2 + \theta_k \nabla^T g(\boldsymbol{U}_k)\boldsymbol{d}_{k-1} < -\|\nabla g(\boldsymbol{U}_k)\|^2$ and it is conducted as $\nabla^T g(\boldsymbol{U}_k)\boldsymbol{d}_k < -\|\nabla g(\boldsymbol{U}_k)\|^2$. Thus, TCFS holds the SDC and proof is compilated.

It can be given from the SDC as $\|\boldsymbol{d}_{k+1}\| < \|\boldsymbol{d}_k\|$. This means that $\|\boldsymbol{d}_k\| = \|\boldsymbol{d}_1\| \prod_{m=1}^{k-1} t_m$, $0 < t_1, t_2, \ldots, t_{k-1} < 1$. Consequently for $k \to \infty$, we have $\lim_{k\to\infty} \prod_{m=1}^{k-1} t_m \approx 0$. Therefore, $\lim_{k\to\infty} \|\boldsymbol{d}_k\| = \lim_{k\to\infty} \|\boldsymbol{d}_1\| \prod_{i=1}^{k-1} t_i \approx 0$ and $\boldsymbol{d}_k \approx 0$ when $k \to \infty$. If $\boldsymbol{d}_k \approx 0$ then new point is fixed on the previous point thus it is obtained stable results without iteration using TCFS.

Figure 7.11 demonstrates a cycle of the TCFS method with CSDVs and its terms. In this figure, schematically shows three values of LSF as $g(\boldsymbol{U}_{k-1})$, $g(\boldsymbol{U}_k)$ and $g(\boldsymbol{U}_{k+1})$ and also plots the three-term conjugate search vector ($\boldsymbol{d}_k$) with normalized sensitivity vector ($\boldsymbol{\alpha}_k$). The search direction applied in TCFS method is improved by term $\eta_k \nabla g(\boldsymbol{U}_{k-1})$, while the other CSDV presented in Chap. 6 is formulated by a CSF. The $\boldsymbol{\alpha}_k$ is not paralleled previous sensitivity vectors by applying the two scalar factors. Therefore, the FORM method iterated by the TCFS formula may provide results without oscillating solutions.

The MPP search and computing reliability index are cited by the following steps for coding the TCFS framework in a computer program,

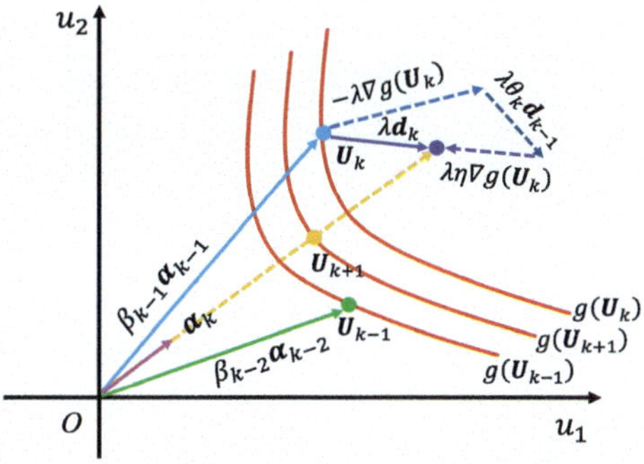

Fig. 7.11 Schematic of iteration for TCFS algorithm

Step 1: Initial parameters as.

(i) Give LSF $g(\boldsymbol{U})$;
(ii) Choose variables and give their distributions, $\boldsymbol{\mu}, \boldsymbol{\sigma}$;
(iii) Set $k = 0$, $\boldsymbol{X}_0 = \boldsymbol{\mu}$, $\boldsymbol{d}_0 = 0$, $c = 1.4$ and $\eta_0 = 0$;
(iv) Choose stopping criterion $\varepsilon << 1$;

Step 2: Transfer variables into $\boldsymbol{U}$-space.

(i) $u = \frac{x-\mu_x^e}{\sigma_x^e}$ where, $\sigma_x^e = \frac{1}{f_X(x)}\varphi\left[\Phi^{-1}\{F_X(x)\}\right]$ and $\mu_x^e = x - \sigma_x^e\Phi^{-1}[F_X(x)]$;

Step 3: Iterative procedure.

(i) Compute $\nabla g(\boldsymbol{U}_k)$ and if $k = 0$ then $\lambda_0 = \frac{100}{\|\nabla g(\boldsymbol{U}_{|U=\mu})\|}$;
(ii) Determine $\eta_k = \min\left\{0.9\|\nabla g(\boldsymbol{U}_k)\|^2/\|\nabla g(\boldsymbol{U}_{k-1})\|^2, \|\nabla g(\boldsymbol{U}_k)\|^2/\|\boldsymbol{d}_{k-1}\|^2\right\}$ and $\theta_k = -\|\nabla g(\boldsymbol{U}_k)\|^2/\|\nabla g(\boldsymbol{U}_{k-1})\|^2$;
(iii) Compute $\boldsymbol{d}_k = -\nabla g(\boldsymbol{U}_k) + \theta_k \boldsymbol{d}_{k-1} + \eta_k \nabla g(\boldsymbol{U}_{k-1})$;
(iv) Compute $\boldsymbol{\alpha}_k = \frac{\boldsymbol{U}_k + \lambda_k \boldsymbol{d}_k}{\|\boldsymbol{U}_k + \lambda_k \boldsymbol{d}_k\|}$;
(v) Determine $\beta_k = \frac{\nabla^T g(\boldsymbol{U}_k)\boldsymbol{U}_k - g(\boldsymbol{U}_k)}{\nabla^T g(\boldsymbol{U}_k)\boldsymbol{\alpha}_k}$ and $\boldsymbol{U}_{k+1} = \beta_k \boldsymbol{\alpha}_k$;

Step 4: Update step length.

(i) Determine $\|\boldsymbol{U}_{k+1} - \boldsymbol{U}_k\|$ and $\|\boldsymbol{U}_k - \boldsymbol{U}_{k-1}\|$ for $k > 1$;
(ii) adjust FSS as $\lambda_k = \begin{cases} \lambda_{k-1} & \|\boldsymbol{U}_{k+1} - \boldsymbol{U}_k\| < \|\boldsymbol{U}_k - \boldsymbol{U}_{k-1}\| \\ \frac{\lambda_{k-1}}{c} & \|\boldsymbol{U}_{k+1} - \boldsymbol{U}_k\| \geq \|\boldsymbol{U}_k - \boldsymbol{U}_{k-1}\| \end{cases}$;

Step 5: Check convergence.

(i) If $\|\boldsymbol{U}_k - \boldsymbol{U}_{k-1}\| \geq \varepsilon$ then $k = k + 1$ and $x = u\sigma_x^e + \mu_x^e$, Go to Step 2 for following Step 3 and 4; else;
(ii) If $\|\boldsymbol{U}_k - \boldsymbol{U}_{k-1}\| < \varepsilon$ then print $\boldsymbol{X}^* = \boldsymbol{X}_{k+1}$, $\boldsymbol{U}^* = \boldsymbol{U}_{k+1}$, and β_k.

The framework of TCFS for coding in a computer program is plotted in Fig. 7.12. It can be shown, that TCFS for the FORM iterative method is established using CG and steepest descent information which are cooperated by two scalar factors. The iterative TCFS method is more complicated formulation compared to FR and the

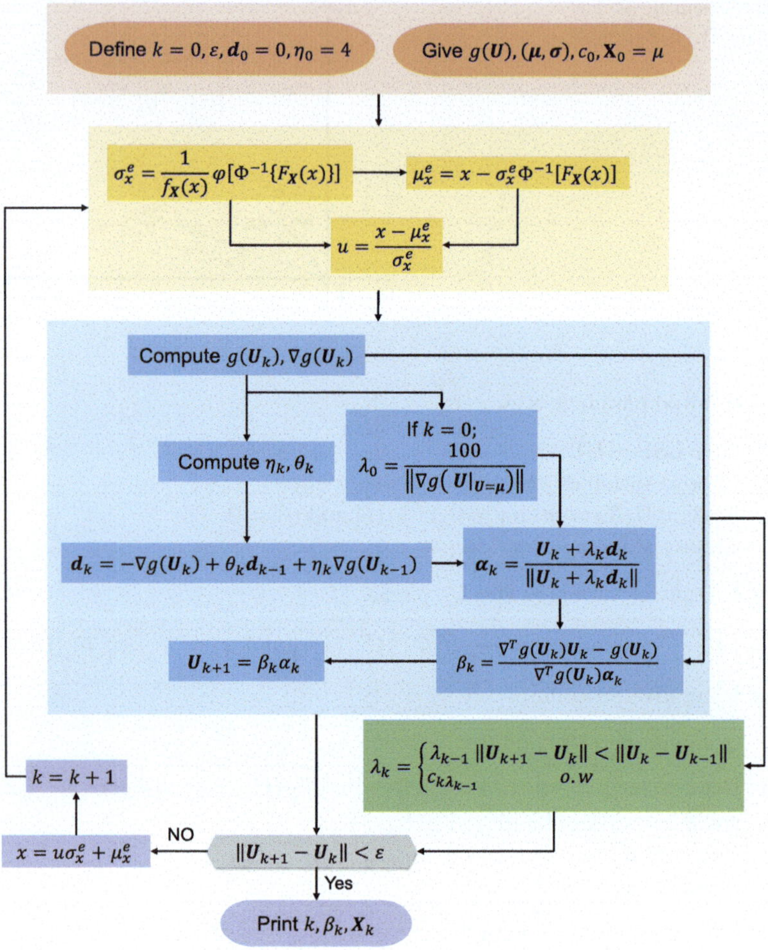

Fig. 7.12 Framework of CFORM-based TCFS for SRA

steepest descent methods such as HL-RF, FSL, STM, and DSTM (i.e. directional stability transformation method) applied to search the MPP. However, the search direction applied in TCFS for computing the sensitivity vector may improve the computational burden of FORM and stabilization of iterative formulation.

The MPP obtained by TCFS is computed using the CSDV, while the traditional FORM formulations such as: STM, HL-RF, and RHL-RF are established by using the SDSD. The iterative CCC formulation is improved using the chaotic CC (i.e. chaos control method) factor, whereas the TCFS method has a CG vector with three-term which is theoretically satisfied SDC. The dynamical step size can be neglected in TCFS method as well as CHL-RF, LFR, HCFS methods which are given by SDC. Satisfying the SDC in TCFS specified a fixed point at final iterations [18].

7.4.2 Examples for Comparative Studies of TCFS

The performances if iterative methods such as robustness and efficiency for FORM-formulas given by HL-RF, DSTM (with parameters of $\lambda = 0.1$), CCC and TCFS are compared using five engineering examples. The FSS initial value of $\lambda_0 = \frac{100}{\|\nabla g(U_{|U=\mu})\|}$ are applied in CCC and TCFS methods that it is adjusted by factor $c = 1.4$ with below relation,

$$\lambda_k = \begin{cases} \lambda_{k-1} & \|\boldsymbol{U}_{k+1} - \boldsymbol{U}_k\| < \|\boldsymbol{U}_k - \boldsymbol{U}_{k-1}\| \\ \frac{\lambda_{k-1}}{c} & \|\boldsymbol{U}_{k+1} - \boldsymbol{U}_k\| \geq \|\boldsymbol{U}_k - \boldsymbol{U}_{k-1}\| \end{cases} \tag{7.35}$$

In CCC method, the chaotic CC factor is determined based on Logistic map with parameters of $a = 4$ and $\xi_{c0} = 0.375$. The condition of $\|\boldsymbol{U}_k - \boldsymbol{U}_{k-1}\| \leq \varepsilon = 10^{-6}$ is applied to stop all studied FORM algorithms. The results such as: β, call function and Iterations are used to compare the convergent performances of studied algorithms.

Example 7.13 Nonlinear function as $g = x_1^3 + 2x_1^2 x_2 + x_2^3 - 18$ with normal x_1 and x_2 variables followed by $\mu_1 = 10$, $\mu_2 = 9.9$ and $\sigma_1 = \sigma_2 = 5$.

The converged reliability results using TCFS are $\beta = 2.34059$ and $\boldsymbol{X}^* = [1.3955,\ 1.9677]$ that they are captured after 12 iterations but, HL-RF produces unstable results while the DSTM and CCC have the robust performance. The $\rho(\boldsymbol{J})$ is shown in Fig. 7.13 for HL-RF, DSTM, CCC, and TCFS methods in Example 7.13. It can be illustrated that the $\rho(\boldsymbol{J})$ for HL-RF is more than 1 while it is computed for other methods formulated by CG vector including CCC and TCFS less than the 1 i.e. $\rho(\boldsymbol{J}) \leq 1$. This means that FORM using SDSD may show the instable solutions for nonlinear LSFs. However, the instability of FORM should be controlled using modification in SDV and step size for achieving $\rho(\boldsymbol{J}) \leq 1$. The TCFS is an efficient method by comparing its iterations with DSTM and CCC methods in Example 7.13. The DSTM is provided a $\rho(\boldsymbol{J})$ more than 1 at the first iterations but its stability is given at final iterations and it is obtained a fixed point by this method due to $\rho(\boldsymbol{J}) \approx 0$.

The bifurcation plot of the reliability index for the HL-RF and the TCFS methods in different mean values of random variable x_2 i.e. μ_2 is shown in Fig. 7.14 for Example 7.13. It can be extracted from Fig. 7.14 that the periodic oscillating, chaos, and bifurcation solutions are obtained based on the HL-RF, while the TCFS provides the stable results for different mean values. The nonlinear mapping using conjugate search direction can be improved the instability of the FORM for highly nonlinear performance functions.

The reliability indices corresponding to different CC factors (ξ) in DSTM and different FFS (λ) in the TCFS method are plotted in Fig. 7.15 for Example 7.13 with variables $x_1 \sim N(10, 5)$ and $x_2 \sim N(5, 2)$. As seen, the TCFS method converges to stable reliability index as $\beta = 2.2433$ for different finite-step lengths (λ). However, the DSTM provides stable results for the chaos control factors less than 0.35 i.e. $\xi < 0.35$. The robustness of the DSTM is strongly dependent on the factor ξ (e.g. $\xi > 0.4$ is provided unstable results).

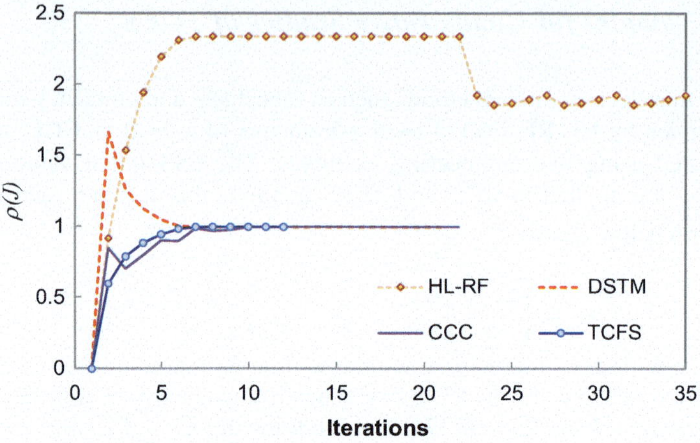

Fig. 7.13 The $\rho(\boldsymbol{J})$ obtained by iterative formulation of reliability methods for Example 7.13

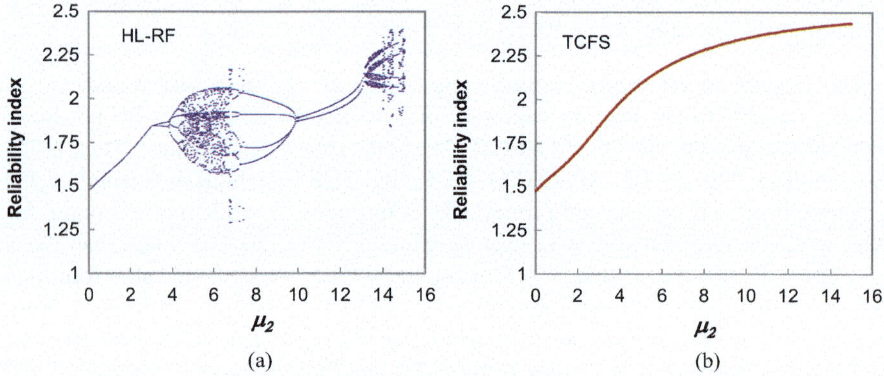

Fig. 7.14 Bifurcation plot of HL-RF and TCFS for Example 7.13. **a** HL-RF; **b** TCFS

Example 7.14 A LSF generated by response-surface function given as below [14],

$$\begin{aligned}g = 1.1 &- 0.00115x_1x_2 + 0.00157x_2^2\\&+ 0.00117x_1^2 + 0.0135x_2x_3 - 0.0705x_2\\&- 0.00534x_1 - 0.0149x_1x_3 - 0.0611x_2x_4\\&+ 0.0717x_1x_4 - 0.226x_3\\&+ 0.0333x_3^2 - 0.558x_3x_4 + 0.998x_4 - 1.339x_4^2\end{aligned}\tag{7.36}$$

where, type-II largest value x_1 with mean of 10 and STD of 5. Normal x_2 with means of 25 and STD of 5, Normal x_3 with mean of 0.8 and STD of 0.2 and lognormal x_4 with mean of 0.0625 and STD of 0.0625.

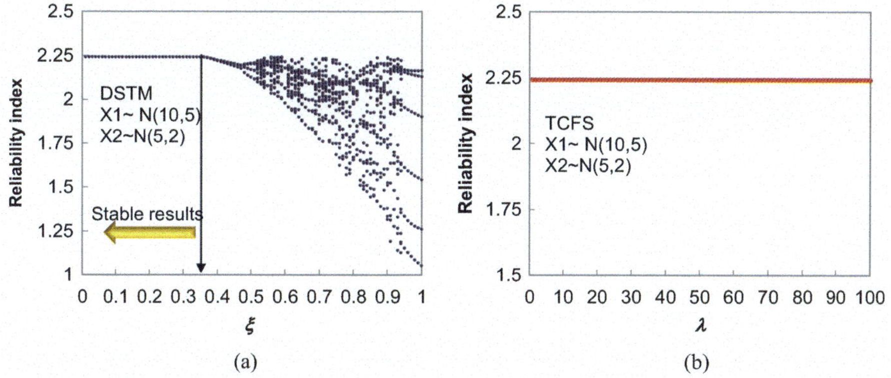

Fig. 7.15 The converged reliability indices for different chaos control factors of DSTM and different finite-step lengths of TCFS for Example 7.13. **a** different chaos control factors of DSTM; **b** different finite-step lengths of TCFS

Using extracted results from Liu and Kiureghian [14], we have $\boldsymbol{X}^* = [15.09, 25.027, 0.8653, 0.03582]$ and $\beta = 1.36$. Using RHL-RF given from Ref. [6], the $\boldsymbol{X}^* = [14.905, 25.067, 0.85956, 0.046061]$ and $\beta = 1.33053$. The TCFS method is converged to $\beta = 1.33179$ and MPP of $\boldsymbol{U}^* = [1.29608, 0.02297, 0.28525, 0.10928]$ and $\boldsymbol{X}^* = [14.9075, 25.1149, 0.8571, 0.048403]$. The converged results based on the TCFS are closely followed the extracted results from literatures.

The reliability index of HL-RF, CCC, DSTM, and TCFS approaches are shown in Fig. 7.16 for Example 7.14. As seen, the DSTM, CCC, and TCFS are respectively converged to stable results after 143, 71, and 47 iterations, while the HL-RF is yielded unstable results as periodic-2 solutions as $\beta = (1.27374, 1.05445)$. The efficiency of FORM is improved based on the TCFS method while the DSTM is a stable method but it is an inefficient approach. As seen, the steepest descent method without line search criterion should be improved for achieving the stable results.

The bifurcation plot of the HL-RF and the proposed TCFS methods with respect to various standard deviations of random variable x_2 are shown in Fig. 7.17. The HL-RF produces unstable results as chaotic and periodic solutions. However, the nonlinear discrete map using a three-term conjugate search direction provides stable results for all standard deviations. The TCFS method is improved the instability of FORM compared to the traditional method.

Example 7.15 A vehicle side-impact performance function is captured by response surface method as follows,

$$
\begin{aligned}
g = {} & 0.489x_3x_7 + 0.843x_5x_6 - 0.0432x_9x_{10} \\
& + 0.0556x_9x_{11} + 0.000786x_{11}^2 - 0.75
\end{aligned}
\tag{7.37}
$$

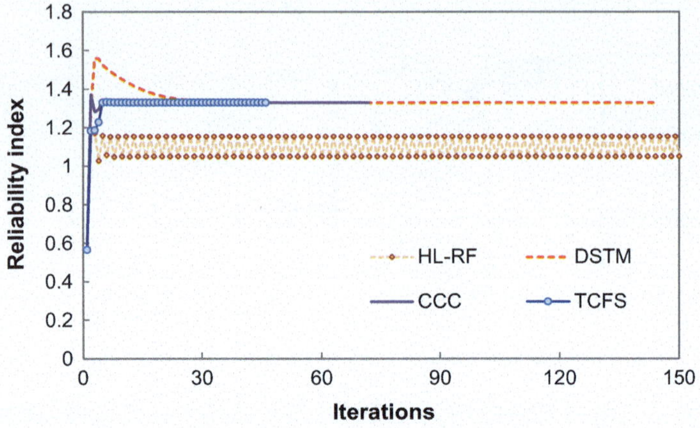

Fig. 7.16 Reliability index of reliability methods for Example 7.14

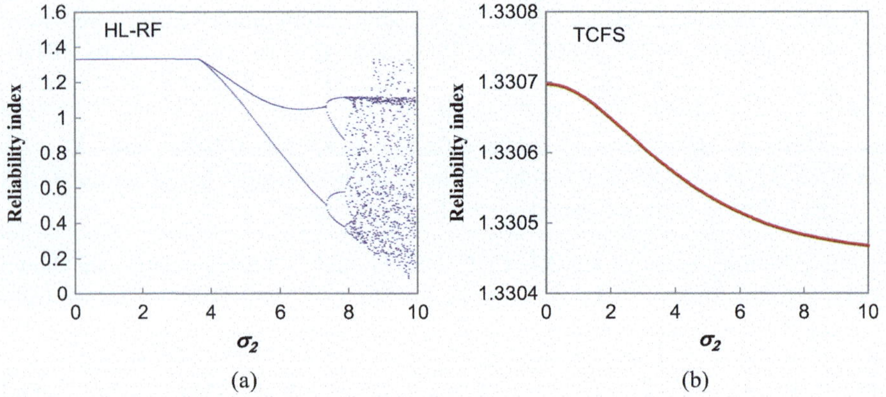

Fig. 7.17 Bifurcation plot for HL-RF and TCFS methods for various standard deviations of random variable x_2 of Example 7.14. **a** HL-RF; **b** TCFS

where $i = 1 \sim 7x_i \sim N(1, 0.005)$, $i = 8 \sim 9x_i \sim N(0.3, 0.006)$, $i = 9 \sim 10x_i \sim N(0, 10.0)$. The reliability index using MCS with 4.8×10^6 samples is obtained as $\beta = 3.71236$ ($P_f = 0.000103$ with COV of 4.5%). The reliability index is obtained using the DSTM, CCC, and TCFS methods after 118, 71 and 42 iterations as 3.49748, 3.49748, and 3.49767, respectively. The TCFS is efficient method among others while the CCC and DSTM are as robust as the TCFS method. The convergence histories of spectral radiuses of the Jacobian matrix for different reliability methods i.e. HL-RF, DSTM, CCC, and TCFS are illustrated in Fig. 7.18 for Example 7.15. As seen, the HL-RF yields unstable results as periodic-4 solutions while the proposed TCFS with discrete conjugate mapping for computing the search direction of FORM provides the stable results more efficient than CCC and DATM methods.

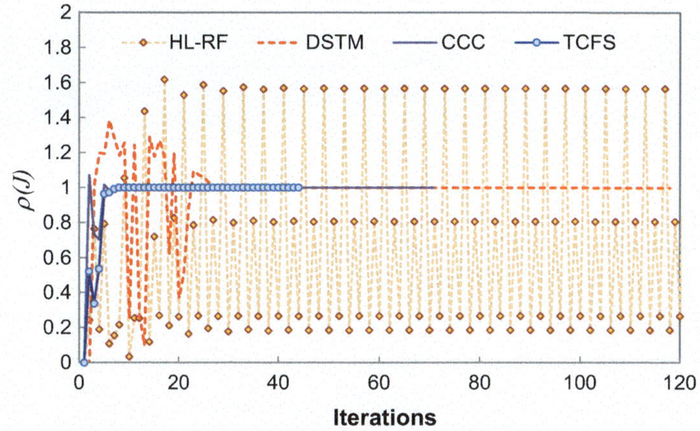

Fig. 7.18 The $\rho(\boldsymbol{J})$ of different reliability methods for Example 7.15

Figure 7.19 illustrates the reliability index with respect to the different chaos control factors of DSTM and finite-step lengths of the TCFS method. It can be conducted from the results of Fig. 7.19 that the robustness of the DSTM is strongly dependent on the chaos control factor. The DSTM with smaller chaos control factor than 0.03 i.e. $\xi < 0.03$ while the DSTM produces the instable periodic oscillating solutions for larger chaos control factor than 0.44 i.e. $\xi > 0.44$. However, the conjugate nonlinear discrete map can be improved the robustness of the FORM formula due to satisfy the SDC.

Example 7.16 The brake disc, friction materials and back plates is shown in Fig. 7.20, that as LSF based on the response surface function for friction loads of the disc system is expressed as follow [20],

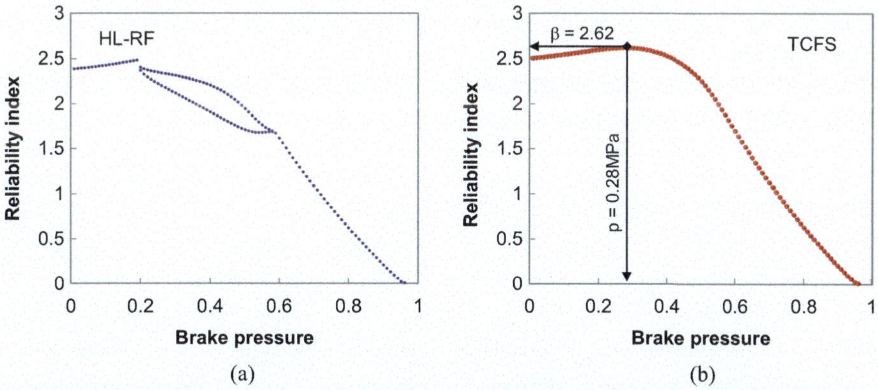

Fig. 7.19 The reliability index for different CC factors in DSTM method and different FFSs in TCFS for Example 7.15. **a** different CC factors in DSTM; **b** different FFSs in TCFS

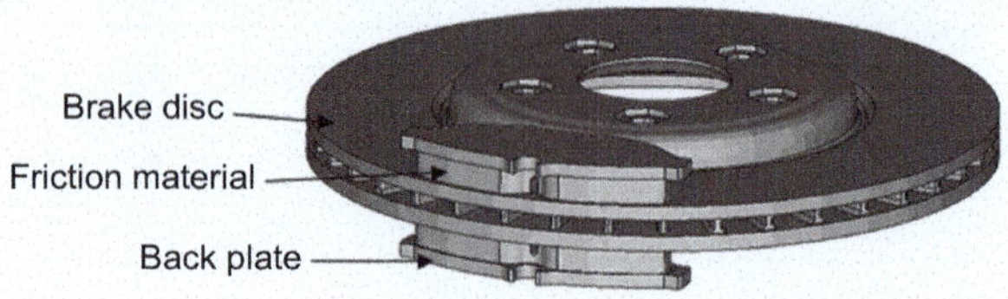

Fig. 7.20 Schismatic view of disc brake system. Reprinted from [20], with permission from Elsevier

$$\begin{aligned} g = {} & 0.046287 + 0.20458\mu - 0.059821p - 0.00036549h_1 - 0.010037h_2 \\ & + 0.013836h_3 + 0.24308\mu p - 0.0037884\mu h_1 + 0.0023358\mu h_2 \\ & - 0.016981\mu h_3 + 0.029287ph_1 - 0.015872ph_2 \\ & - 0.0028333ph_3 + 0.0007175h_1h_2 - 0.00046158h_1h_3 \\ & - 0.00036486h_2h_3 - 0.39076\mu^2 - 0.015986p^2 \\ & - 0.0011936h_1^2 + 0.000269h_2^2 + 0.00062638h_3^2 - 0.01 \end{aligned} \tag{7.38}$$

The basic variables applied for computing the LSF of Eq. (7.38) are listed in Table 7.12.

The converged results such as the failure probability (P_f), reliability index (β), number of call functions (Call function), the number of evaluating the gradient vector (Iterations) and CPU (i.e. Central Processing Unit) run-times (CPU-times) are tabulated in Table 7.13 for different reliability methods. It can be conducted that the HL-RF shows the periodic-2 solutions, although other methods converge, more robustly. The TCFS method is respectively iterated about 6-times and 25-times faster than CCC and DSTM. The obtained results using TCFS are $\beta = 2.251418$, $\boldsymbol{X}^* = [\mu^* = 0.25163, p^* = 0.61806$ MPa, $h_1^* = 13.3265$ mm, $h_2^* = 18.9771$ mm, $h_3^* = 10.9791$ mm] for the disc brake system example. The search direction vector obtained by the TCFS method provides strongly efficient results compared to CCC and DSTM FORM algorithms. By comparing the CCC and DSTM, the CCC is more efficient than the DSTM for this problem. the DSTM is robust method but it is loosed its efficiency for this moderately nonlinear LSF which is computed based on the response surface method.

Figure 7.21 presents the reliability index given from iterative schemes of HL-RF and TCFS for various brake pressures of vehicle discs. As seen, the unstable

Table 7.12 Variables for disc brake system applied in Example 7.16

Variable	p (MPa)	μ	h_1 (mm)	h_2 (mm)	h_3 (mm)
Mean	0.5	0.4	15	20	10
CoV	0.25	0.35	0.1	0.1	0.1
PDF	Lognormal	Normal	Gumbel	Gumbel	Gumbel
Description	Brake pressure	Friction coefficient	Friction material thickness	Disc thickness	Back plate thickness

Table 7.13 The results of the disc brake system for different reliability methods

Method	P_f	β	Evaluating LSF	Iterations	CPU-time (sec)
HL-RF	Periodic-2 as $\beta = (1.7112, 1.9448)$				
DSTM	0.016	2.144414	5526	460	274.4
CCC	0.015996	2.144522	781	71	45.69
TCFS	0.01218	2.251418	176	16	10.95

results are attained by HL-RF that this instability showed by priodic-2 reliability index in the range from 0.2 to 0.6 MPa. However, stable results are captured by TCFS as presented reliability index in Fig. 7.21. The maximum reliability index is 2.62 with related brake pressure of 0.28 MPa. Almost the engineering problems are described with a meta-model using the response surface method by using the nonlinear function with second-order polynomial form that the traditional reliability method i.e. HL-RF and the modified FORM-based SDSD i.e. DSTM may provide unstable solutions. However, the CSDV applied in CCC and TCFS improved both robustness and efficiency in comparison with DSTM (see Examples 7.14, 7.15, and 7.16). Consequently, the TCFS method can be implemented for the real engineering problems that their LSF are expressed from the response surface method with robust and efficient results.

Example 7.17 A steel frame as presented is given for this example that its LSF is defined as $g = 0.05 - \Delta$ [3], where, the statistical Δ is the horizontal displacement computed using finite element model with linear analysis in geometry and materials. Towel basic random variables are applied for computing the displacement that their properties are listed in Table 7.14.

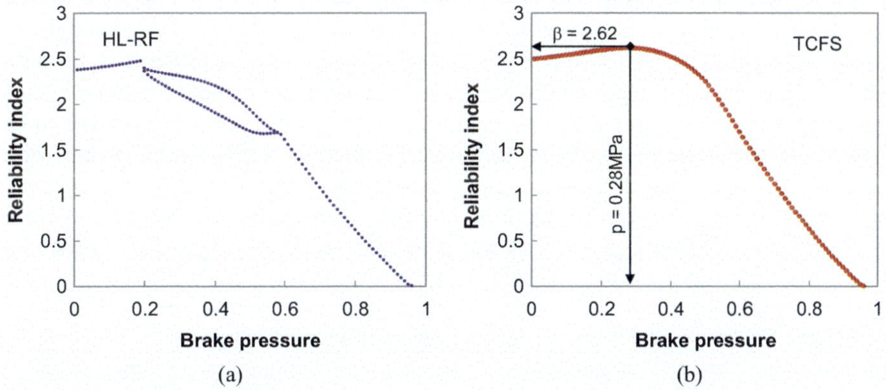

Fig. 7.21 The reliability indices of HL-RF and TCFC for various brake pressures of disc system. **a** HL-RF; **b** TCFC

Table 7.14 Variables for steel frame applied in Example 7.17

Variable	Mean	STD	PDF	Variable	Mean	STD	PDF
A_1 (m^2)	91×10^{-4}	9.1×10^{-4}	Lognormal	E_1 (kg/m^2)	2.38×10^{10}	1.9×10^9	Normal
A_2 (m^2)	53.8×10^{-4}	5.38×10^{-4}	Lognormal	E_2 (kg/m^2)	2.17×10^{10}	1.74×10^9	Normal
A_3 (m^2)	106×10^{-4}	10.6×10^{-4}	Lognormal	H (m)	4.5	0.36	Normal
I_1 (m^4)	80.9×10^{-6}	8.09×10^{-6}	Lognormal	L (m)	6	0.48	Normal
I_2 (m^4)	83.6×10^{-6}	8.36×10^{-6}	Lognormal	P (kg)	2000	350	Gumbel
I_3 (m^4)	112.6×10^{-6}	11.26×10^{-6}	Lognormal	W (kg/m)	4500	630	Gumbel

The different reliability methods of FORM are applied to search the reliability index that the reliability results of this example are listed in Table 7.15 and reliability index iterations are plotted in Fig. 7.22 for steel frame structure. It can be given as,

(i) The HL-RF and DSTM methods are produced unstable iterative solutions with chaotic results;
(ii) The CSDV applied in CCC and TCFS formulations are enhanced the stability of FORM thus we have robust convergence;
(iii) The TCFS method is an efficient method compared to CCC method by using a CG vector less than FR approach that it is obtained that TCFS is about twice faster than CCC with as stable reliability index of 3.152635.

The conjugate sensitivity vector applied in the CFORM is much effective factor to provide the stable and efficient results. It is shown that the CFORM method can be applied to find the MPP that it is given a stable result for FORM iterative formula when the CG vector is satisfied the SDC. An idea using conjugate sensitivity vector holed by SDC i.e. $\nabla^T g(\boldsymbol{U}_k)\boldsymbol{d}_k < -c_1\|\nabla g(\boldsymbol{U}_k)\|^2$ called TCFS is established in FORM formula. The main efforts in the TCFS method are to find a formulation for FORM which it has an efficient and robust iterative relation. By comparing the TCFS formulation, we derived that it is used for establishing the FORM by (1) new gradient vector, (2) conjugate search direction and (3) previous gradient vector that

Table 7.15 The Results of different analytical methods for frame structure

Method	P_f	β	Call function	Iteration	CPU-times (sec)	Rel-error%
HL-RF	Chaotic solutions					Failed
DSTM	Chaotic solutions					Failed
TCFS	0.000809	3.152635	2902	35	41.1	2.84
CCC	0.000766	3.168373	5833	71	75.5	3.35

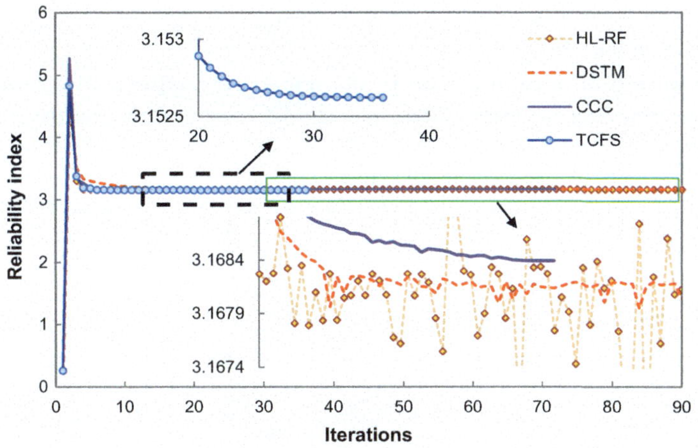

Fig. 7.22 The iterations for the reliability index obtained by analytical methods for the steel frame example

the previous CG and previous gradient vector are combined with a dynamical related scalar factor. The results validated that TCFS method improved computational burden and provided better performance than other studied FORM algorithms.

7.5 Discussion

As mentioned, the FORM formulas-based CSDVs for abilities of efficiency and robustness are strongly dependent on the conjugate line search including step size and CG vectors. These two factors are applied for computing the normalized sensitivity vectors. The conjugate line search applied for computing the sensitivity vector which is determined by two main factors named (i) CSF utilizing the CG vector and (ii) step length as a cooperative factor for considering CG vector effect are main factors for determining the normalized conjugate sensitivity vector. If the step size selects a smaller value which is tended to zero than the effect of CG is neglected for computing the sensitivity vector while it is tended to an infinite value (larger factor) then it is considered the maximum effect for computing the sensitivity vector with respect to CG vector. As seen from the methods established by the CG vector in this chapter, different types of scalar factors with the FR method are presented. The gradient methods developed in this section are different in CSFs or line search related to the CG vector. In this section, due to satisfying the sufficient descent condition, almost methods of CFORM provided the stable results without appalling the step length which is given for controlling the stability of the FORM formula using the sufficient descent and the Armijo-type conditions. As showed that the intimal step length is computed with an Armijo-type condition by using the gradient vector at the mean point of random variables.

In this section as well as Chap. 6, the parameters applied in computing the line search including (i) step length and (ii) CG vectors are discussed through several examples. In this comparative study, the different CFORM iterative formulations presented in this section are also discussed with two formulations applied for computing the reliability index that the reliability index is computed by CG vector ($\boldsymbol{d}$) as below,

$$\beta_{k+1} = \frac{\boldsymbol{d}_k^T \boldsymbol{U}_k - g(\boldsymbol{U}_k)}{\boldsymbol{d}_k^T \boldsymbol{\alpha}_k^c} \tag{7.39}$$

$$\beta_{k+1} = \frac{\nabla^T g(\boldsymbol{U}_k)\boldsymbol{U}_k - g(\boldsymbol{U}_k)}{\nabla^T g(\boldsymbol{U}_k)\boldsymbol{\alpha}_k^c} \tag{7.40}$$

In Eq. (7.39), the reliability index is computed by both the conjugate sensitivity vector and CG. Although in Eq. (7.40), the steepest descent vector and sensitivity are applied for determining the reliability index. After computing the reliability index using Eqs. (7.39) and (7.40), the new point is computed as $\boldsymbol{U}_{k+1} = \beta_{k+1}\boldsymbol{\alpha}_k^c$. The normalized conjugate sensitivity vector is determined computed as $\boldsymbol{\alpha}_k^c = \frac{\boldsymbol{U}_k + \lambda_k \boldsymbol{d}_k}{\|\boldsymbol{U}_k + \lambda_k \boldsymbol{d}_k\|}$ in above relation.

In computing $\boldsymbol{\alpha}_k^c$ obvious, $\boldsymbol{\alpha}_k^c$ is computed using three factors of (i) CG vector, (ii) previous convergence point and (iii) finite step length. The CG vector is the main different between FORM formulas presented in current chapter. Step length λ_k is as important as the CG vector $\boldsymbol{d}_k$. Generally, its initial value is determined by using Armijo condition as follows,

$$\lambda_0 = \frac{M}{\left\|\nabla g\left(\boldsymbol{U}|_{\boldsymbol{U}=\mu}\right)\right\|} \tag{7.41}$$

in which, $10 \leq M \leq 100$. In this section, we given $M = 30$ and the step length is dynamically adjusted by sufficient descent condition as Eq. (7.42),

$$\lambda_k = \begin{cases} \lambda_{k-1} & \|\boldsymbol{U}_{k+1} - \boldsymbol{U}_k\| < \|\boldsymbol{U}_k - \boldsymbol{U}_{k-1}\| \\ \frac{\lambda_{k-1}}{c} & \|\boldsymbol{U}_{k+1} - \boldsymbol{U}_k\| \geq \|\boldsymbol{U}_k - \boldsymbol{U}_{k-1}\| \end{cases} \tag{7.42}$$

where, $c = 1.4$. Using above relations for computing the sensitivity vector applied for searching the MPP and reliability index, it is designed three scenarios for comparative results of different CFORM formulated in this section as below,

Scenario 1: named as SCE-I.

(i) Reliability index computed as $\beta_{k+1} = \frac{\nabla^T g(\boldsymbol{U}_k)\boldsymbol{U}_k - g(\boldsymbol{U}_k)}{\nabla^T g(\boldsymbol{U}_k)\boldsymbol{\alpha}_k^c}$;

(ii) λ_k is adjusted as $\lambda_k = \begin{cases} \lambda_{k-1} & \|U_{k+1} - U_k\| < \|U_k - U_{k-1}\| \\ \frac{\lambda_{k-1}}{1.4} & \|U_{k+1} - U_k\| \geq \|U_k - U_{k-1}\| \end{cases}$ with $\lambda_0 = \frac{30}{\|\nabla g(U|_{U=\mu})\|}$;

Scenario 2: named as SCE-II.

(i) Reliability index computed as $\beta_{k+1} = \frac{\nabla^T g(U_k)U_k - g(U_k)}{\nabla^T g(U_k)\alpha_k^c}$;
(ii) λ_k is given a constant value as $\lambda_k = 1$;

Scenario 3: named as SCE-III.

(i) Reliability index computed as $\beta_{k+1} = \frac{d_k^T U_k - g(U_k)}{d_k^T \alpha_k^c}$;
(ii) λ_k is adjusted b as $\lambda_k = \begin{cases} \lambda_{k-1} & \|U_{k+1} - U_k\| < \|U_k - U_{k-1}\| \\ \frac{\lambda_{k-1}}{1.4} & \|U_{k+1} - U_k\| \geq \|U_k - U_{k-1}\| \end{cases}$ with $\lambda_0 = \frac{30}{\|\nabla g(U|_{U=\mu})\|}$.

As seen from SEC-I, SEC-II and SEC-III, the CG vectors are the main difference between various CG methods. The new point in iterative FORM formulations is computed based on CG which are presented for different methods in Table 7.16. It's worth noting that the parameters of CG methods are different, thus, the different CG vectors are computed for MPP search.

The formulation in Table 7.16 for CFORM showed that the methods of SAC, HSAC, and TCFS methods have been formulated based on a CSF less than the conjugate FR factor. Finally, a different CSF is considered with the FR method in CD, RMIL, and HCD. By giving the main formulation of FORM using FR, it can be given three main categories for CFORM which are formulated by different methods including,

(i) variant type of CG vector-based FR which is presented in TCFS CG vector;
(ii) modified FR using controlling the CSF in SAC and HSAC methods;

Table 7.16 Methods of CFORM for computing CG vector $\boldsymbol{d}$

Method	CG vector	Conjugate parameters
CD	$d_k = -\nabla g(U_k) - \theta_k d_{k-1}$	$\theta_k^{CD} = \frac{\|\nabla g(U_k)\|^2}{d_{k-1}^T \nabla g(U_{k-1})}$
RMIL	$d_k = -\nabla g(U_k) - \theta_k d_{k-1}$	$\theta_k^{RMIL} = \frac{\nabla^T g(U_k)[\nabla g(U_k) - \nabla g(U_{k-1})]}{\|d_{k-1}\|^2}$
HCD	$d_k = -\nabla g(U_k) - \theta_k d_{k-1}$	$\theta_k^{HCD} = \eta\theta_k^{CD} + (1-\eta)\theta_k^{RMIL}$ $\eta = -\frac{\|\nabla g(U_{k-1})\|^2}{4\nabla^T g(U_k) d_{k-1}}$
SAC	$d_k = -\nabla g(U_k) - \theta_k d_{k-1}$	$\theta_k^{SA} = \theta_k^{FR} - \frac{\sqrt{\theta_k^{FR}}\|\nabla^T g(U_k)\nabla g(U_{k-1})\|}{\|d_{k-1}\|^2}$
HSAC	$d_k = -\nabla g(U_k) - \theta_k d_{k-1}$	$\theta_k^{HS} = \min\{\theta_k^{FR}, \|\theta_k^{SA}\|\}$
TCFS	$d_k = -\nabla g(U_k) - \theta_k^{FR} d_{k-1} + \eta_k \nabla g(U_{k-1})$	$\eta_k = \min\left\{0.9\frac{\|\nabla g(U_k)\|^2}{\|\nabla g(U_{k-1})\|^2}, \frac{\|\nabla g(U_k)\|^2}{\|d_{k-1}\|^2}\right\}$

(iii) non-general FR CFORM which is presented in CD, RMIL, and HCD.

The robust results are given based on all CFORM algorithms when it is applied an adjusting step size but efficiency as an essential challenge should be disused that it is an effort to develop of CFORM. As presented in Chap. 6, the efficiency of CFORM is compared by six reliability problems with stopping error of $\varepsilon = 10^{-4}$. The problems including LSFs and their statistical properties are given in Examples 7.18, 7.19, 7.20, 7.21, 7.22, and 7.23 as below,

Example 7.18 $g = x_1^3 + 2x_1^2x_2 + x_2^3 - 18$ where x_1 and x_2 are normal with $\mu_1 = 10$, $\mu_2 = 9.9$ and $\sigma_1 = \sigma_2 = 5$;

Example 7.19 $g(X) = x_1^4 + x_2^2 - 50$ where, x_1: Lognormal (5,1) and x_2: Gumbel (10,10);

Example 7.20 $$\begin{aligned} g = 1.1 - 0.00115x_1x_2 + 0.00157x_2^2 + 0.00117x_1^2 + 0.0135x_2x_3 \\ -0.0705x_2 - 0.00534x_1 - 0.0149x_1x_3 - 0.0611x_2x_4 \\ +0.0717x_1x_4 - 0.226x_3 + 0.0333x_3^2 - 0.558x_3x_4 \\ +0.998x_4 - 1.339x_4^2 \end{aligned},$$
where, x_1: extreme value II (10,5), x_2: N (25,5), x_3: N (0.8,0.2), x_4: LN (0.0625,0.0625);

Example 7.21 A conical structure as indicated in Fig. 7.9 by LSF as $g = 1 - \frac{\sqrt{3(1-v^2)}}{\pi E t^2 \cos^2 \alpha} \times \left(\frac{P}{0.66} + \frac{M}{0.41 r_1} \right)$ where, $v = 0.3$ and data of Table 7.10 is the random variables of conical example;

Example 7.22 A dynamic system as presented with LSF of $g = Fs - Ks \times P(E\left[x_s^2\right])^{1/2}$, where, $P = 3$, and $E\left[x_s^2\right] = \frac{\pi S_0}{4\xi_s\omega_s}\left[\frac{\xi_a\xi_s}{\xi_p\xi_s(4\xi_a^2+\theta^2)+\gamma\xi_a^2} \times \frac{(\xi_p\omega_p^3+\xi_s\omega_s^3)\omega_p}{4\xi_a\omega_a^4}\right]$, in which $\gamma = \frac{M_s}{M_p}$, $\omega_a = \frac{\omega_p+\omega_s}{2}$, $\xi_a = \frac{\xi_p+\xi_s}{2}$ and $\theta = \frac{\omega_p-\omega_s}{\omega_a}$ that this factors are computed by using Lognormal variables presented in Table 7.17.

Example 7.23 A planar ten-bar structure presented in Fig. 7.5 by LSF of $g = 0.004 - \left|\Delta_3^V\right|$ that this LSF is determined by variables listed in Table 7.18.

Three categories of CFORM as (i) variant-type FR, (ii) modified FR and (iii) non-general FR are applied for Example 7.18 in SEC-I that their convergences in normal standard space are plotted in Figs. 7.23, 7.24, and 7.25, respectively. It can be conducted that all CFORM with different formulations with FR are converged to

Table 7.17 Random variables of Example 7.22

Variable	M_s	M_p	K_s	K_p	ξ_s	ξ_p	F_s	S_0
Mean	0.01	1	0.01	1	0.02	0.05	15	100
STD	0.001	0.1	0.001	0.2	0.01	0.02	1.5	10

Table 7.18 Variables applied in Example 7.23

Variable	A_i (m^2)	E (Gpa)	L (m)	P_1 (kN)	P_2 (kN)	P_3 (kN)
Mean	0.001	100	1	80	10	10
COV	0.15	0.05	0.05	0.1	0.1	0.1
PDF	Normal	Normal	Log-Normal	Gumbel	Gumbel	Gumbel

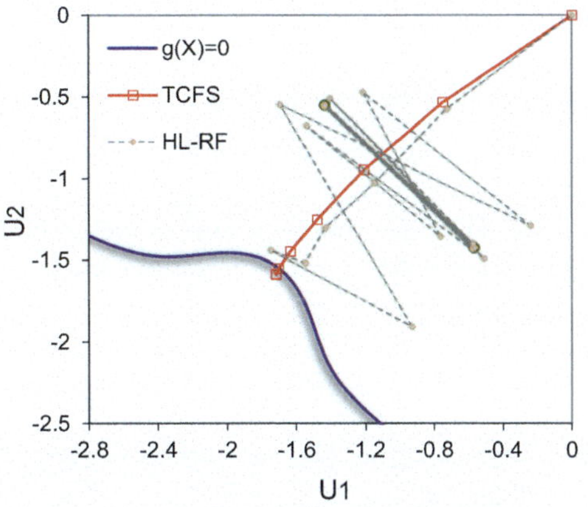

Fig. 7.23 Iterations of variant type FR-based TCFS methods for Example 7.18

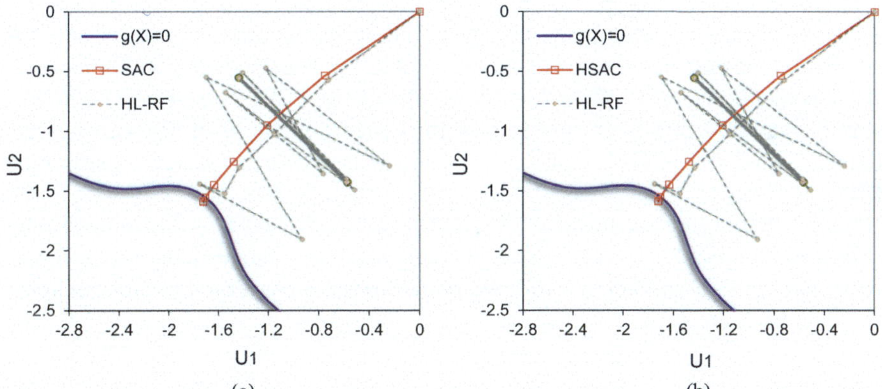

Fig. 7.24 Iterations of modified FR-based SAC and HSAC methods for Example 7.18. **a** SAC; **b** HSAC

stable MPP, more robustly. By comparing the initial iterations of FORM formulated by steepest descent search direction and CG search direction vector, the iteration is not to a same line search. The CFORM provides a nonlinear line search to find MPP as seen in these methods. Consequently, the parallel risk of the search direction

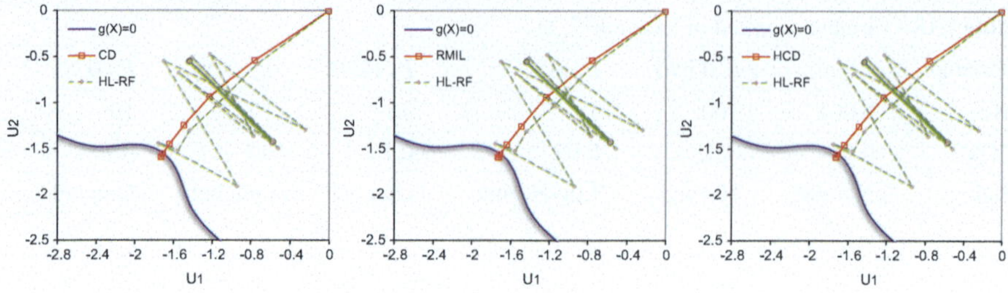

Fig. 7.25 Iterations of non-general FR-based CD, RMIL and HCD methods for Example 7.18. **a** CD; **b** RMIL; **c** HCD

vector is strongly reduced. The lines search of almost method are obtained a similar curve that it is strongly tended to MPP at first iterations but at final iteration tries to find MPP with more than initial iterations.

The reliability index and iteration for convergence as β (iter) are computed for CFORM in three categories of CG vectors applied in three scenarios,

(i) SCE I as $\beta_{k+1} = \frac{\nabla^T g(\boldsymbol{U}_k)\boldsymbol{U}_k - g(\boldsymbol{U}_k)}{\nabla^T g(\boldsymbol{U}_k)\boldsymbol{\alpha}_k^c}$ and λ_k with $\lambda_0 = \frac{30}{\|\nabla g(U|_{U=\mu})\|}$;

(ii) SCE II as $\beta_{k+1} = \frac{\nabla^T g(\boldsymbol{U}_k)\boldsymbol{U}_k - g(\boldsymbol{U}_k)}{\nabla^T g(\boldsymbol{U}_k)\boldsymbol{\alpha}_k^c}$ and $\lambda_k = 1$;

(iii) SCE III as $\beta_{k+1} = \frac{\boldsymbol{d}_k^T \boldsymbol{U}_k - g(\boldsymbol{U}_k)}{\boldsymbol{d}_k^T \boldsymbol{\alpha}_k^c}$ and λ_k with $\lambda_0 = \frac{30}{\|\nabla g(U|_{U=\mu})\|}$.

The results of variant-types of FR formulation using TCFS are given in Table 7.19. All scenarios for this method are converged. Therefore, the variant formulation of FR just applied for the reliability index based on SCE I or II extended by TCFS method. The step size is strongly affected on the results of the TCFS compared. For SCE I, the efficiency of TCFS is acceptable and it is better than the other SECs. Adjusting gradient vector in the CG relation can be provided a suitable line search for searching MPP.

Table 7.20 presents the result of CFORM-based modified FR as the limited scalar factor for Examples 7.18, 7.19, 7.20, 7.21, 7.22, and 7.23. The step length is an important factor for SAC and HSAC while these methods are converged for all scenarios. These methods are generally efficient in SCE I compared to SCEII and SCEIII. The

Table 7.19 Results of different variant FR using TCFS reliability methods for SCE I-III

Examples	SCE I	SCE II	SCE III
7.18	2.3406 (14)	2.3406 (14)	2.3398 (16)
7.19	3.2593 (15)	3.2593 (17)	3.2581 (19)
7.20	1.3305 (45)	1.3309 (315)	1.3289 (56)
7.21	4.6281 (86)	4.6279 (312)	4.6337 (101)
7.22	2.0165 (82)	2.0164 (86)	2.0116 (108)
7.23	1.7446 (99)	1.7401 (468)	1.7385 (75)

Table 7.20 Results of different modified FR including SAC and HSAC for SCE I-III

Examples	SAC			HSAC		
	SCE I	SCE II	SCE III	SCE I	SCE II	SCE III
7.18	2.3406 (10)	2.3406 (18)	2.3405 (10)	2.3406 (10)	2.3406 (16)	2.3405 (9)
7.19	3.2593 (17)	3.2593 (16)	3.25788 (25)	3.2593 (17)	3.2593 (14)	3.25788 (25)
7.20	1.3306 (36)	1.3308 (274)	1.3298 (40)	1.3306 (36)	1.3308 (276)	1.32988 (40)
7.21	4.6279 (55)	4.6279 (279)	4.6279 (58)	4.6279 (55)	4.6279 (279)	4.6279 (58)
7.22	2.0165 (71)	2.0164 (63)	2.0125 (129)	2.0165 (68)	2.0164 (63)	2.0125 (129)
7.23	1.7412 (87)	1.7393 (424)	1.7332 (96)	1.7473 (80)	1.7412 (456)	1.7462 (83)

Table 7.21 Results of different non-general FR including CD, RMIL, HCD for SCE I to III

Examples	CD			RMIL			HCD		
	SCE I	SCE II	SCE III	SCE I	SCE II	SCE III	SCE I	SCE II	SCE III
7.18	2.3406 (12)	2-periodic	2.3405 (18)	2.3406 (10)	Chaos	2.3405 (11)	2.3406 (9)	Chaos	2.3405 (12)
7.19	3.2593 (28)	2-periodic	3.2593 (27)	3.2593 (29)	Chaos	3.2593 (29)	3.2593 (23)	Chaos	3.2593 (25)
7.20	1.3591 (17)	1.3305 (101)	1.3305 (17)	1.3305 (846)	1.3305 (173)	1.3305 (866)	1.3305 (503)	1.3305 (149)	1.3305 (508)
7.21	4.6278 (61)	4.6278 (165)	4.6279 (63)	4.6281 (69)	4.6277 (877)	4.6281 (73)	4.6278 (66)	4.6277 (294)	4.6279 (59)
7.22	2.0164 (78)	2-periodic	2.0164 (71)	2.0164 (88)	2-periodic	2.0164 (82)	2.0164 (102)	2-periodic	2.0164 (103)
7.23	1.7414 (96)	Chaos	1.7521 (89)	1.7443 (95)	Slowly	1.7616 (89)	1.7366 (84)	Slowly	1.7366 (84)

HSAC method provides the better ability for efficiency in almost examples. However, HSAC is not introduced as a method with significant computational burden in SCE I and SCE III. Thus, the search direction for computing the reliability index using gradient vector in SCE I and CG vector in SCEII are not affected in the SAC and HSCA methods. these methods can be adjusted line search for finding MPP with CG vectors with limited scalar factor which is a modification of FR method. The SAC and HSAC are better in the SEC I and SCE III compared to SCE II.

The results of non-general FR methods for Examples 7.18, 7.19, 7.20, 7.21, 7.22, and 7.23 are presented in Table 7.21 in SCE I-III. The step size in this category of CFORM is strongly important for convergence. If the step size is given 1, then the instabilities and slow convergence of the FORM formula using non-FR methods are provided. A good convergence rate for componential effort is obtained based on the CD method but RMIL is the inefficient method I Example 7.20. SCE I and SCE III are similar for computing iterations in relative methods of CD, RMIL and HCD.

Consequently, the best method is CD but its formulation should be controlled based on a step size adjusted by sufficient descent conditions as applied in SCE I and SCE II.

The results in Tables 7.19, 7.20, and 7.21 showed that an acceptable formulation for CFORM formulas is developed. There are two types of CFORM algorithms including modified FR and variant-type FR methods that these methods are a robust formulation which is obtained without line search rules as Armijo or sufficient descent condition. However, the FSS should be given in the no-gradient FR methods.

By comparing the compactional burden, it can be given for the first category as variant FR method, then modified FR and finally non-general FR method can be introduced for future applicable reliability problems. However, the nonlinearity degree and concavity phenomena are directly affected on the efficiency of FORM formulation which is expected by the conjugate sensitivity vector.

The HSAC in modified FR and CD in non-general FR categories are the best methods but it is suggested that the FSS with adjusting procedure is applied for better computational burden. The FSS with adjusting procedure given by sufficient descent conditions can improve the efficiency of CFORM but the FR-type families have the stable results.

7.6 Summary

In this chapter, the extended different types of conjugate FORM are presented, which are formulated with different scalar factor by comparing with FR. Several conjugate FORM formulations are established based on novel conjugate gradient vectors in this chapter. Three formulations of analytical conjugate FORM are presented as (1) hybrid conjugate finite-step length method; (2) hybrid self-adaptive directional conjugate method and (3) three-term conjugate gradient method. In the hybrid conjugate finite-step length method, two conjugate scalar factors are presented and these factors are dynamically combined, while, in the hybrid self-adaptive directional conjugate method, the FR formulation is enhanced by using a dynamical conjugate scalar factor as a hybrid conjugate formulation given from FR and adjusted FR scalar factors. The conjugate gradient vector is dynamically determined based on three terms given from gradient vector, previous conjugate gradient vector, and previous gradient vector, in the three-term conjugate gradient method. These three methods are investigated for convergence performances for both accuracy and efficiency in this chapter. The parameters applied in computing the line search including (i) step length and (ii) conjugate gradient vectors are discussed through several examples. By this comparative study, the two different conjugate FORM iterative formulations applied for computing the reliability index is also discussed. By comparing the compactional burden, it can be given, the nonlinearity degree and concavity phenomena are directly affected on the efficiency of FORM formulation which is expected by improved conjugate sensitivity vector.

References

1. B. Keshtegar, Enriched FR conjugate search directions for robust and efficient structural reliability analysis. Eng. Comp. **34**, 117–128 (2018)
2. B. Keshtegar, Limited conjugate gradient method for structural reliability analysis. Eng. Comp. **33**, 621–629 (2017)
3. B. Keshtegar, Chaotic conjugate stability transformation method for structural reliability analysis. Comp. Methods Appl. Mech. Eng. **310**, 866–885 (2016)
4. M. Rivaie, M. Mamat, L.W. June, I. Mohd, A new class of nonlinear conjugate gradient coefficients with global convergence properties. Appl. Math. Comp. **218**(22), 11323–11332 (2012)
5. B. Keshtegar, A hybrid conjugate finite-step length method for robust and efficient reliability analysis. Appl. Math. Model. **45**, 226–237 (2017)
6. B. Keshtegar, M. Miri, An enhanced HL-RF Method for the computation of structural failure probability based on relaxed approach. Civil Eng. Infrastructures J. **1**(1), 69–80 (2013)
7. B. Keshtegar, S. Chakraborty, An efficient-robust structural reliability method by adaptive finite-step length based on Armijo line search. Reliability Eng. Syst. Safety **172**, 195–206 (2018)
8. X. Huang, Y. Zhang, Reliability–sensitivity analysis using dimension reduction methods and saddlepoint approximations. Int. J. Numer. Meth. Eng. **93**(8), 857–886 (2013)
9. L. Zhang, Z. Lu, P. Wang, Efficient structural reliability analysis method based on advanced Kriging model. Appl. Math. Model. **39**(2), 781–793 (2015)
10. Z. Meng, G. Li, D. Yang, L. Zhan, A new directional stability transformation method of chaos control for first order reliability analysis. Struct. Multidiscipl. Optim. **55**, 601–612 (2017)
11. J.X. Gong, P. Yi, A robust iterative algorithm for structural reliability analysis. Struct. Multidiscipl. Optim. **43**, 519–527 (2011)
12. B. Keshtegar, Stability iterative method for structural reliability analysis using a chaotic conjugate map. Nonlinear Dyn. **84**, 2161–2174 (2016)
13. B. Keshtegar, Z. Meng, A hybrid relaxed first-order reliability method for efficient structural reliability analysis. Struct. Saf. **66**, 84–93 (2017)
14. P.L. Liu, A. Der Kiureghian, Optimization algorithms for structural reliability. Struct. Saf. **9**(3), 161–177 (1991)
15. S.R.d. Santos, L.C. Matioli, A.T. Beck, New optimization algorithms for structural reliability analysis. Comp. Model. Eng. Sci. **83**(1), 23–55 (2012)
16. B. Keshtegar, S. Chakraborty, A hybrid self-adaptive conjugate first order reliability method for robust structural reliability analysis. Appl. Math. Model. **53**, 319–332 (2018)
17. C. Luo, S.P. Zhu, B. Keshtegar, X. Niu, O. Taylan, An enhanced uniform simulation approach coupled with SVR for efficient structural reliability analysis. Reli. Eng. Syst. Safety **237**, 109377 (2023)
18. D. Yang, Chaos control for numerical instability of first order reliability method. Comm. Nonlinear Sci. Numerical Simul. **15**(10), 3131–3141 (2010)
19. B. Keshtegar, S.P. Zhu, Three-term conjugate approach for structural reliability analysis. Appl. Math. Model. **76**, 428–442 (2019)
20. B. Xia, H. Lü, D. Yu, C. Jiang, Reliability-based design optimization of structural systems under hybrid probabilistic and interval model. Comp. Struct. **160**, 126–134 (2015)

Chapter 8
Original-Space Search for FORM

Abstract In this chapter, the original probabilistic model (OPM) is proposed by a novel probabilistic optimization model for most probable point MPP search in original space for interval basic random variables. The novel MPP search approach is formulated by using joint cumulative distribution function (CDF) of basic random variables in X-space. In novel searching MPP, an individual MPP can be provided for multi-MPP problems. Then, four harmony search (HS) methods as a music-inspired formulated by human-based optimization methods named HS, global-best HS, improved HS gaussian global-best HS are applied for searching MPP computed by the proposed novel original probabilistic model.

Keywords Original probabilistic model · Harmony search · MPP · Cumulative distribution function

8.1 Introduction

In structural reliability analysis (SRA), the iterative process for searching the most probable point (MPP) is one of the reliability methods which are applied to compute the safety index. These iterative methods can be categorized as the dimension reduction [1], saddle point approximation [2], first-order reliability method (FORM) [3] and second-order reliability method [4]. Due to computational efforts give from simulation methods-based reliability analysis for application in real engineering problems, FORM is also a popular SRA by providing efficient commutation for reliability-based design optimization [5]. As mentioned in FORM algorithms, the search of MPP is the primary effort for approximating failure probability using analytical iterative approaches as presented in Chaps. 3–7. Generally, the MPP in FORM is evaluated by transferring random variables into $\boldsymbol{U}$-space defined as normal standard space in an iterative process [6]. The transformation of real (original) i.e. $\boldsymbol{X}$ normal variables into $\boldsymbol{U}$-space is simple but nonlinear map should be applied for non-normal variables that this map is computed by using statistical information of variables and new points obtained by iterative approach. The reliability method as an analytical-based

S.-P. Zhu and B. Keshtegar, *Structural Reliability Analysis*, Structural Integrity 31,
https://doi.org/10.1007/978-3-031-75197-4_8

first-order saddle point approximation method (FOSAM) [2] and simulation-based weighted uniform (WUS) [7] are extended by using PDF (i.e. probability density function) of variables to approximate the MPP without transferring variables from $\boldsymbol{X}$ to $\boldsymbol{U}$. Nevertheless, in FOSAM, the probability model to search MPP is similar to FORM. Consequently, FOSAM and FORM have shown the same abilities for problems with several MPPs, because these two methods are sensitive to the initial point in the iteration process for searching MPP. Thus, they may provide a fake MPP. The maximum weighted index applied in WUS is used for choosing the MPP where the number of sampling points is directly affected on its accuracy and it has the same issue as well as FORM for flinging MPP of problems with multi-MPP. Therefore, it is a vital challenge for finding individual MPP of LSFs (i.e. limit state functions) which have several points with the same minimum distance from the origin.

A novel probabilistic model is introduced which is applied for MPP using joint cumulative distribution function (CDF). In this method, the MPP is searched by using CDF of random variables in $\boldsymbol{X}$-space. A probabilistic optimization model is introduced for MPP search that this optimization model is objected to solve maximize probabilistic model which is defined by a joint CDF as objective function and limit state function (LSF) as subjective function, that is named as original probabilistic model (OPM). In OPM-based MPP search, the proposed probabilistic model can provide an individual MPP for problems under multi-MPP. The OPM is applied for MPP by using SQP optimizer as an analytical approach at the first stage and then an optimization-based population approach as harmony search with three improved versions are applied to search MPP using OPM at the next stage.

8.2 Original Probabilistic Model (OPM) for FORM

The main challenge of FORM for multi-MPP reliability problems is shown with a simple function in this section. Finally, a probabilistic model is presented which does not need to search MPP in normal standard space and also this model may solve an issue of FORM-based MPP search.

8.2.1 MPP Search Using FORM

In FORM algorithm, the shortest distance a point on LSF from origin is defined as MPP which the distance is computed in standard normal space with uncorrelated random variables [6]. Thus, by using this definition of FORM, it can be given as optimum value of a constrain optimization problem for MPP which is obtained by solving the probabilistic model as follows,

$$\mathit{find}\ \boldsymbol{U}^*,\ \min F(\boldsymbol{U}) = (\boldsymbol{U}^T\boldsymbol{U})^{1/2} = \left[\left(\frac{\boldsymbol{X}-\mu_x}{\sigma_x}\right)^T\left(\frac{\boldsymbol{X}-\mu_x}{\sigma_x}\right)\right]^{1/2}$$
$$\mathit{subjected\ to}\ g(\boldsymbol{U}) = 0 \tag{8.1}$$

where, $\boldsymbol{U} = (u_1, u_2, \ldots, u_n)^T$ is standard normal variables which are given as below,

$$u = \frac{x-\mu_x}{\sigma_x} \tag{8.2}$$

As seen in FORM, original variables i.e. $\boldsymbol{X}$-space should be transformed into standard normal variables i.e. $\boldsymbol{U}$-space. However, in FOSAM, the PDF of variables is applied to define the optimum probabilistic model which is applied to search MPP by the following model,

$$\mathit{find}\ \boldsymbol{X}^*,\ \max F(\boldsymbol{X}) = \prod\nolimits_{i=1}^{n} f_X(x_i)$$
$$\mathit{subjected\ to}\ g(\boldsymbol{X}) = 0 \tag{8.3}$$

where, $f_X(x_i)$ is PDF of variable x_i. Figure 8.1 illustrates a LSF which is plotted in $\boldsymbol{X}$ and $\boldsymbol{U}$-space. The optimal point can be obtained by solving Eq. (8.1) using FORM and Eq. (8.3) using FOSAM formula, respectively. The MPP as optimum points of optimization models in Eqs. (8.1) and (8.2) are shown as $\boldsymbol{X}^*$ (in $\boldsymbol{X}$-space) and $\boldsymbol{U}^*$ (in $\boldsymbol{U}$-space) that they are related to maximum likelihood for failure of LSF.

The optimization problem can be simply applied for MPP search of LSF with normal variables using Eq. (8.1) that its MPP is also related to MPP search in $\boldsymbol{X}$-space by Eq. (8.3) without transferring process. However, in real engineering problems, we have non-normal variables, thus it should be computed the equivalent mean (i.e. μ_x^e) and STD (i.e. σ_x^e) which is applied for transferring variables into $\boldsymbol{U}$-space as presented in FORM optimum model as follows,

$$\sigma_x^e = \frac{1}{f_X(x)}\varphi\left[\Phi^{-1}\{F_X(x)\}\right] \tag{8.4}$$

$$\mu_x^e = x - \sigma_x^e\Phi^{-1}[F_X(x)] \tag{8.5}$$

where, $F_X(x)$ represents CDF of variable x. φ and Φ^{-1} are respectively standard normal PDF and inverse standard normal CDF. As presented, several improved FORM-based steepest descent and conjugate search direction algorithms have been extended for MPP search that some of them have been named as HL-RF (Hasofer-Lind and Rackwitz-Fiessler) [8], the STM (i.e. stability transformation method) [3], accelerated CC (i.e. chaos control method) [5] and CCC (i.e. chaotic conjugate chaos control method) [6]. There is presented that failure probability (P_f) is approximated by reliability index ($\beta = \boldsymbol{U}^*$) as follows,

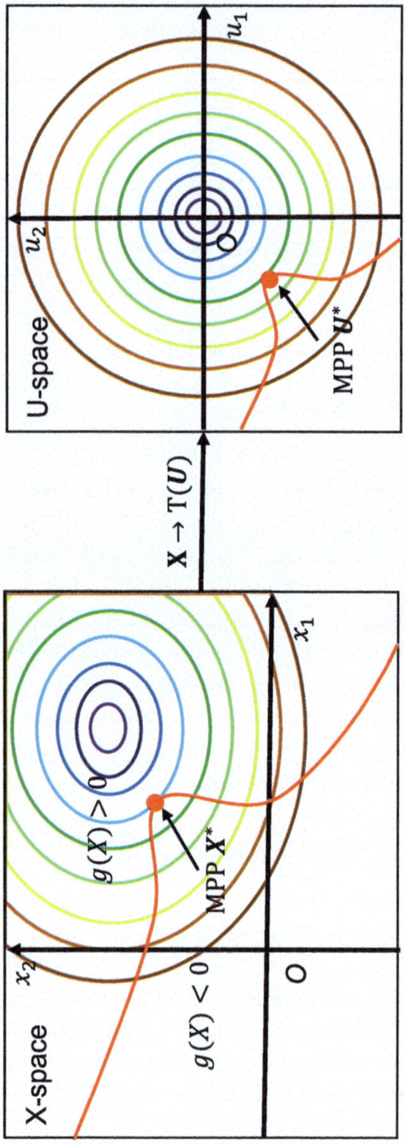

Fig. 8.1 FOSAM and FORM with MPP in $\boldsymbol{X}$-space and $\boldsymbol{U}$-space

$$P_f \approx 1 - \Phi(\beta) = \Phi\left(-\boldsymbol{U}^*\right) \tag{8.6}$$

The MPP is directly applied for determining the reliability index that MPP iteratively computed by nonlinear map i.e. $u_{k+1} = f\left(x_k, \mu_x^e, \sigma_x^e\right)$ in FORM. As seen in the MPP search process as well as formulation of FORM, iterative FORM approach is also dependent on (i) point x_k (ii) distribution function which applied for determining μ_x^e and σ_x^e as presented in Eqs. (8.2–8.4). However, in FOSAM using Eq. (8.3), the PDFs of variables are directly utilized to search MPP. $\boldsymbol{U}$-space transformation in FORM may increase nonlinearity effects for FORM compared to FOSAM in a LSF with non-normal variables. On the other hand, the MPP obtained by FOSAM and FORM are the same thus these two MPP approaches cannot converge to an individual MPP when it is selected a LSF with several optimal points which are located on LSF with the same distance from the origin. For better observation of this issue, a one-dimension LSF with normal standard variable x as $N(0, 1)$ are selected as $g(X) = 4 - x^2$, where, MPP is computed by FOSAM as below probabilistic model,

$$\begin{aligned} &\max \tfrac{1}{\sqrt{2\pi}} \exp\left(-\tfrac{(x)^2}{2}\right) \\ &s.t\, 4 - x^2 = 0 \end{aligned} \tag{8.7}$$

In $g(X) = 4 - x^2$ using FORM, $g = 0$ at two point 2 and -2 that these located two points on LSF have minimum distance. As presented in Fig. 8.2, the normal standard PDF showed two maximum values for MPP in point 2 and -2 with related objective of 0.05399. Consequently, the multi-MPP is provided by FOSAM and FORM. The objective function for MPP search using the probabilistic model applied in FORM and FOSAM is a hyper-sphere surface in $\boldsymbol{U}$-space that it may locate several points on this hyper-sphere surface with minimum distance. Consequently, the probabilistic model applied in FORM and FOSAM has a main issue as several MPPs that it is inherently provided by FORM in Eq. (8.1) and FOSAM in Eq. (8.3) optimization models.

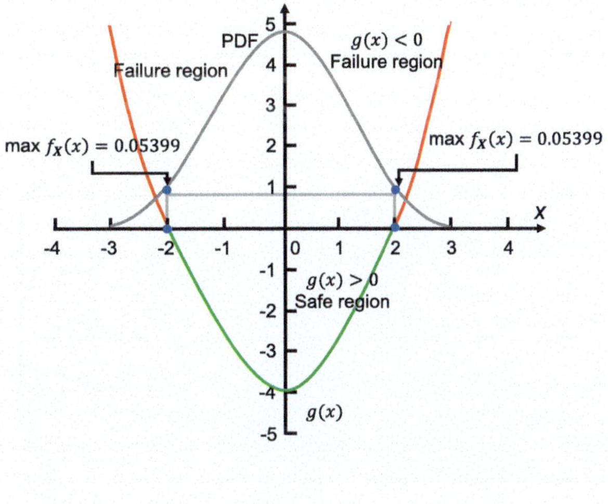

Fig. 8.2 The FOSAM results for LSF of $g(X) = 4 - x^2$

8.2.2 MPP Search in Original Space

It is showed that a nonlinear map applied for the transformation of variables into $\boldsymbol{U}$-space that it directly affected on performances of analytical reliability methods. By eliminating the transformation mapping process in $\boldsymbol{U}$-space, it may reduce risk of instability of analytical iterative method and also inaccuracy of MPP. In this section, we try to establish an optimum model which is searched MPP without mapping in $\boldsymbol{U}$-space. As presented in Sect. 8.2.1, the FORM and FOSAM models provided several optimums as MPP. Thus, the main effort of this section is to introduce a probabilistic model for searching individual MPP. For this propose, a joint CDF objective function is defined in this section.

For better description of opposed OPM, it is schematically plotted a one-dimensional problem by using MPP and failure probability in $\boldsymbol{X}$-space in Fig. 8.3. By view to Fig. 8.3, we have a new definition for MPP ($x*$). It is defined MPP that it is a point which has the maximum value of its CDF (joint CDF) i.e. $F_X(x)$ on failure domain i.e. $g(X) \leq 0$. Consequently, as presented in Fig. 8.3 MPP should be searched in failure domain for achieving the maximum CDF. Therefore, a novel probabilistic model is defined for MPP search without transformation process. In this idea, we try to find the maximum CDF in the $\boldsymbol{X}$-failure domain.

Therefore, the new formulation is established by searching MPP as $\boldsymbol{X}$-physical point in failure domain with maximum CDF that MPP is extracted from optimization model as follows,

$$\begin{array}{c} \textit{find}\ \boldsymbol{X}^*,\ \max F(\boldsymbol{X}) = \prod_{i=1}^{n} F_X(x_i) \\ \textit{subjected to}\ g(\boldsymbol{X}) = 0 \end{array} \tag{8.8}$$

where, $F_X(x_i)$ represents CDF for x_i, and $F(\boldsymbol{X})$ defines as joint CDF as objective and $g(\boldsymbol{X})$ denotes LSF in $\boldsymbol{X}$-space.

For illustrating the advantages of proposed OPM, a LSF of $g(X) = 4 - x^2$, $x \sim N(0, 1)$ is considered that its probabilistic optimization model is defined as Eq. (8.9),

$$\begin{array}{c} \max \Phi(x) \\ s.t\, 4 - x^2 = 0 \end{array} \tag{8.9}$$

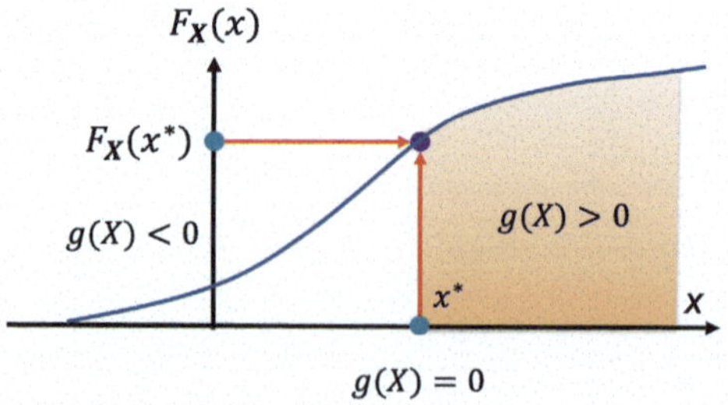

Fig. 8.3 A view for one-dimension LSF, its MPP and CSF using OPM

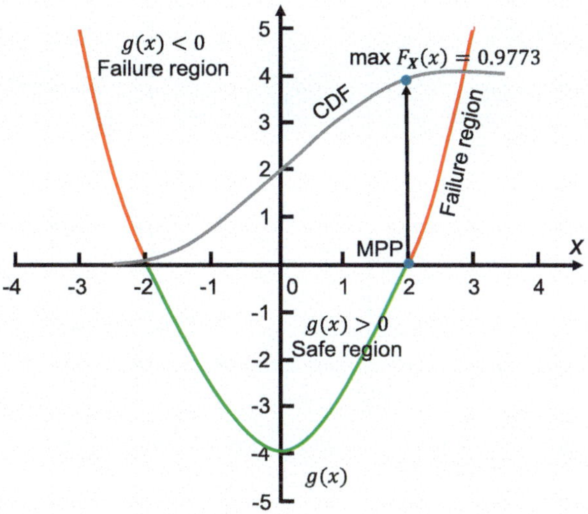

Fig. 8.4 The OPM results for LSF of $g(X) = 4 - x^2$

Figure 8.4 illustrates the results of the above OPM for random variable x. By applying the new definition for MPP, the optimum value for model in Eq. (8.9) is 2 that this point provided maximum $\Phi(x)$. Therefore, it is captured one MPP using OPM while two MPPs are obtained by FORM and the FOSAM for this LSF i.e. $g(X) = 4 - x^2$. as presented in Sect. 8.2.1. We know that $F(\boldsymbol{X})$ is increased by increasing X in failure domain form 0–1. Consequently, this function CDF just at one point can be crossed with LSF that CDF have maximum value. This means by using OPM that one of main challenge of FORM to analyze the multi-MPPs is avoided.

As a schematic view of OPM to search MPP, $\boldsymbol{X}$-MPP space, LSF defined by two variables is plotted in Fig. 8.5. The joint CDF is maximized by OPM for finding a point on LSF that this definition is called MPP. As seen from this figure, it is obtained one individual MPP as optimal point using OPM. The difference of FORM and OPM models for searching MPP is plotted in Fig. 8.6. In OPM unlike FORM, it is omitted the transformation process of variables in $\boldsymbol{U}$-space.

8.2.3 Illustrative Examples for Performance of OPM

The performances of prosed OPM are examined based on several LSFs. The MPPs captured by OPM in $\boldsymbol{X}$-space are compared MPP obtained with FOSAM and FORM. The MATLAB software is applied as an optimization solver given from SQP applied for MPP search in FORM, FOSAM, and OPM. The HL-RF and CFORM (i.e. conjugate first-order reliability method)-based FR (i.e. Fletcher-Reeves) with an initial finite step size of 50 and adjusting factor of 1.4 are utilized in the compression phase.

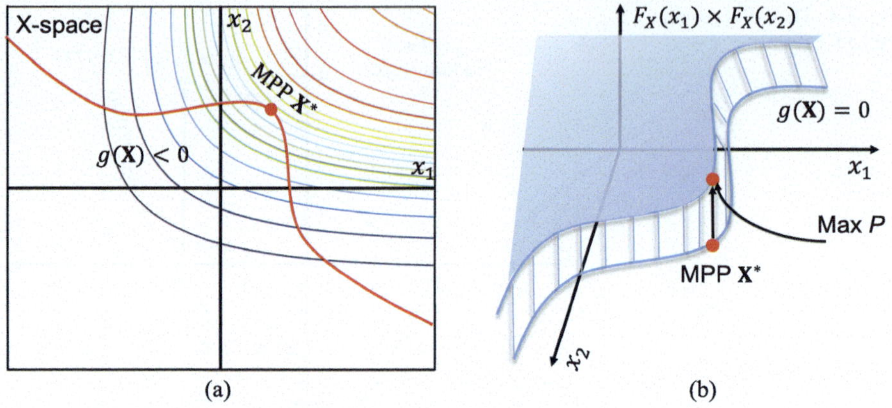

Fig. 8.5 X-MPP space and LSF for demonstrating OPM

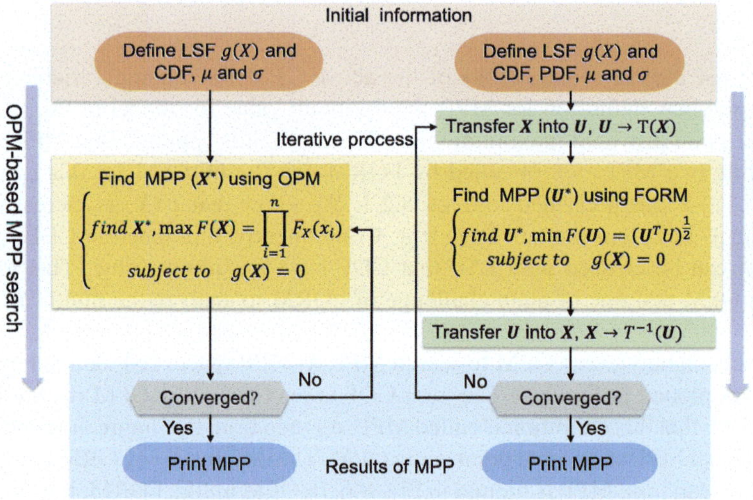

Fig. 8.6 MPP search framework using probabilistic models of FORM and OPM

Example 8.1 An LSF is considered as below,

$$g = 3 - 0.5x_1^2 - x_2 \tag{8.10}$$

where, normal x_1 and x_2 defined as $x_1, x_2 \sim N(0, 1)$. For this example, obtained MPP procedure are plotted for FOSAM, FORM and OPM in Fig. 8.7 where two initial points for the iterative process of MPP search methods are shown as $x_0 = (-1,\ 1)$ and $(1,\ 1)$.. As seen, the FOSAM as well as FORM probabilistic model provide two MPPs as (1) $X^* = (2,\ 1)$ and (2) $X^* = (-2,\ 1)$. Whereas, the MPP search using OPM provided one point as $X^* = (1.5508,\ 1.7957)$ for these two initial points.

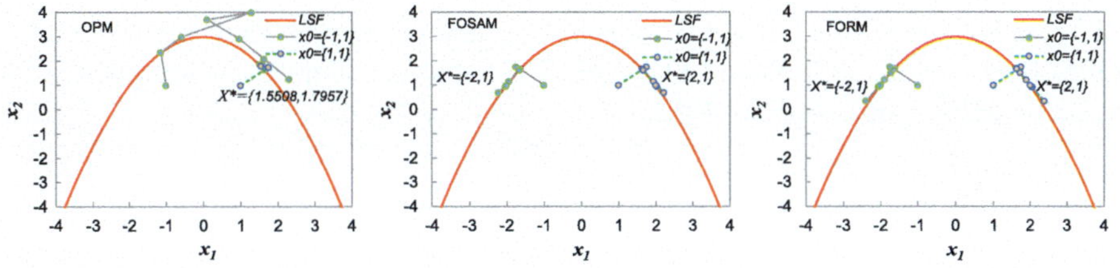

Fig. 8.7 The MPP search process of OPM, FOSAM and FORM for $g = 3 - 0.5x_1^2 - x_2$ obtained by two initial points. **a** OPM; **b** FOSAM; **c** FORM

Therefore, it is presented a probabilistic model and describe a new definition for MPP that it provided an individual point, unlike FORM and FOSAM.

Example 8.2 The nonlinear LSF is defined as,

$$g = x_1^4 + x_2^4 - 20 \tag{8.11}$$

where, normal x_1 and x_2 have $\mu_1 = \mu_2 = 10$ and $\sigma_1 = \sigma_2 = 5$ that using 10^6 simulation data point applied in MCS (i.e. Monte Carlo simulation), we have $\beta_{MCS} = 2.8861$.

The iterations (Iter), MPP, function calls, and β captured by FOSAM, FORM, and OPM-based SQP optimization solver and two methods of HL-RF and FR are listed in Table 8.1. HL-RF periodically provides unstable solutions, while FR as well as FOSAM, FORM and OPM are robustly provided the same MPP. It is obvious that the MPP using OPM is accurately computed.

As presented, the analytical approaches are iterated by a mathematical framework which is depended on a descript map computed by mean and STD (i.e. standard deviation) random variables. To illustrate the effects of mean and STD on MPP results of OPM mode, parameters of variable x_1 is changed and then FORM, FOSAM and OPM are applied for reliability analysis that their captured results in the ***X***-space

Table 8.1 Results of different MPP search models for Example 8.2

Method	***X****	***U****	β	Calls	Iter
OPM	(1.819216, 1.458369)	(−1.636157, −1.708326)	2.365457	42	12
FOSAM	(1.815783, 1.461680)	(−1.636843, −1.707664)	2.365454	46	14
FORM	(1.815783, 1.461680)	(−1.636843, −1.707664)	2.365454	79	25
CFORM	(1.928665, 1.552691)	(−1.614267, −1.689462)	2.336694	871	49
HL-RF	Periodic-2 as $\boldsymbol{X}_1 = (5.403079,\ 8.215047)$ and $\boldsymbol{X}_2 = (9.347417,\ 5.412545)$				

($\boldsymbol{X}^*$) are listed in Table 8.2. As seen from this table, faster convergence is given by OPM compared to FORM and FOSAM. The superior computational burden with accurate MPP is obtained by OPM.

Example 8.3 A non-normal LSF as follows [9],

$$g(X) = x_1^4 + x_2^2 - 50 \tag{8.12}$$

where, x_1 and x_2 respectively follow Lognormal ($\mu = 5$ and $\sigma = 1$) and Gumbel ($\mu = 10$ and $\sigma = 10$). The extracted reliability index from Ref. [9] as a FORM result is $\beta = 3.259315$.

The periodic-4 $\beta = 2.5118, 1.8550, 2.3857$, and 1.5966 is extracted by the HL-RF method. The converged results of FOSAM, FORM, and OPM are listed as follows,

(1) OPM; $\boldsymbol{U}^* = (-3.2587, -0.7377)$, $\boldsymbol{X}^* = (2.5714, 2.5057)$ and $\beta = 3.34118$ with iterations = 4;
(2) FORM; $\boldsymbol{U}^* = (-3.1115, -0.9706)$, $\boldsymbol{X}^* = (2.6475, 0.9315)$ and $\beta = 3.25933$ with iterations = 11;
(3) FOSAM; $\boldsymbol{U}^* = (-3.0962, -1.0344)$, $\boldsymbol{X}^* = (2.6556, 0.5197)$ and $\beta = 3.264419$ with $iterations = 8$. By comparing with MCS as $\beta = 3.5009$, the

Table 8.2 Results of different statistical properties of random variable x_1 for Example 8.2

Parameters	Proposed OPM		FOSAM		FORM	
	$\boldsymbol{X}^*$	β (Iter)	$\boldsymbol{X}^*$	β (Iter)	$\boldsymbol{X}^*$	β (Iter)
$\mu_1 = 5, \mu_2 = 10$ $\sigma_1 = \sigma_2 = 5$	(1.57109, 1.67198)	1.8124 (12)	(1.53822, 1.63811)	1.8101 (12)	(1.53811,1.63822)	1.8100 (30)
$\mu_1 = 15, \mu_2 = 10$ $\sigma_1 = \sigma_2 = 5$	(1.90459, 1.35998)	3.1378 (13)	(1.92424, 1.33169)	3.1376 (22)	(1.92424, 1.33169)	3.1376 (22)
$\mu_1 = 20, \mu_2 = 10$ $\sigma_1 = \sigma_2 = 5$	(1.95675, 1.27827)	4.0081 (14)	(1.97913, 1.23533)	4.0078 (29)	(1.97912, 1.23533)	4.0078 (22)
$\mu_1 = \mu_2 = 10$ $\sigma_1 = 1, \sigma_2 = 5$	(2.11134, 0.50344)	8.1141 (4)	(1.74078, −1.52501)	8.5748 (5)	(2.10756, 0.60627)	8.1130 (14)
$\mu_1 = \mu_2 = 10$ $\sigma_1 = 2, \sigma_2 = 5$	(2.05647, 1.01407)	4.3595 (5)	(2.07326, 0.93423)	4.3584 (19)	(2.07326, 0.93423)	4.3584 (12)
$\mu_1 = \mu_2 = 10$ $\sigma_1 = 10, \sigma_2 = 5$	(1.46325, 1.66623)	1.8727 (11)	(1.36371, 1.69585)	1.8720 (30)	(1.36371, 1.69585)	1.8720 (33)

OPM is an accurate formulation. Searching MPP in $\boldsymbol{X}$-space is reduced the iterations of OPM and FOSAM compared to FORM. The LSF and iterative process to find MPP using FORM, FOASM, HL-RF and proposed OPM are presented in Fig. 8.8. It is worth noting from Fig. 8.8 that unstable formulations captured by HL-RF while others are robust.

Now, various distributions for variables x_1. and x_2. are examined for Example 8.3 as follows,

Case 1: $x_1 \sim Normal(\mu = 5\, and\, \sigma{=}1)$ and $x_2 \sim Normal(\mu = 10\, and\, 10)$;
Case 2: $x_1 \sim Gumbel(\mu = 5\, and\, \sigma = 1)\, and\, x_2 \sim Gumbel(\mu = 10\, and\, \sigma = 10)$;
Case 3: $x_1 \sim Lognormal(\mu = 5\, and\, \sigma = 1)$ and $x_2 \sim Lognormal(\mu = 10$ and $\sigma = 10)$;
Case 4: $x_1 \sim Lognormal(\mu = 5\, and\, \sigma = 1)$ and $x_2 \sim Normal(\mu = 10\, and\, \sigma = 10)$;
Case 5: $x_1 \sim Normal(\mu = 5\, and\, \sigma = 1)$ and $x_2 \sim Gumbel(\mu = 10\, and\, \sigma = 10)$;

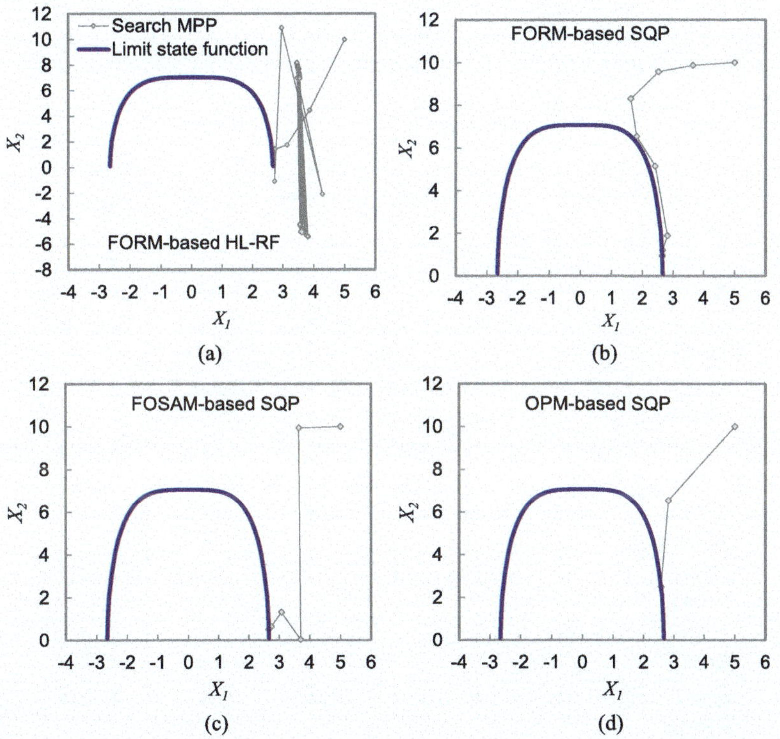

Fig. 8.8 Iterations of different analytical MPP search Example 8.3. **a** FORM-based HL-RF; **b** FORM-based SQP; **c** FOSAM-based SQP; **d** OPM-based SQP

Case 6: $x_1 \sim Uniform(\mu = 5\, and\, \sigma = 2)$ and $x_2 \sim Gumbel(\mu = 10\, and\, \sigma = 10)$;
Case 7: $x_1 \sim Uniform(\mu = 5\, and\, \sigma = 2)$ and $x_2 \sim Uniform(\mu = 10\, and\, \sigma = 10)$;
Case 8: $x_1 \sim Lognormal(\mu = 5\, and\, \sigma = 1)$ and $x_2 \sim Uniform(\mu = 10\, and\, \sigma = 10)$;
Case 9: $x_1 \sim Normal(\mu = 5\, and\, \sigma = 1)$ and $x_2 \sim Uniform(\mu = 10\, and\, \sigma = 10)$;
Case 10: $x_1 \sim Normal(\mu = 5\, and\, \sigma = 1)$ and $x_2 \sim Lognormal(\mu = 10\, and\, \sigma = 10)$.

The $\boldsymbol{X}^*$, β, function calls and MCS reliability index as β computed by $1.0 \sim 1.8 \times 10^6$ samples are tabulated in Table 8.3 for different distributions presented in Cases 1–10. Results of this table indicated that the MPP is provided by OPmethod for interval random variables. As seen in Case 7, applied interval variables x_1 and x_2 as followed by uniform distributions, the MPP is not captured by FOASM for while OPM provides MPP, more efficiently. In Cases 8 and 9 using OPM by applying joint CDF in probabilistic optimum model applfor MPP search, the reliability indexes are tended to MCS results by comparison with FORM.

Table 8.3 Results FORM, FOSAM and OPMor different distributions applied in cases 1–10 for Example 8.3

Cases	OPM		FOSAM		FORM		MCS
	$\boldsymbol{X}^*$	β/Iter	$\boldsymbol{X}^*$	β/Iter	$\boldsymbol{X}^*$	β/Iter	β
Case 1	(2.605, 1.981)	2.519/21	(2.635, 1.340)	2.519/32	(2.635, 1.340)	2.519/37	2.902
Case 2	(2.645, 0.744)	4.371/16	(2.658, 0.3)	4.355/43	(2.657, 0.445)	4.354/22	4.573
Case 3	0.602, 2.047)	3.529/16	(2.598, 2.111)	3.521/48	(2.558, 2.685)	3.486/44	3.654
Case 4	(2.622, 0.947)	3.297/16	(2.654, 0.634)	3.237/36	(2.654, 0.597)	3.238/38	3.691
Case 5	0.602, 2.047)	2.550/15	(2.634, 1.379)	2.533/31	(2.613, 1.836)	2.529/24	2.808
Case 6	(2.367, 4.316)	1.160/33	(2.108, 5.4)	1.428/24	(2.482, 3.476)	1.251/67	1.028
Case 7	(2.534, 3.071)	1.157/18	No-converged	–	(2.5689, 3.539)	1.186/21	1.131
Case 8	(2.403, 4.084)	3.629/15	(2.659,0.000)	3.192/62	(2.656, 0.464)	3.186/33	3.599
Case 9	(2.261, 4.884)	2.765/21	(2.659, 0.002)	2.474/79	(2.645, 1.025)	2.458/44	2.854
Case 10	(2.426, 3.721)	2.670/23	(2.568, 2.555)	2.722/66	(2.479, 3.501)	2.659/34	2.835

Table 8.4 Variables applied for determining LSF of Example 8.4

Variable	Mean	STD	PDF
x_1	4	0.1	Weibull
x_2	25,000	2000	Log-normal
x_3	0.875	0.1	Gumbel
x_4	20	1	Uniform
x_5	100	100	Exponential
x_6	150	10	Normal

Example 8.4 An LSF with normal/non-normal variables [10] is defined as below,

$$g(X) = x_1x_2x_3x_4 - x_5x_6^2/8 \tag{8.13}$$

The means, STD and distributions of variables are shown in Table 8.4.

The $\beta = 2.67$ using MCS with 1×10^6 has been extracted from Ref. [11] for this example. The reliability results including iterations, β, MPP as $\boldsymbol{X}^*$ or $\boldsymbol{U}^*$, and call LSF as *Eval* for studied reliability methods are arranged in Table 8.5. The proposed OPM for searching MPP is efficiently provided an accurate MPP as well as FOSAM, and FORM. By comparing the errors of the reliability index with MCS as Errors, OPM and FOSAM improve the accuracy of reliability analysis due to the elimination of terraformed $\boldsymbol{U}$-space process in iterative procedure.

Example 8.5 A dynamic structure as plotted in Fig. 8.9 is defined using LSF by force capacity (Fs) as below [6],

$$g = Fs - Ks \times 3\left(E\left[x_s^2\right]\right)^{1/2} \tag{8.14}$$

where,

$$E\left[x_s^2\right] = \frac{\pi S_0}{4\xi_s\omega_s}\left[\frac{\xi_a\xi_s}{\xi_p\xi_s\left(4\xi_a^2 + \theta^2\right) + \gamma\xi_a^2} \times \frac{\left(\xi_p\omega_p^3 + \xi_s\omega_s^3\right)\omega_p}{4\xi_a\omega_a^4}\right] \tag{8.15}$$

and $\gamma = \frac{M_s}{M_p}$, $\omega_a = \frac{\omega_p+\omega_s}{2}$, $\xi_a = \frac{\xi_p+\xi_s}{2}$, and $\theta = \frac{\omega_p-\omega_s}{\omega_a}$ that these factors for determining LSF is computed by lognormal variables as listed in Table 8.6.

Extracted reliability indexes from Ref. [6] for FORM and MCS (with 5.8×10^5 point for samples) are respectively 2.0165 and 2.5874. Unstable periodic-2 $\beta = 4.21724, 4.98089$ solutions obtained by HL-RF. The FOSAM, FORM and OPM provide stable results as presented in Table 8.7. As seen from the results of Table 8.7, the OPM and FOSAM by $\boldsymbol{X}$-space MPP search converged, more accurately but, OPM is more efficient than others. The robust formulation of SQP is provided in FORM, OPM and FOSAM.

Table 8.5 Results reliability for different MPP search methods in Example 8.4

Method	HL-RF	FORM	CFORM	FOSAM	OPM	Melchers and Ahammed [10]	Zhao et al. [11]
u_1^*	−0.1015	−0.1014	−0.1045	−0.1374	−0.1378	–	–
u_2^*	−0.3649	−0.3646	−0.3724	−0.2131	−0.2131	–	–
u_3^*	−0.4300	−0.4296	−0.4353	−0.7235	−0.7233	–	–
u_4^*	−0.3075	−0.3073	−0.3059	−0.5461	−0.5461	–	–
u_5^*	2.5222	2.5222	2.5836	2.4859	2.4859	–	–
u_6^*	0.5862	0.5857	0.6125	0.5020	0.5020	–	–
x_1^*	4.0057	4.0057	4.0054	4.0025	4.0025	4.01	4.006
x_2^*	24,204.6	24,205.2	24,204.4	24,499.9	24,499.9	24,193	24,234
x_3^*	0.8227	0.8228	0.8225	0.8011	0.8011	0.822	0.824
x_4^*	19.581	19.582	19.578	19.281	19.281	19.58	19.59
x_5^*	514.34	514.46	531.98	504.20	504.20	509.02	517.4
x_6^*	155.86	155.86	156.13	155.02	155.02	157.01	155.68
Eval	127	426	1352	90	42	–	–
Iter	8	69	68	11	6	–	–
β	2.6699	2.6696	2.7355	2.7051	2.7051	2.68	2.67
Error (%)	0.89	0.90	1.54	0.43	0.43	0.50	0.88

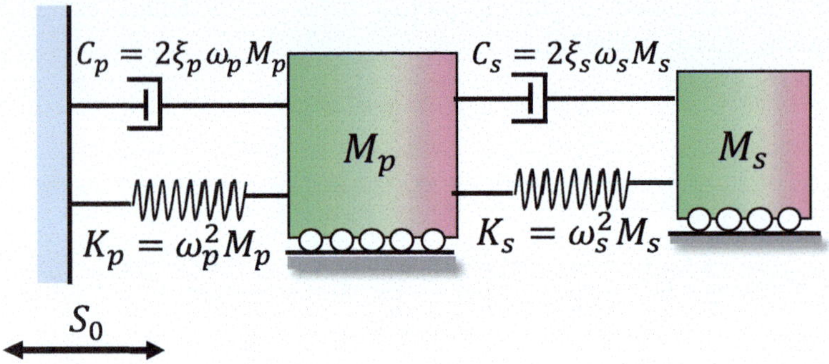

Fig. 8.9 Dynamic system of Example 8.5. Reprinted from [12], with permission from Elsevier

Table 8.6 Random variables applied for computing LSF in Example 8.5

Variable	Mean	STD	Variable	Mean	STD
M_s	0.01	0.001	ξ_s	0.02	0.01
M_p	1	0.1	ξ_p	0.05	0.02
K_s	0.01	0.002	F_s	15	1.5
K_p	1	0.2	S_0	100	10

Table 8.7 Results of MPP search method by different probabilistic models in Example 8.5

Method	M_p	M_s	K_s	K_p	ξ_s	ξ_p	S_0	F_s	β	Iter
FORM	1.0019	0.0101	0.0112	1.1022	0.0121	0.0280	13.737	103.723	2.01644	45
OPM	1.0013	0.0101	0.0102	1.0362	0.0100	0.0210	12.973	104.996	2.25961	12
FOSAM	0.9916	0.0100	0.0105	1.0421	0.0105	0.0261	13.770	102.062	2.10435	55
CFORM	1.0017	0.0101	0.0112	1.1044	0.0120	0.0278	13.721	103.777	2.04413	37

Example 8.6 A conical shell as plotted in Fig. 8.10 [9] under buckling failure mode is computed with below LSF,

$$g = 1 - \frac{\sqrt{3(1-v^2)}}{\pi E t^2 \cos^2 \alpha} \times \left(\frac{P}{0.66} + \frac{M}{0.41 r_1} \right) \tag{8.16}$$

in which, $v = 0.3$, Table 8.8 presents variables as normal /non-normal distributions for computing LSF in Example 8.6. COV is the coefficient of variation of each variable.

The 10^7 samples given in MCS for approximating reliability index that by MCS we have $\beta = 4.4163$. MPP, β and iterations for different probabilistic models applied to search MPP are presented in Table 8.9. The chaotic solutions are given by the HL-RF scheme, while SQP optimizer applied in FORM, OPM and FOSAM and analytical method of CFORM are converged as robust results. MPP extracted from OPM is as same as MPP computed by Ref. [9] and FORM.

A probabilistic model named OPM is established for MPP search in ***X***-space by avoiding the ***U***-transformation. The proposed OPM is improved the accuracy of FORM by searching MPP in ***X***-space. The OPM is an optimum model which is used

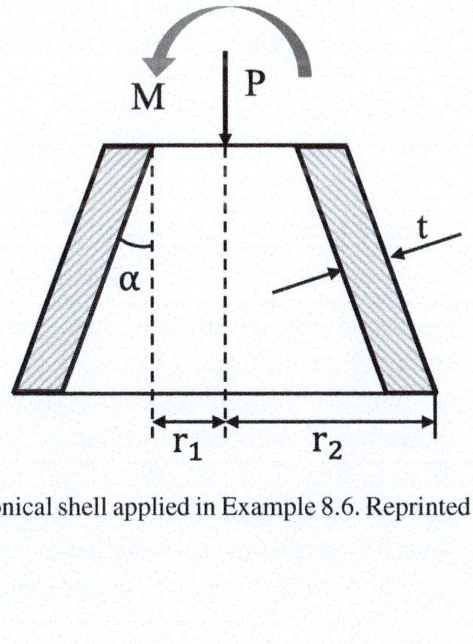

Fig. 8.10 A view of the conical shell applied in Example 8.6. Reprinted from [12], with permission from Elsevier

Table 8.8 Variables for conical structure applied in Example 8.6

Variable	Description	PDF	Mean	COV
E	Young's modulus (MPa)	Lognormal	70,000	0.05
α	Slop angle (rad)	Normal	0.524	0.02
t	Thickness (m)	Normal	0.0025	0.05
r_1	Internal radius (m)	Normal	0.9	0.025
P	Axial load (N)	Gumbel	70,000	0.08
M	Bending moment (N-m)	Gumbel	80,000	0.08

Table 8.9 Results of MPP for different probabilistic models for Example 8.6

Method	E^*	α^*	t^*	r_1*	P^*	M^*	Iter	β
Ref. [9]	65,442.63	0.5275	0.00208	0.8887	73,112.66	106,149.59	43	4.632489
FORM	65,131.86	0.5275	0.0021	0.8888	72,673.82	108,053.51	75	4.628886
OPM	65,383.38	0.5239	0.0021	0.8849	72,391.45	109,145.01	17	4.616935
FOSAM	64,856.21	0.5276	0.00208	0.8884	70,331.55	101,964.91	46	4.711865
CFORM	65,393.61	0.5275	0.0021	0.8887	72,694.41	108,498.89	23	4.627782

to maximizes joint CDF as objective function. The joint CDF is computed by information distribution of random variables the probabilistic constraint of this optimization model is LSF. Unlike FOSAM and FORM, the individual MPP is computed by OPM. Consequently, OPM improves the major issue of FORM for LSF with several MPPs.

The numerical results presented for studied examples clearly demonstrate that OPM provided an ***X***-MPP space. As well as FORM, the MPP of this method can be applied for reliability analysis of engineering structures in the next research. The reliability analysis by correlated random variables is a main issue of OPM that it is needed to be investigate by researchers for further exploration. In future work, the MPP point obtained by OPM can be applied for implementation in reliability-based design optimization.

8.3 Metaheuristic Optimization Methods

In this section, we introduce a constrained optimization method that its objective is to search MPP in real space by CDF information. The analytical optimization method are efficient approaches but they are not efficient methods with low-computational burden in reliability problems with multimodal properties, non-continuous performance function or non-differentiable limit state function at the design point. Therefore, the metaheuristic optimization methods can be applied for optimal designating the large verities of engineering problems such as SRA-based MPP search. The metaheuristic methods as an alternative optimizer have a simple implementation for both coding in a computer program by easy formulation and analysis process with on gradient iterative procedure. In addition, the metaheuristic algorithms have been categorized into two main types as (i) single solution based given by simulated results and (ii) population-based optimization methods. In this section, a population-based metaheuristic method is applied for searching the MPP in real-space of basic random variables.

The metaheuristic population-based techniques are given from the natural phenomena of the human-based schemes, animal-based swarms, physical-informed methods and evolutionary algorithms. In these algorithms, the optimization process is given based on generating of a population as design variables which are detected the solutions of the objective function. The population of swarm/human/particles is updated based on stochastic operators given from the results of objective-based previous populations. In metaheuristic optimization methods, the stopping criteria is commonly given based on the maximum number of iterations, thus, it is an optimization result by these methods. The biological behaviors including recombination, mutation-based random behavior and selection-based best population are generally applied in evolutionary-based optimization methods. The physical-informed methods are inspired using physical information of population such as gravitational, central force big-bang theory. The population-metaheuristics using human behaviors is also impressed based on the human relations together for a social works such as learning and musical implementation. The last category of metaheuristics algorithms is the swarm-based methods which is indicated the social behavior of animal swarm to find the foods or organisms of each animal for living in swarms [13]. In the population-metaheuristics algorithm, the searching optimum position is mainly two phases including exploration and exploitation. In exploration, an optimization algorithm should be promoted its operators for giving the random selection of particles in the rage from possible design domains. Thus, exploratory property of an optimization method can provide a well-designed solver with enough random nature to promote the position by randomly-generated solutions in different areas of the design domain. Usually, the exploitation process is executed after the exploration of particles that the main effort of this phase is to search the neighborhood of particles in the feasible domain to improve the quality of the objective function. In a practical optimization method, searching optimum point should be capable for a balance between global and local regions by using exploration and exploitation.

The harmony search (HS) optimization is a music-inspired using human-based optimization method that can be applied for searching MPP by OPM presented in this chapter. The HS method has a suitable random selection from the domain and current positions while it is provided a local search based on a tuning factor named as bandwidth. The MPP of LSFs is approximated based on four HS methods that are given based on the dynamical random selection and adjustment. The original HS and its three modifications are introduced for structural reliability analysis that they are used for approximating the MPP is $\boldsymbol{X}$-space at first propose.

8.3.1 Harmony Search Optimization Methods

In this section, an original probabilistic model named as OPM is presented to find MPP in the original space. In OPM, a new definition for MPP is established as a point which is corresponded to a physical-informed point on LSF with maximum joint CDF-based random variables. The OPM in optimization form is given as below probabilistic model,

$$\begin{aligned} &\textit{find}\ \boldsymbol{X}^*,\ \max F(\boldsymbol{X}) = \prod\nolimits_{i=1}^{n} F_X(x_i) \\ &\textit{subjected to}\ g(\boldsymbol{X}) = 0 \end{aligned} \tag{8.17}$$

in which, n is the number of random variables x followed by cumulative distribution function $F_X(x)$. As it can be seen, this model is a multiplier function i.e. $F(\boldsymbol{X})$ of CDF of random variable with a constraint of LSF $g(\boldsymbol{X})$. The main effort of the above OPM is to find the MPP which is related to its optimum design point. In this work, we have a new definition for MPP as a point on the LSF in $\boldsymbol{X}$-space that provided maximum joint CDF which is defined by multiplier function as $F(\boldsymbol{X}) = \prod_{i=1}^{n} F_X(x_i)$ in this current work.

The metaheuristics-based optimization algorithms given by the HS optimization method can be evaluated the following optimization model as an unconstructed optimization model.

$$\max f = \prod_{i=1}^{n} F_X(x_i) - \eta \max\{0, g(\boldsymbol{X})^2 - \varepsilon\} \tag{8.18}$$

where η denotes the penalty factor computed as $\eta = \frac{500}{|g(\boldsymbol{\mu})|}$ and $\max\{0, g(\boldsymbol{X})^2 - \varepsilon\}$ related to the penalty function which is given from the LSF at point $\boldsymbol{X}$ and ε denotes the tolerance of LSF to zero which is given a smaller value as $\varepsilon = |g(\boldsymbol{\mu})| \times 10^{-4}$. It can be conducted that, the unconstraint optimization model for searching the design point named MPP can be directly applied in the human-based behavior of metaheuristic optimization algorithms without computing the sensitivity vector. In the HS optimization method, the probabilistic objective function is directly applied to update positions of new iteration thus it is a method free gradient in FORM-based OPM. Thus, it is saved the computation efforts applied for determining the sensitivity vectors in HS algorithms.

8.3.1.1 Harmony Search Optimization

The HS is established based on the musical-inspired process that it is formulated by a human-based jazz population to find a perfect harmony of musicians. In HS optimization, a best condition of music related to global optimization which is randomly selected based on searching musical process to obtain a perfect state as global optimum [14]. The musical team has several players (particles) that they try to search perfect state of harmony improvisation (iterations). The HS algorithm was proposed with a random search for optimum using a harmony memory involved several members related to design variables and the number of musical notes related to harmony memory size (particles). In every improvisation, every players-applied in jazz team has several adjusting notes in their memory that the adjusting notes are related to the local search of HS for achieving the perfect music. In HS optimization method, the local search is randomly adjusted based on a smaller value named bandwidth (*bw*). Therefore, for a search process using HS, it should be given several parameters including number of iteration (*NI*), memory size (*MS*), bandwidth (*bw*) and constant parameters which are applied for adjusting new harmony memory named as harmony memory considering rate (*HMCR*) and pitch adjusting rate (*PAR*).

The HMCR and *PAR* are two parameters which are applied for a random generation of the new HM (i.e. harmony memory) as below relations,

$$x_{i,j}^{n} = \begin{cases} x_{i,j}^{o} + (2r - 1) \times bw\, r_1 \leq HMCR \\ x_i^L + r \times \left(x_i^U - x_i^L\right) r_1 > HMCR \end{cases} \tag{8.19}$$

where, $r, r_1 \in [0, 1]$ relates a random number and x_i^U,x_i^L respectively relate the maximum and minimum design variable $x_i \in \left[x_i^L, x_i^U\right]$. Using HMCR, the new components of harmony memory $x_{i,j}^n$ are given form old harmony memory with the probability of HMCR while it is computed from design domain with probability of 1-HMCR in HS method. The probability of considering the old harmony memory $x_{i,j}^o$ is evaluated based on a random number between 0 and 1. The new harmony memory is adjusted based on bandwidth (*bw*) with probability of *PAR* as follows,

$$x_{i,j}^{n} = \begin{cases} x_{i,j}^{n} + (2r - 1) \times bw r_2 \leq PAR \\ x_{i,j}^{n} r_2 > PAR \end{cases} \tag{8.20}$$

where, $r, r_2 \in [0, 1]$. The new HM (i.e. harmony memory) elements given form the previous step is adjusted with probability of PAR × HMCR by using *bw*. The bandwidth is computed as $bw_i = 0.001\left(x_i^U - x_i^L\right)$ and $i = 1, 2, \ldots, RV$ (numbers of random variable) and $j = 1, 2, \ldots, MS$ (harmony memory size). Based on this random generated the HS method, it is cited the steps of HS process for coding a computer program as below,

Step 1: Give performance function, statistical properties of basic random variables including CDF, σ and μ. Compute feasible design domain of random variables as i) $x_i^U = \mu_i + 4\sigma_i$ and ii) $x_i^L = \mu_i - 4\sigma_i$ with $i = 1, 2, \ldots, RV$;

Step 2: Set HS parameters including *NI* representing the number of iterations, *MS* representing the memory size. *HMCR* is the harmony memory considering rate, *PAR* is the pitch adjusting rate and *bw* indicates the bandwidth as $bw_i = 0.001\left(x_i^U - x_i^L\right)$. Compute the penalty factor as $\eta = \frac{500}{|g(\boldsymbol{\mu})|}$ and $\varepsilon = |g(\boldsymbol{\mu})| \times 10^{-4}$;

Step 3: Generate initial HS in feasible design domain for random variables as $x^o_{i,j} = x^L_i + r_{i,j} \times \left(x^U_i - x^L_i\right) i = 1, 2, ...RVj = 1, 2, ..., MS$, where, $r_{i,j} \in [0, 1]$ is a random uniform number and also $x^L_i = \mu_i - 4\sigma_i$ and $x^U_i = \mu_i + 4\sigma_i$ are respectively lower and upper bounds of variable x_i with mean of μ_i and STD of σ_i. *MS* and *RV* respectively represent number of random variables and harmony memory size. Based on this initial HM, the optimization model is computed as $f^o_j = \prod_{i=1}^{RV} F_X\left(x^o_{i,j}\right) - \eta \max\{0, g(\boldsymbol{X}^o_j)^2 - \varepsilon\}$;

Step 4: Create a new HM by using three factors of *HMCR*, *PAR*, and *bw* that new HM is updated based on the old HM as below,

(i) Generate uniform random values as $r, r_1, r_2, r_3 \in [0, 1]$;

(ii) If $r_1 \leq HMCR$ then, $x^n_{i,j} = x^o_{i,j}$ and IF $r_2 \leq PAR$ then $x^n_{i,j} = x^n_{i,j} + (2r_3 - 1) \times bw$, end if, Else;

(iii) give new HM from design domain as $x^n_{i,j} = x^L_i + r \times \left(x^U_i - x^L_i\right)$, end if;

Step 5: Update HM by the new HM elements which is given in Step 4 as below,

(i) compute the new objective function as $f^n_j = \prod_{i=1}^{RV} F_X\left(x^n_{i,j}\right) - \eta \max\{0, g(\boldsymbol{X}^n_j)^2 - \varepsilon\}$;

(ii) If $f^n_j > f^0_j$ then $x^o_{i,j} = x^n_{i,j}$, end if;

Step 6: Done Steps 4 and 5 until k less than *NI* or $g(\boldsymbol{X}_{best}) < 0.1\varepsilon$.

As presented sub-step in Step 3, the new HM is randomly generated using three random phases as (1) selecting from the previous HM, (2) adjusting by *bw* of several HM, and (3) computing from the design domain. By using six steps for HS, the framework of HS applied in a computer program for MPP search is plotted in Fig. 8.11. The HS is simply applied to find the optimum results as MPP using OPM.

In addition, the HS has shown some deficiencies for slowly convergent rate and it is sensitive to its parameters [15]. Some modifications in adjusting the parameters of HS have been establish for enhancing its performances. That several of improved versions of HS are presented as below sections.

8.3.1.2 Improved Harmony Search Optimization

Mahdavi et al. [16] improved HS by dynamical parameters in adjusting phase. All most iterative scheme of HS and improved HS (IHS) are same but in Step 4, IHM algorithm is determined new HM elements using dynamical *PAR* and *bw* which are computed as below formulation,

$$PAR(k) = PAR_{\min} + \frac{PAR_{\max} - PAR_{\min}}{NI}k \tag{8.21}$$

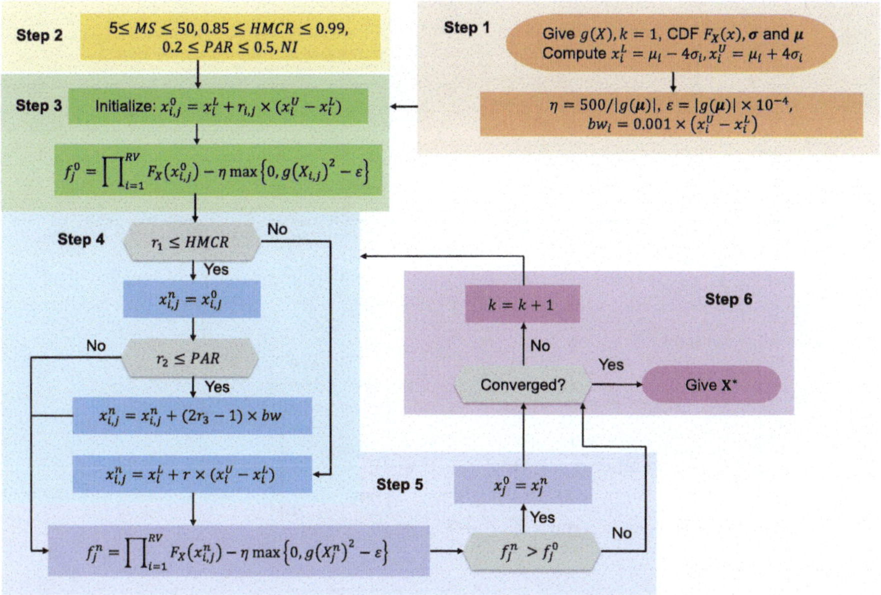

Fig. 8.11 Framework of HS algorithm for MPP search in $\boldsymbol{X}$-space

$$bw_i(k) = bw_{imax}\exp\left[\frac{Ln(bw_{imin}) - Ln(bw_{imax})}{NI}k\right] \tag{8.22}$$

where, $PAR_{\min}$ is the minimum value and $PAR_{\max}$ denotes the maximum value which are given for PAR. minimum and maximum bw are shown by $bw_{\min}$ and $bw_{\max}$, respectively. Ln is the logarithmic operator and k denotes current iteration that it is kNI. Thus, for IHS, the Step 4 of HS is rewritten as below,

Step 4: Create a new HM by using three factors of $HMCR$, PAR, and bw as below,

(i) Generate uniform random numbers between 0 and 1 for $r, r_1, r_2, r_3 \in [0, 1]$ and computed $bw(k)$ and $PAR(k)$;

(ii) If $r_1 \leq HMCR$ then, $x_{i,j}^n = x_{i,j}^o$ and If $r_2 \leq PAR(k)$ then $x_{i,j}^n = x_{i,j}^n + (2r_3 - 1) \times bw_i(k)$, end If; else;

(iii) Give the new HM from design domain as $x_{i,j}^n = x_i^L + r \times \left(x_i^U - x_i^L\right)$, end If.

As seen, the IHS has the extra parameters which is applied for determining $PAR(k)$ and $bw(k)$ of basic random variables. The $PAR(k)$ is linearly changed $PAR_{\min} = 0.01$ form to $PAR_{\max} = 0.99$ while $bw(k)$ is changed as nonlinear form as logarithmic between $bw_{imin} = 0.0001\left(x_i^U - x_i^L\right)$ and $bw_{imax} = 0.005\left(x_i^U - x_i^L\right)$. These four initial parameters are also given in the IHS algorithm which is applied in MPP search. The framework of IHS for MPP search-based original space is presented in Fig. 8.12. Two HS and IHS are similar to the same adjusting procedure but the dynamical

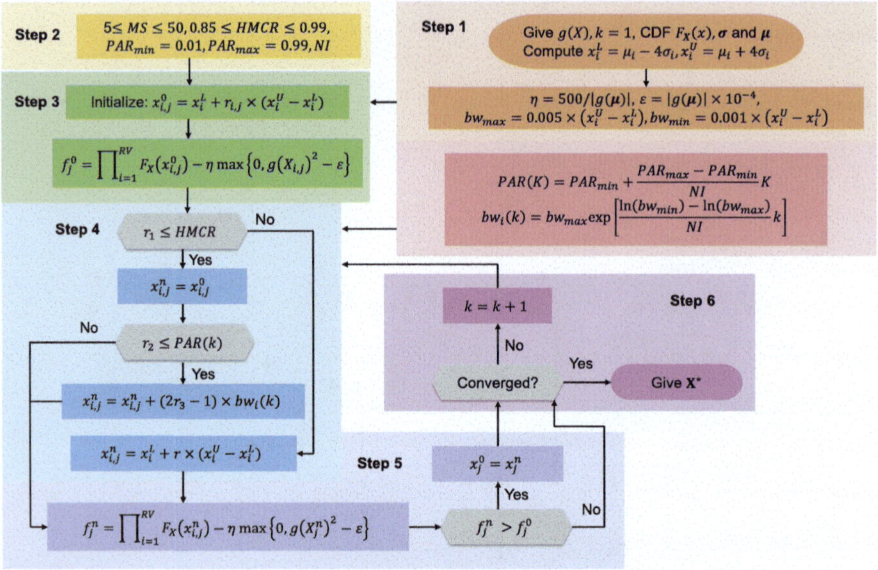

Fig. 8.12 Framework of IHS algorithm for MPP search in $\boldsymbol{X}$-space

parameters are the major different in the local search which is applied for adjusting new harmony memory.

8.3.1.3 Global-Best Harmony Search

Mahdavi and Omran [17] intruded global-best HS (GHS). In GHS, *PAR* is confirmed by parameters applied in IHS by Eq. (8.21) while adjustment of particles in new HM is considered based on global-best elements as best position among other form HM. The Step 4 in the GHS is similar to the IHS by this difference that new HM is replaced by global-best HM with probability of $HMCR \times PAP(k)$, thus $x_{i,j}^n = x_i^{best}$ when $r_2 \le PAR(k)$; where, x_i^{best} is best harmony for *i*-th variable. The flowchart of the GHS for coding in a computer program is plotted in Fig. 8.13. It is shown that the new elements of harmony are replaced by the best HM by probability of $HMCR \times PAP(k)$. The random discrete search is obtained based on this approach and it has a good exploration for global search of optimum while the exploitation is poor without any tuning the position for local search.

8.3.1.4 Gaussian Global-Best Harmony Search (GGHS)

The global convergence with fast convergence rate is the main efforts in improved version of HS algorithms. The GGHS has been established by Keshtegar et al. [15,

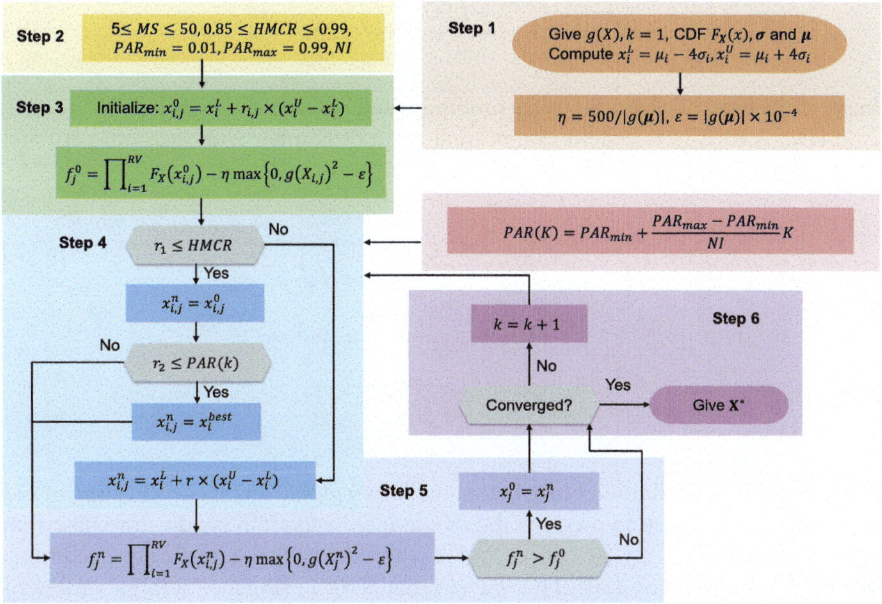

Fig. 8.13 Framework of GHS algorithm for MPP search in $\boldsymbol{X}$-space

18] which has been proposed as novel improved HS algorithm for a global optimum result. The GGHS can be applied for performance functions with nonlinear and discrete problems. The GGHS can be utilized for searching MPP by using of probabilistic model of max$f = \prod_{i=1}^{n} F_X(x_i) - \eta \max\{0, g(\boldsymbol{X})^2 - \varepsilon\}$ which is a probabilistic optimization model. In this GGHS, two adjusting processes is applied for new *HM* which are adjusted with probability of *HMCR* and *HMCR* × *PAR*(k). In first stage, new *HM* is computed a global search using dynamical bandwidth with probability of *HMCR* by the following relation,

$$x_{i,j}^n = \begin{cases} x_{i,j}^o + N(0, 1) \times bw_i(k) r_1 \le HMCR(k) \\ x_i^L + r \times \left(x_i^U - x_i^L\right) r_1 > \text{HMCR}(k) \end{cases} \tag{8.23}$$

where, $N(0, 1)$ relates a random normal generated number by standard normal PDF. On the other hand, in the second process, new HM is pitched using global-best position with probability of *HMCR* × *PAR*(k) as follows,

$$x_{i,j}^n = \begin{cases} x_i^{best} + \gamma \times N(0, S_i(k)) \;\; r_2 \le PAR(k) \\ x_{i,j}^n \;\; r_2 > PAR(k) \end{cases} \tag{8.24}$$

where, $bw_i(k)$ defines as dynamical bandwidth for variable x_i computed by,

$$bw_i(k) = \frac{|x_i^{\max} - 1.001x_i^{\min}|}{10} \exp[-\frac{10k}{NI}] \quad (8.25)$$

where, $x_i^{\min}$ and $x_i^{\max}$ are respectively minimum and maximum values for *i-th* variable in old HM. $N(0, S_i(k))$ generates a normal random number with mean of zero and dynamical STD of $S_i(k)$, which is determined as below relation,

$$S_i(k) = |x_i^{\max} - 1.001x_i^{\min}| \times \left(1 - \frac{k}{NI}\right)^{|x_i^{\max} - 1.001x_i^{\min}|} \quad (8.26)$$

And γ is controlling factor which is computed as below,

$$\gamma = \sqrt{1 - \frac{k}{NI}} \quad (8.27)$$

The Gaussian distribution is applied in these two global and local adjusting of old HM. In the local adjusting process, the dynamical bandwidth is adjusting based on a random factor given by $\gamma \times N(0, S_i(k))$ while the global adjusting procedure is randomly determined by $N(0, 1) \times bw_i(k)$ that $N(0, 1)$ provided a large variety of random number. It is shown that these two parameters as HMCR and PAR of HS are dynamically computed in this method as below,

$$HMCR(k) = 0.95 + 0.1 \times \sqrt{\frac{k}{NI} - (\frac{k}{NI})^2} \quad (8.28)$$

$$PAR(k) = 0.3 + 0.6 \times \left[1 - \sqrt{1 - k/NI}\right] \quad (8.29)$$

The best HM of the variable is adjusted by using the harmony memory information in GGHS with dynamical parameters. These modifications of GGHS for HS optimization framework are the major differences with HS, IHS, and GHS. The statistical bandwidth computed by normal distribution with STD of $S_i(k)$ in local adjusting process is controlled by a dynamical power function which is determined based on the statistical properties of HM design variables. Based on a large difference between the maximum and minimum value of variables in HM, larger values are determined for $S_i(k)$ at the beginning iteration thus, it is provided large verities to search of MPP domain using GGHS. At final iterations, we have $1 - \frac{k}{NI} \approx 0$ thus $\gamma \times N(0, S_i(k))$ is tended to zero. This means that global optimal is a fixed point and the adjusting process is searched a smaller domain to find the best results. The main different in Step 4 of HS for GGHS is shown in the below step,

Step 4: Create a new HM by using three factors of *HMCR*, *PAR*, and *bw* as below,

(i) Generate uniform random numbers between 0 and 1 for $r, r_1, r_2, r_3 \in [0, 1]$ and computed *bw(k)*, $S_i(k)$, γ, *HMCR (k)* and *PAR(k)* and give x_i^{best};

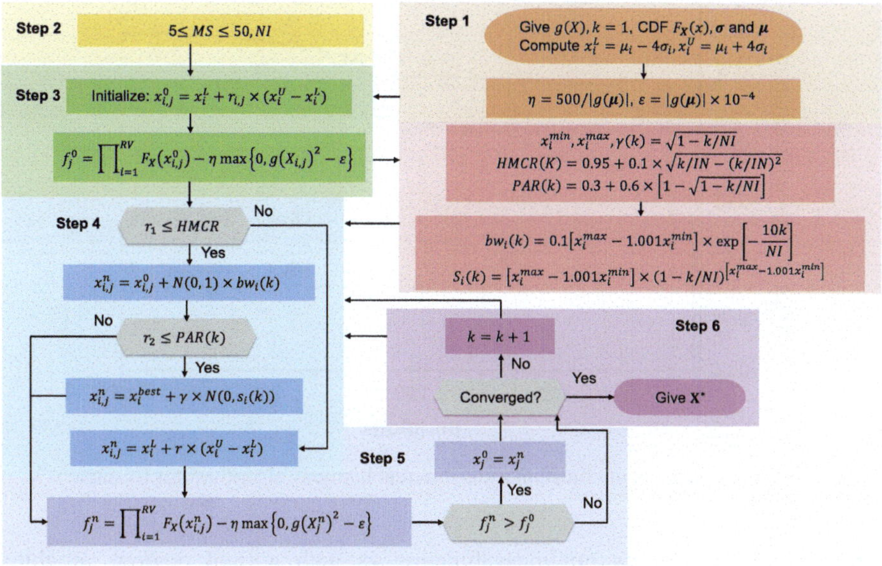

Fig. 8.14 Framework of GGHS algorithm for MPP search in $\boldsymbol{X}$-space

(ii) If $r_1 \leq HMCR$ then, $x_{i,j}^n = x_{i,j}^o + N(0,1) \times bw_i(k)$ and If $r_2 \leq PAR(k)$, then $x_{i,j}^o = x_i^{best} + \gamma \times N(0, S_i(k))$, end If, else;

(iii) Give the new HM from design domain as $x_{i,j}^n = x_i^L + r \times \left(x_i^U - x_i^L\right)$, end If.

The flowchart for a computer program of GGHS is presented in Fig. 8.14 for MPP search using OPM-based FORM. It can be seen that the parameters of HS are self-adaptively adjusted with dynamical relation thus the GGHS is a self-adaptive dynamical method that is free-parameters unlike HS.

8.3.2 Illustrative Examples and Discussion

In this section, all HS frameworks are coded in a MATLAB programming computer for MPP search of reliability problems. Generally, harmony memory size (MS) is given as 10 but it is investigated as $MS = 5, 10, 20, 40$ for some reliability problems. The NI is depended on the number of random variables that it is increased by increasing random variables that it is generally selected as $NI = 200RV$. In some cases, for evaluating the performances of different HS algorithms, the NI is given as $NI = 50, 100, 300, 500RV$. Other parameters of HS methods are presented in related frameworks in Figs. 8.11, 8.12, 8.13, 8.14 and $HMCR = 0.95$ and $PAR = 0.35$ for HS and also $HMCR = 0.95$ for GHS and IHS. The examples are given similarly to the Sect. 8.2.2 for better comparison of accuracy for searching MPP. The selected examples are summarized as below,

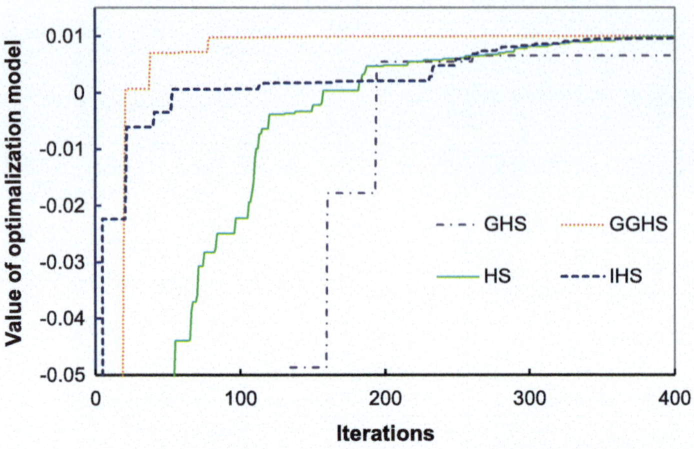

Fig. 8.15 Iteration for optimum function using different harmony algorithms for Example 8.7

Example 8.7 The highly nonlinear function as with normal random variables followed by $\mu_1 = \mu_2 = 10$ and $\sigma_1 = \sigma_2 = 5$.

The convergence iterations for optimum unconstrained function of $f_j^{best} = \prod_{i=1}^{RV} F_X\left(x_{i,j}^{best}\right) - \eta \max\{0, g(\mathbf{X}_{.j}^{best})^2 - \varepsilon\}$ with $NI = 400$ and $MS = 10$ for HS, IHS, GHS and GGHS is presented in Fig. 8.15. It can be seen that the convergence rate of the GGHS is strongly better than others. The IHS at the final iteration tended to the optimum. Nevertheless, it is given that the HS and GHS are provided an optimum with random selection of design variables. Consequently, the results of these two methods are evaluated with a low-convergence rate compared to GGHS. The adjusting procedure of new HM is strongly affected on performances of HS methods.

Figure 8.16 illustrates the error bars for the reliability index with respect to various NI and $MS = 10$ which are computed based on 10-time run of HS optimization algorithms. As seen, by increasing the NI, the robustness of the IHS, GHS and HS algorithms are increased but the HS and GHS are instable algorithms compared to GGHS and IHS. Generally, the GGHS is a robust method for $NI > 100$ iterations. The robustness of convergence for GGHS is obtained based on $NI = 400, 600, 1000$. Thus, the proposed method is strongly more robust than other methods.

The error bars as mean and STD for reliability index are plotted for different methods of HS with various harmony memory size (MS) of 5, 10, 20, 40 and $NI = 200$ for Example 8.8. It can be conducted that the GGHS is not depended on the MS compared to GHS and HS. The robust results are obtained using GGHS in all MSs and IHS in $MS = 40$. The random search using GHS is shown by respect to different MSs while the HS is shown better robustness for $MS = 20$ and 40.

By comparing the results of Figs. 8.16 and 8.17, the feasible domain for search is increased based on the NI and MS that it is affected on the robustness of the HS optimization method while the GGHS showed better robustness with increments

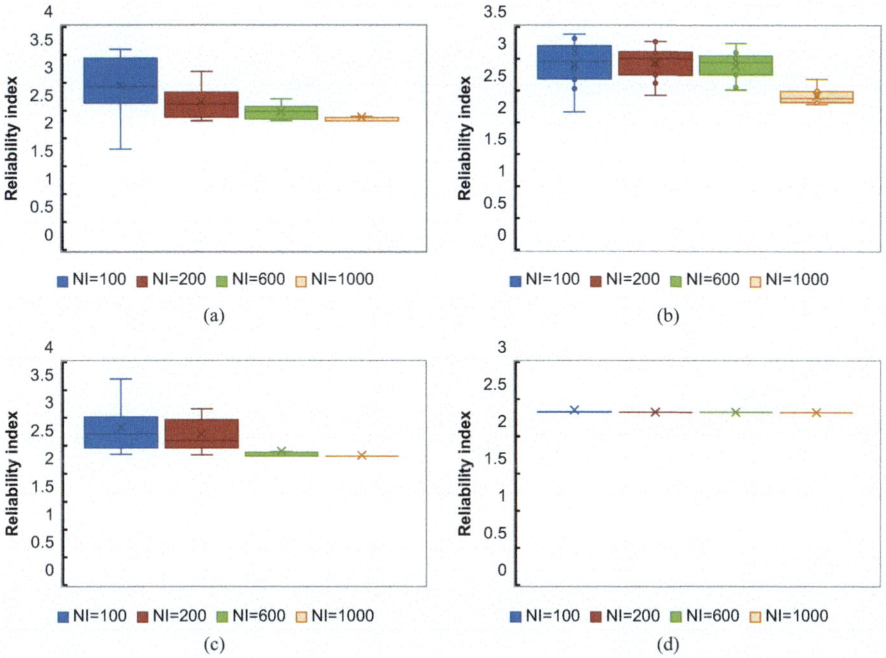

Fig. 8.16 Bar diagram of reliability index for various total iterations of different harmony algorithms for Example 8.7 **a** HS; **b** GHS; **c** HIS; **d** GGHS

of NI compared to MS. The dynamical adjusting HM in IHS can provide a robust result when its parameters including NI and MS are selected for an acceptable search in feasible design domain. However, the adjusting formulation with dynamical self-adaptive parameters can improve the robustness of HS applied in GGHS. This means that this method has a global condition with high-conversant rate and it can provide stable results as robust solutions with minimum search in feasible domain for this example.

The reliability index is $\beta_{MCS} = 2.8861$ using MCS with 10^6 samples and by using GGHS with $NI = 200$ and $MS = 10$, the reliability index computed as $\beta = 2.3246$. This reliability index, is given based on the MPP converged results at last iterations while it is not to transfer form $\boldsymbol{X}$-space to $\boldsymbol{U}$-space during run-process of GGHS. The converged results of GGHS as MPP and reliability index with other HS methods are compared in Table 8.10. It is shown that, the reliability index using the HS optimization method has a value less than the OPM-SQP and FSL (i.e. finite step length) approach. The IHS results are closely followed to the results of CFORM given by the MPP-based transformation in $\boldsymbol{U}$-space. It can be conducted that the metaheuristic optimization algorithm with balancing in global and local searches for the best point as GGHS performed better than the analytical method such as SQP for searching optimum. The reliability index is defined as shorten distance LSF with origin is given by the GGHS, while the LSF value at MPP is 0.131 which is small value.

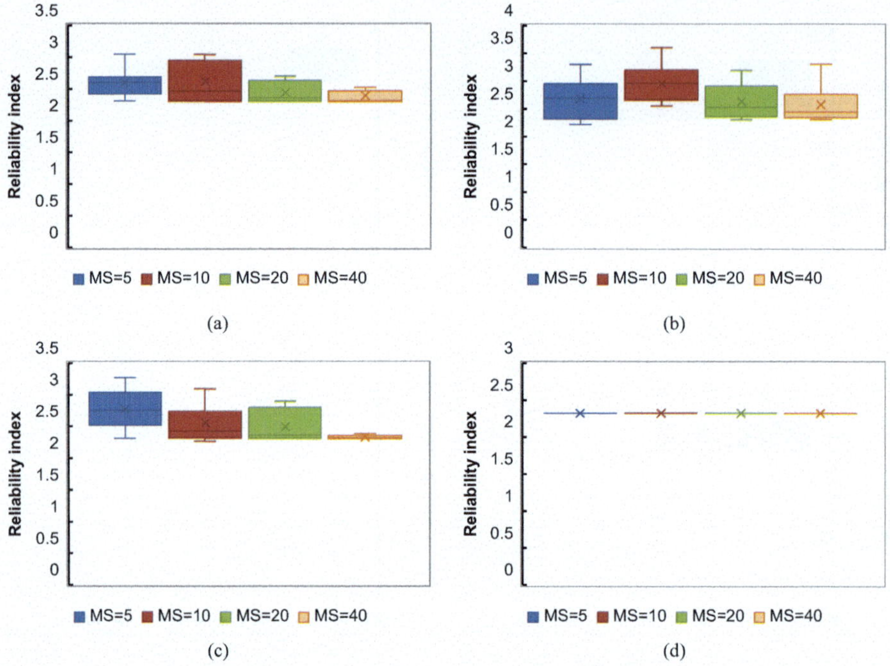

Fig. 8.17 Bar diagram of reliability index for various HM size (MS) of different harmony algorithms for Example 8.7. **a** HS; **b** GHS; **c** HIS; **d** GGHS

Table 8.10 Results of different OPM given by different HS method for Example 8.7

Method	$\boldsymbol{X}^*$	$\boldsymbol{U}^*$	β
OPM-SQP	(1.819216, 1.458369)	(−1.636157, −1.708326)	2.365457
FSL	(1.825913, 1.451805)	(−1.634817, −1.709639)	2.365479
CFORM	(1.928665, 1.552691)	(−1.614267, −1.689462)	2.336694
HS	(1.623052, 1.968477)	(−1.675389, −1.606304)	2.327024
IHS	(2.028156, 1.459093)	(−1.594368, −1.708181)	2.336641
GHS	(2.248609, 1.203419)	(−1.550278, −1.759316)	2.3448998
GGHS	(1.785544, 1.777182)	(−1.642891, −1.644563)	2.324581

Example 8.8 A non-normal LSF with relation of $g(X) = x_1^4 + x_2^2 - 50$, where, variables x_1 and x_2 respectively are Lognormal ($\mu = 5$ and $\sigma = 1$) and Gumbel ($\mu = 10$ and $\sigma = 10$).

For this nonlinear problem with non-normal variables, the HS, IHS, GHS and GGHS are applied that the iteration for different $MS = 5, 10, 20, 40$ of optimum function is presented in Fig. 8.18. It is obvious that the MS is an affective parameter for IHS and HS methods. By increasing the MS, the globality of HS methods is also increased. The GGHS provides stable results with the same optimum function at the

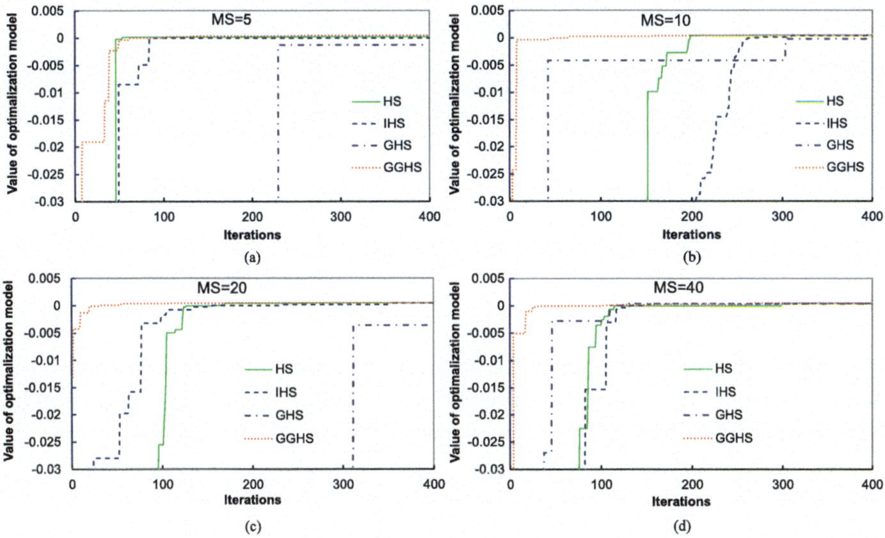

Fig. 8.18 Iteration for optimum function using different harmony algorithms for Example 8.8. **a** $MS = 5$; **b** $MS = 10$; **c** $MS = 20$; **d** $MS = 40$

final iteration in almost memory sizes. In this figure can be extracted, the GGHS has a top convergence rate for all harmony memory sizes compared to other methods. The GHS is performed with local results and it cited in last method for globality and robustness. HS and IHS performed with the same robustness by increasing memory size while the globality of IHS is better than the HS. By comparing with optimal results and convergence rate, the best to worst methods are GGHS, IHS, HS and GHS.

The extracted reliability index by GGHS and MCS (with 1.2×10^6 datapoints) are respectively $\beta = 3.26201$ and $\beta = 3.5009$. The OPM using SQP provided $\boldsymbol{U}^* = (-3.2587, \ -0.7377), \boldsymbol{X}^* = (2.5714, \ 2.5057)$, and $\beta = 3.34118$. FORM has $\boldsymbol{U}^* = (-3.1115, \ -0.9706)$, $\boldsymbol{X}^* = (2.6475, \ 0.9315)$, and $\beta = 3.25933$ however, the GGHS provides results as $\boldsymbol{X}^* = (2.6381, \ 1.2602)$, $\boldsymbol{U}^* = (-3.1294, \ -0.9206)$ with $\beta = 3.26201$. These results indicated that the accurate convergence is obtained by the proposed OPM-based GGHS for this example. The reliability index of OPM-based GGHS is closely agreement with results of FORM-based SQP.

The results of MPP search real (x)-space by OPM using metaheuristic-based musical human behavior studied in the current chapter with $NI = 200$ and $MS = 10$ are presented in Table 8.11 for Example 8.8. It can be the driver for the results of Table 8.1, the GGHS provided the global results with minimum reliability index compared to other methods. The worst method is GHS with a large different with GGHS. The LSF at MPP is computed a small value as 0.03 which is obtained by GGHS. This indicated that the GGHS can be searched the MPP on the LSF when the joint CDF of basic random variables have a maximum value.

Table 8.11 Results of different OPM given by different HS method for Example 8.8

Method	X^*	U^*	β
OPM	(2.5714, 2.5057)	(−3.2587, −0.7377)	3.34118
FORM	(2.6475, 0.9315)	(−3.1115, −0.9706)	3.25933
HS	(2.6346, 1.3793)	(−3.1359, −0.9026)	3.26331
IHS	(2.6157, 1.7924)	(−3.1724, −0.8411)	3.28207
GHS	(2.6229, 1.1966)	(−3.1585, −0.9302)	3.29267
GGHS	(2.6381, 1.2602)	(−3.1294, −0.9206)	3.26201

The FORM is very sensitive to the uniform distribution for these cases of basic random variables, the uniform distribution for variables x_1 and x_2 are discussed by the following four cases.

Case 1: $x_1 \sim Uniform(\mu = 5\ and\ \sigma = 2)$ and $x_2 \sim Uniform(\mu = 10\ and\ \sigma = 10)$;
Case 2: $x_1 \sim Normal(\mu = 5\ and\ \sigma = 1)$ and $x_2 \sim Uniform(\mu = 10\ and\ \sigma = 10)$;
Case 3: $x_1 \sim Lognormal(\mu = 5\ and\ \sigma = 1)$ and $x_2 \sim Uniform(\mu = 10\ and\ \sigma = 10)$;
Case 4: $x_1 \sim Uniform(\mu = 5\ and\ \sigma = 2)$ and $x_2 \sim Gumbel(\mu = 10\ and\ \sigma = 10)$.

As given results from OPM-based SQP in Sect. 8.2, the results of GGHS including MPP in $\boldsymbol{X}$-space ($\boldsymbol{X}^*$) and β computed by OPM are listed in Table 8.12 for above cases. As seen, the OPM-given by GGHS is globally converged to an accurate MPP without any transformation mapping from $\boldsymbol{X}$-pace to $\boldsymbol{U}$-space. The results of OPM-based GGHS are similar to the FORM-based SQP. Consequently, the MPP can be search based on the information of CDF for reliability analysis of problems with interval variables.

Table 8.12 Results of Cases 1–4 for different MPP search methods for Example 8.8

Cases	OPM-SQP		OPM-GGHS		FORM-SQP		MCS
	X^*	β/ (error%)	X^*	β/ (error%)	X^*	β/ (error%)	β
Case 1	(2.534, 3.071)	1.157/ (2.3)	(2.534, 3.112)	1.183/ (4.5)	(2.5689,3.539)	1.186/ (4.8)	1.131
Case 2	(2.261, 4.884)	2.765/ (3.1)	(2.629, 1.520)	2.460/ (13.7)	(2.645, 1.025)	2.458/ (13.9)	2.854
Case 3	(2.403, 4.084)	3.629/ (0.8)	(2.653, 0.722)	3.187/ (11.5)	(2.656, 0.464)	3.186/ (11.5)	3.599
Case 4	(2.367, 4.316)	1.160/ (12.8)	(2.431, 3.997)	1.249/ 21.3	(2.482, 3.476)	1.251/ (21.6)	1.028

Table 8.13 Random variables for Example 8.9

Variable	Mean	STD	Distribution
x_1	4	0.1	Weibull
x_2	25,000	2000	Log-normal
x_3	0.875	0.1	Gumbel
x_4	20	1	Uniform
x_5	100	100	Exponential
x_6	150	10	Normal

Example 8.9 A LSF as $g(X) = x_1x_2x_3x_4 - x_5x_6^2/8$ is given by random variables listed in Table 8.13.

Extracted from Ref. [11] is $\beta = 2.67$ that it is given by MCS using 1×10^6 samples as the reliability index of $\beta = 2.6935$. This example is solved by using the HS applied in OPM that the iterations for determining the optimum function are presented in Fig. 8.16 for HS, GHS, IHS, and GGHS with 1200 iterations. The reliability indexes for different HS methods are computed as $\beta_{HS} = 2.6719$, $\beta_{IHS} = 2.6339$, $\beta_{GHS} = 2.6502$, and $\beta_{GGHS} = 2.6871$. These results are agreement with the results of literature and computed by MCS. Figure 8.19 showed that the GGHS strongly showed the best globality compared to others while the optimum of HS, IHS and GHS are the same for this problem. At the final iteration, the IHS performed better than the HS and GHS methods. However, the GGHS is strongly performed with superior abilities for convergence rate and globality.

The reliability index (β), iterations, MPP ($\boldsymbol{X}^*$ and $\boldsymbol{U}^*$), and function calls ($Eval$) of OPM-based SQP, FORM-based SQP and HL-RF, OPM-GGHS and CFORM are tabulated in Table 8.14. As seen, the GGHS results were followed by the results of various methods. However, this method is time consuming approach with efficient

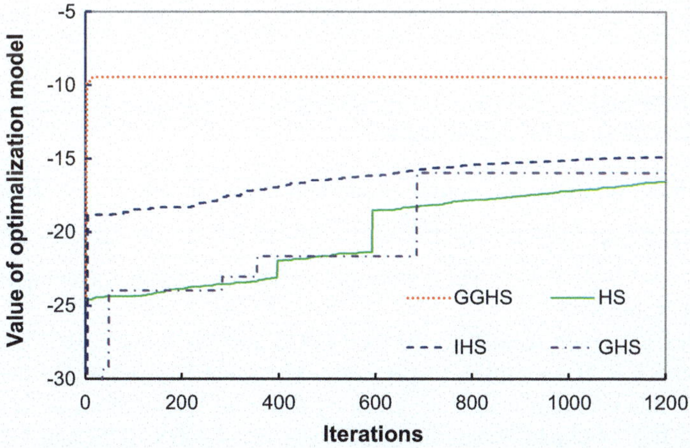

Fig. 8.19 Iteration for optimum function using different harmony algorithms for Example 8.9

Table 8.14 Results for different probabilistic models for Example 8.9

Method	HL-RF	OPM-GGHS	CFORM	FORM	OPM-SQP	Melchers and Ahammed [10]
x_1^*	4.0057	3.985	4.0054	4.0057	4.0025	4.01
x_2^*	24,204.6	24,700	24,204.4	24,205.2	24,499.9	24,193
x_3^*	0.8227	0.86	0.8225	0.8228	0.8011	0.822
x_4^*	19.581	19.0146	19.578	19.582	19.281	19.58
x_5^*	514.34	520	531.98	514.46	504.20	509.02
x_6^*	155.86	151.5	156.13	155.86	155.02	157.01
u_1^*	−0.1015	−0.32553	−0.1045	−0.1014	−0.1378	–
u_2^*	−0.3649	−0.11121	−0.3724	−0.3646	−0.2131	–
u_3^*	−0.4300	0.015857	−0.4353	−0.4296	−0.7233	–
u_4^*	−0.3075	−0.78735	−0.3059	−0.3073	−0.5461	–
u_5^*	2.5222	2.541648	2.5836	2.5222	2.4859	–
u_6^*	0.5862	0.15	0.6125	0.5857	0.5020	–
Iter	8	1200	68	69	6	–
Eval	127	12,000	1352	426	42	–
β	2.6699	2.6871	2.7355	2.6696	2.7051	2.68
Error (%)	0.89	0.6	1.54	0.90	0.43	0.50

Table 8.15 Statistical properties for random variables in Example 8.10

Variable	Mean	STD	Variable	Mean	STD
M_s	0.01	0.001	ξ_s	0.02	0.01
M_p	1	0.1	ξ_p	0.05	0.02
K_s	0.01	0.002	F_s	15	1.5
K_p	1	0.2	S_0	100	10

results for MPP search. It provides a suitable ability of globality for searching MPP but the GGHS is an inefficient method.

Example 8.10 A dynamic system described by LSF of $g = Fs - Ks \times 3(\frac{\pi S_0}{4\xi_s\omega_s}\left[\frac{\xi_a\xi_s}{\xi_p\xi_s(4\xi_a^2+\theta^2)+\gamma\xi_a^2} \times \frac{(\xi_p\omega_p^3+\xi_s\omega_s^3)\omega_p}{4\xi_a\omega_a^4}\right])^{1/2}$, in which, $\gamma = \frac{M_s}{M_p}$, $\omega_a = \frac{\omega_p+\omega_s}{2}$, $\xi_a = \frac{\xi_p+\xi_s}{2}$ and $\theta = \frac{\omega_p-\omega_s}{\omega_a}$. The lognormal variables applied for computing LSF are listed in Table 8.15.

The extracted result from Ref. [6] for FORM and MCS are 2.0165 and 2.5874 obtained by 5.8×10^5 samples, respectively. The reliability index and the MPP in ***X***-space for this problem which are given by different HS methods as tabulated in Table 8.16. It is obtained inaccurate MPP by using the HS, IHS, and GHS while the GGHS provides the accurate reliability index more closely to MCS. The two adjusting processes for this problem can be enhanced the global and local updating

Table 8.16 Results of OPM using different reliability methods for Example 8.10

Method	M_p	M_s	K_p	K_s	ξ_p	ξ_s	F_s	S_0	β
HS	0.974	0.009	1.190	0.012	0.036	0.011	103.034	13.552	2.1989
IHS	1.071	0.010	1.072	0.010	0.027	0.010	105.898	14.923	2.1756
GHS	1.037	0.010	1.077	0.010	0.055	0.022	103.635	15.585	1.1639
GGHS	1.085	0.011	1.17	0.0117	0.033	0.012	108.5	14.043	2.4240

HM in HS framework, thus it can be applied for real applicable optimization problems to search optimum conditions.

Example 8.11 A structure with LSF $g = 1 - \frac{\sqrt{3(0.91)}}{\pi E t^2 \cos^2 \alpha} \times \left(\frac{P}{0.66} + \frac{M}{0.41 r_1}\right)$ computed for buckling failure mode and random variables listed in Table 8.17.

The reliability index and MPP for Example 8.11 obtained by HS, IHS, GHS, and GGHS are presented in Table 8.18. It is can be given that the GGHS and IHS provided accurate results which is confirmed the MPP obtained by FORM method. The GGHS provided the results of the reliability index followed by the result of MCS with reliability index of 4.4163 obtained by ten million samples. The convergence of optimum function for different methods of HS is plotted in Fig. 8.17 for Example 8.11. It can be seen that the optimum value of the GGHS is more efficient than other methods. The maximum value for joint CDF of basic random variables is obtained while other methods have the same convergence rate. As seen from Fig. 8.20, the convergence rate of GGHS is highly performance level that it is a good ability for this method.

The optimization method using the HS methods are time-consuming approach and the efficiency of computational burden of these methods are increased by increasing the random variables. However, for application of this method given by publication-based optimization, it should be used the hybrid method of optimization and machine learning approaches that the machine learning methods applied for improving the efficiency of computational burden as a nonlinear approximated function for LSF while the optimization method is applied for MPP search of LSF approximated by machine learning [19, 20].

Table 8.17 The variables applied for computing LSF in Example 8.11

Variable	Description	PDF	Mean	COV
E	Young's modulus (MPa)	Lognormal	70,000	0.05
α	Slop angle (rad)	Normal	0.524	0.02
t	Thickness (m)	Normal	0.0025	0.05
r_1	Internal radius (m)	Normal	0.9	0.025
P	Axial load (N)	Gumbel	70,000	0.08
M	Bending moment (N-m)	Gumbel	80,000	0.08

Table 8.18 Results of different reliability methods using FORM and OPM for Example 8.11

Method	E^*	t^*	$\alpha*$	r_1*	M^*	P^*	β
Ref. [9]	65,442.63	0.00208	0.5275	0.8887	106,149.59	73,112.66	4.632489
FORM	65,131.86	0.0021	0.5275	0.8888	108,053.51	72,673.82	4.628886
HS	6299.534	0.00223	0.5388	0.8561	87,836.10904	69,365.12018	4.0405
IHS	6195.467	0.00221	0.5474	0.9122	92,787.83789	78,731.1192	4.6322
GHS	6213.800	0.00224	0.5270	0.8527	91,931.56646	72,189.91162	4.1723
GGHS	7412.527	0.00279	0.5502	0.9563	95,352.26706	78,220.68568	4.6155

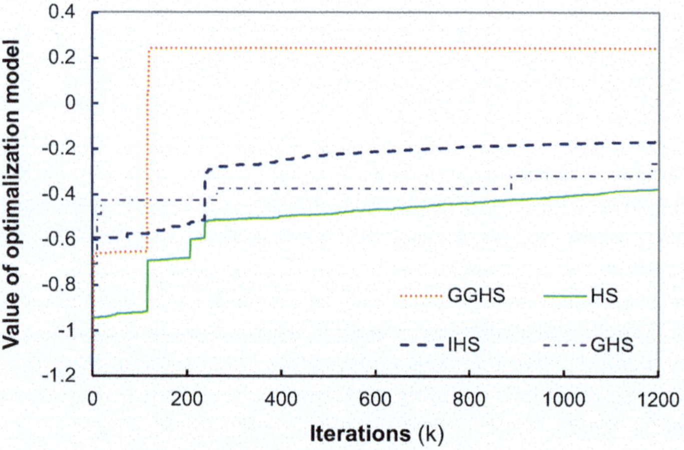

Fig. 8.20 Iteration for optimum function using different harmony algorithms for Example 8.11

For future works, it just presented an improved version of optimization approach but the other methods of population-based optimization algorithms can be applied for MPP search of reliability problems using novel OPM.

The authors well-known that this formulation can be improved based on a novel joint cumulative function which is more stable for a high-dimensional performance function. The joint CDF may converge to zero, thus the optimization method is applied to search a point on LSF without considering the maximization of joint CDF.

8.4 Summary

In this chapter, a novel optimization model for MPP search in original space for interval basic random variables is developed named the original probabilistic model (OPM). The novel MPP search approach is formulated by using the joint cumulative distribution function (CDF) of basic random variables in ***X***-space. The MPP is determined by solving a probabilistic optimization problem as the objective of multiplier CDF named joint CDF under probabilistic constraint of limit state function. In novel searching MPP, an individual MPP can be provided for multi-MPP problems. The performances of OPM are compared with FORM and FOSAM through several reliability problems. The formulation for searching MPP presented in this chapter can be applied for reliability problems with interval basic random variables. Then, harmony search (HS) optimization as a music-inspired using human-based optimization method is applied for searching MPP by presented OPM in this chapter. The MPP of LSFs is approximated based on four HS methods that they are given based on the dynamical random selection and adjustment. The original HS and three modifications of this are introduced for structural reliability analysis that they are used for approximating the MPP is $\boldsymbol{X}$-space. The modified versions which are extended based on the dynamical parameters of HS optimization provided the accurate MPP results. The optimization method using the HS methods are time-consuming approach and the efficiency of computational burden of these methods are increased by increasing the random variables.

References

1. I. Lee, K. Choi, L. Du, D. Gorsich, Dimension reduction method for reliability-based robust design optimization. Comp. Struct. **86**(13–14), 1550–1562 (2008)
2. X. Du, A. Sudjianto, First order saddlepoint approximation for reliability analysis. AIAA J. **42**(6), 1199–1207 (2004)
3. Z. Meng, G. Li, D. Yang, L. Zhan, A new directional stability transformation method of chaos control for first order reliability analysis. Struct. Multidiscipl. Optim. **55**, 601–612 (2017)
4. M. Hohenbichler, R. Rackwitz, Improvement of second-order reliability estimates by importance sampling. J. Eng. Mech. **114**(12), 2195–2199 (1988)
5. P. Hao, Y. Wang, R. Ma, H. Liu, B. Wang, G. Li, A new reliability-based design optimization framework using isogeometric analysis. Comp. Methods Appl. Mech. Eng. **345**, 476–501 (2019)
6. B. Keshtegar, Chaotic conjugate stability transformation method for structural reliability analysis. Comp. Methods Appl. Mech. Eng. **310**, 866–885 (2016)
7. M. Rashki, M. Miri, M.A. Moghaddam, A new efficient simulation method to approximate the probability of failure and most probable point. Struct. Saf. **39**, 22–29 (2012)
8. P.L. Liu, A. Der Kiureghian, Optimization algorithms for structural reliability. Struct. Saf. **9**(3), 161–177 (1991)
9. B. Keshtegar, Stability iterative method for structural reliability analysis using a chaotic conjugate map. Nonlinear Dyn. **84**, 2161–2174 (2016)
10. R. Melchers, M. Ahammed, A fast approximate method for parameter sensitivity estimation in Monte Carlo structural reliability. Comp. Struct. **82**(1), 55–61 (2004)
11. W. Zhao, J. Liu, J. Ye, A new method for parameter sensitivity estimation in structural reliability analysis. Appl. Math. Comp. **217**(12), 5298–5306 (2011)
12. C. Luo, S.P. Zhu, B. Keshtegar, X. Niu, O. Taylan, An enhanced uniform simulation approach coupled with SVR for efficient structural reliability analysis. Reliab. Eng. Syst. Safe. **237**, 109377 (2023)
13. A. Baykasoğlu, F. Ozsoydan, Evolutionary and population-based methods versus constructive search strategies in dynamic combinatorial optimization. Inf. Sci. **420**, 159–183 (2017)

14. K.S. Lee, Z.W. Geem, A new structural optimization method based on the harmony search algorithm. Comp. Struct. **82**(9–10), 781–798 (2004)
15. B. Keshtegar, P. Hao, Y. Wang, Y. Li, Optimum design of aircraft panels based on adaptive dynamic harmony search. Thin-Walled Struct. **118**, 37–45 (2017)
16. M. Mahdavi, M. Fesanghary, E. Damangir, An improved harmony search algorithm for solving optimization problems. Appl. Math. Comp. **188**(2), 1567–1579 (2007)
17. M.G. Omran, M. Mahdavi, Global-best harmony search. Appl. Math. Comp. **198**(2), 643–656 (2008)
18. B. Keshtegar, T. Ozbakkaloglu, A. Gholampour, Modeling the behavior of FRP-confined concrete using dynamic harmony search algorithm. Eng. Comp. **33**(3), 415–430 (2017)
19. C. Luo, S.P. Zhu, B. Keshtegar, W. Macek, R. Branco, D. Meng, Active Kriging-based conjugate first-order reliability method for highly efficient structural reliability analysis using resample strategy. Comp. Methods Appl. Mech. Eng. **423**, 116863 (2024)
20. C. Luo, B. Keshtegar, S.P. Zhu, X. Niu, EMCS-SVR: hybrid efficient and accurate enhanced simulation approach coupled with adaptive SVR for structural reliability analysis. Comp. Methods Appl. Mech. Eng. **400**, 115499 (2022)

The manufacturer's authorised representative in the EU is Springer Nature Customer Service Centre GmbH, Europaplatz 3, 69115 Heidelberg, Germany. If you have any concerns regarding our products, please contact ProductSafety@springernature.com

Printed and bound by CPI Group (UK) Ltd, Croydon, CR0 4YY
01/10/2026
02225958-0008